Aerospace Materials

Series in Materials Science and Engineering

Series Editors: **B Cantor**, Department of Materials, University of Oxford, UK
M J Goringe, School of Mechanical and Materials Engineering, University of Surrey, UK

Other titles in the series

Microelectronic Materials
C R M Grovenor
Department of Materials, University of Oxford, UK

Computer Modelling of Heat, Fluid Flow and Mass Transfer in Materials Processing
C-P Hong, Yonsei University, Korea

Physical Methods for Materials Characterisation
P E J Flewitt
Magnox Electric, Berkeley, UK
and
R K Wild
University of Bristol, UK

Forthcoming titles in the series

Topics in the Theory of Solid Materials
J M Vail
University of Manitoba, Canada

Physical Methods for Materials Characterisation, Second Edition
P E J Flewitt
Magnox Electric, Berkeley, UK
and
R K Wild
University of Bristol, UK

Fundamentals of Fibre Reinforced Composite Materials
A R Bunsell, J Renard and M-H Berger
Centre des Materiaux, Pierre-Marie Fourt, France

Series in Materials Science and Engineering

Aerospace Materials

An Oxford–Kobe Materials Text

Edited by

Brian Cantor, Hazel Assender and Patrick Grant
Department of Materials, University of Oxford, UK

CRC Press
Taylor & Francis Group
Boca Raton London New York

CRC Press is an imprint of the
Taylor & Francis Group, an **informa** business

First published 2001 by IOP Publishing Ltd

Published 2023 by CRC Press
Taylor & Francis Group
6000 Broken Sound Parkway NW, Suite 300
Boca Raton, FL 33487-2742

ISBN 13: 978-0-7503-0742-0 (hbk)

Visit the Taylor & Francis Web site at
http://www.taylorandfrancis.com

and the CRC Press Web site at
http://www.crcpress.com

British Library Cataloguing-in-Publication Data
A catalogue record for this book is available from the British Library.
Library of Congress Cataloging-in-Publication Data are available

Series Editors: **B Cantor and M J Goringe**

Commissioning Editor: Tom Spicer
Production Editor: Simon Laurenson
Production Control: Sarah Plenty
Cover Design: Victoria Le Billon
Marketing Executive: Laura Serratrice

Typeset by Academic + Technical, Bristol

Contents

Preface

This book is a text on aerospace materials, rising out of presentations given at the first Oxford–Kobe Materials Seminar, held at the Kobe Institute, on 22–25 September 1998.

The Kobe Institute is an independent non profit-making organization. It was established by donations from Kobe City, Hyogo Prefecture and more than 100 companies all over Japan. It is based in Kobe City, Japan and is operated in collaboration with St Catherine's College, Oxford University, UK. The Chairman of the Kobe Institute Committee in the UK is Sir Peter Williams, Master of St Catherine's College; the Director of the Kobe Institute Board is Dr Yasutomi Nishizuka; the Academic Director is Dr Helen Mardon, Oxford University; and the Bursar is Dr Kaizaburo Saito. The Kobe Institute was established with the objectives of promoting the pursuit of education and research that furthers mutual understanding between Japan and other nations, and to contribute to collaboration and exchange between academics and industrial partners.

The Oxford–Kobe Seminars are research workshops which aim to promote international academic exchanges between the UK/Europe and Japan. A key feature of the seminars is to provide a world-class forum focused on strengthening connections between academics and industry in both Japan and the UK/Europe, and fostering collaborative research on timely problems of mutual interest.

The first Oxford–Kobe Materials Seminar was on aerospace materials, concentrating on developments in science and technology over the next 10 years. The co-chairs of the Seminar were Professor Akihisa Inoue of Tohoku University, Professor Brian Cantor, Dr Hazel Assender and Dr Patrick Grant of Oxford University, and Dr Kaizaburo Saito of the Kobe Institute. The Seminar co-ordinator was Ms Pippa Gordon of Oxford University. The Seminar was sponsored by the Kobe Institute, St Catherine's College, the Oxford Centre for Advanced Materials and Composites, ONERA, Dowty Aerospace Propellers, Ishikawajima-Harima Heavy Industries and Kobe Steel. Following the Seminar itself, all of the speakers prepared extended manuscripts in order to compile a text suitable for graduates and for researchers entering the field. The contributions are compiled into three sections: industrial applications, lightweight materials, and high temperature materials.

Acknowledgments

Brian Cantor, Hazel Assender and Patrick Grant

The editors would like to thank the following: the Kobe Institute Committee at St Catherine's College, Oxford University for agreeing to support the Oxford–Kobe Materials Seminar on Aerospace Materials; Lord Plant, Sir Peter Williams, Professor Akihisa Inoue and Drs Lincoln Wallen, Helen Mardon and Kaizaburo Saito for help in organizing the Seminar, and Ms Pippa Gordon and Ms Sophie Briant for help with preparing the manuscripts.

Individual authors would like to make additional acknowledgments as follows:

Patrick Grant

The author would like to thank F Lavers and P Pedro for help with figures, and Rolls-Royce plc, The UK Engineering and Physical Sciences Research Council, the DTI Teaching Company Directorate and The Royal Society for financial support.

Mike Loretto

I would like to acknowledge the many colleagues who have allowed me to quote from their work, some of which is as yet unpublished. In addition the support of EPSRC through the IRC core grant and through support for COST 513 is gratefully acknowledged.

Malcolm McLean

The author thanks the following colleagues for their contributions to the paper: M G Ardakani, B A Basoalto, R N Ghosh and B A Shollock. The work was made possible by support from the Engineering and Physical Science Research Council (Grant Numbers GR/J02667, GR/K19358; Visiting Fellowship GR/L67042), the Defence Evaluation and Research Agency and BRITE EURAM III Project BE 96-3911.

Tetsuya Ohashi

The author would like to express his great thanks to Dr K Hidaka, Mr S Imano and M Saito in Hitachi research laboratory for their support in experimental verifications and helpful discussions on the present study.

Chris Peel

The author acknowledges the support of DERA in the preparation of this paper and the contributions of his colleagues. The opinions expressed are those of the author and are offered in the interest of discussion.

Alan Smith

The author extends his thanks to Mr R F Williams (Chief Materials Engineer), Mr G M Byham (Chief Engineer Vehicles) and Mr P N Goddard (Engineering Director) for allowing Westland Helicopter support of this seminar.

SECTION 1

INDUSTRIAL APPLICATIONS

Aerospace materials technology is moving very rapidly, to keep up with the momentum of change in the air transport sector as a whole. Developments in aircraft, spacecraft, engines, landing gear etc. continue to produce improvements in air transport operations, building on the development of new materials and manufacturing processes. The industrial scene, and the key design drivers and materials needs, are covered in detail in this section.

Chapters 1 and 2 set the overall scene for materials needs in commercial aircraft, and chapters 3 to 5 follow with specific additional needs for specialized supersonic, helicopter and spacecraft applications. Chapter 6 describes the manufacture of a composite wing propeller, and chapters 7 and 8 cover major materials uses and developments in aeroengines.

Chapter 1

Aerospace materials and manufacturing processes at the millennium

Norman Barrington and Malcolm Black

Introduction

The aerospace industry worldwide is experiencing a restructuring process in order to achieve the efficiency and productivity levels required to meet the challenges of the market demands in the first quarter of the twenty-first century. Figure 1.1 shows the planned A3XX large body airliner.

It is essential that the supply chain in the industry faces the prime drivers of cost (initial and life cycle), time-to-market and platform performance. Aerospace materials and the associated design and manufacturing processes must be optimized in an integrated manner so as to deliver product competitiveness. The demands of increasing environmental constraints enforced by local and worldwide legislation must also be addressed. These challenges will only be met if the platform suppliers and material and semi-product manufacturers collaborate effectively in strong partnerships to deliver the goals of minimum cost and maximum performance in the required time-scales.

The past 20 years have shown significant improvements in the joint activities of materials suppliers and airframe manufacturers, enabling joint research programmes to be established which address the requirements in a comprehensive manner, with all of the participants being fully aware of the future needs and constraints during the design and manufacture of new and developed products. The optimization of manufacturing processes and production time efficiency is dependent on the initial quality of the materials and semi-products utilized, particularly if the required design properties are influenced or, in some cases, determined by the processes employed by the component manufacturer. Typical examples of this are heat treatments of aluminium alloys after mechanical forming or stretching, or during creep

Figure 1.1. The planned A3XX large body aircraft.

forming, and the development of the required polymer composite component properties through the resin, fibre mix and laminate lay up and final cure processes.

Airframe design and manufacturing processes have developed rapidly over the past 50 years. The mandatory requirements have been strengthened in both the test demonstration of structural integrity, and the necessity for airframe manufacturers to agree with the airlines and publish the maintenance requirements prior to certification, demonstrating the durability and damage-tolerance characteristics of the primary structure through the inspection programmes related to fleet and individual aircraft schedules. These developments have necessitated the definition of minimum requirements for new material characteristics such as fatigue and crack propagation rates under constant amplitude and spectrum loading, and fracture toughness under plane stress and plane strain conditions. Attention has also been diverted to the material environmental performance, with clearer requirements for stress corrosion cracking, exfoliation and intergranular attack. All of these properties must be evaluated against the expected time base and environmental exposure of the structure for the predicted in-service utilization.

To meet the future requirements of the airframe manufacturers the aluminium alloy material suppliers are being asked to extend their capabilities in casting volumes, rolling mill sizes (length, width and thickness), stretching capability, static material properties, durability, damage tolerance and corrosion resistance. Figure 1.2 shows the increased wing size requirements for modern large bodied aircraft. The manufacturers have responded with significant improvements in their production capabilities through statistical process control, and with careful development to extend their range of products whilst maintaining volume and quality.

This chapter considers the developments required by the platform suppliers so as to optimize their production processes and the efficiency of their products in service.

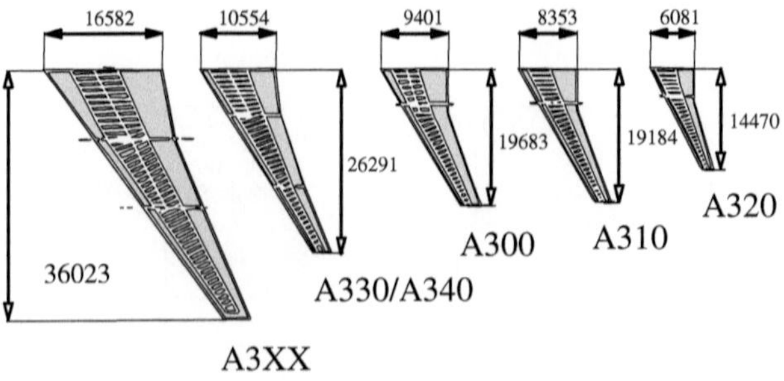

Figure 1.2. Wing sizes for modern large body aircraft.

Global market forecast highlights

Civil

Over the next 20 years great potential exists for an increase in demand for air travel as economic development continues in the populous less-developed nations. Current forecasts indicate that the number of air travellers will more than double over that period of time. The number of new aircraft required to meet this demand is estimated at between 13 500 and 17 000, depending on the seat range and aircraft size distributions used by the forecasters in their predictions. This represents average yearly deliveries of between 675 and 850 aircraft, as shown in figure 1.3. The value of this market is

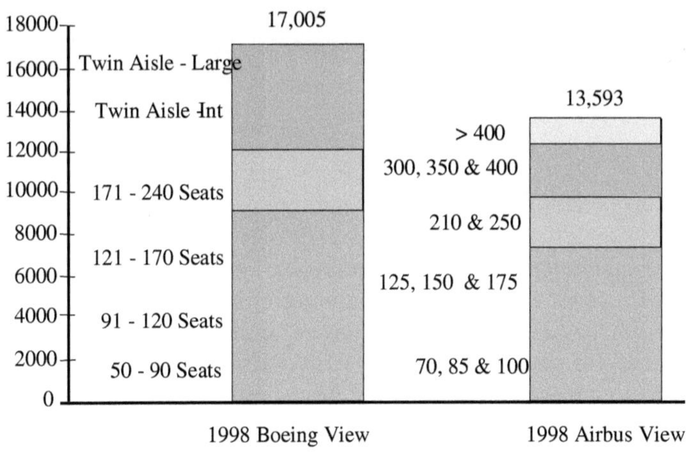

Figure 1.3. Boeing and Airbus estimates of aircraft requirements over the next 20 years.

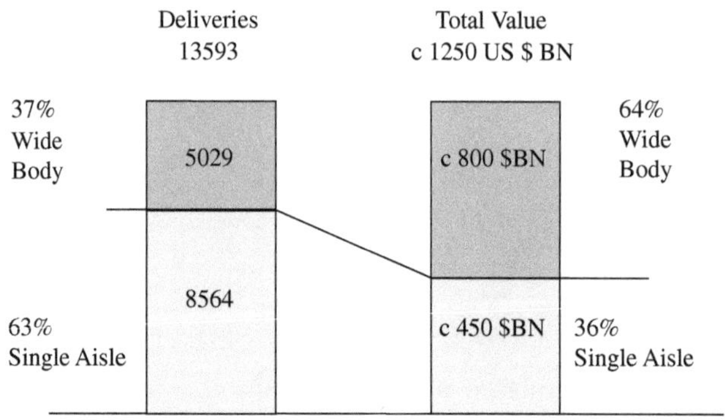

Figure 1.4. Breakdown of aircraft requirements over the next 20 years by size, number and value.

approximately 1.25 trillion dollars at present-day prices, which covers both the replacement of older aircraft and the market growth demands in the ratio of approximately 1 : 3, as shown in figure 1.4. It is anticipated that the world's airport and air traffic infrastructure limitations, despite restructuring and development, will result in the airlines requiring a new type of aircraft which will be larger and more economical than any current model in production today, to meet the demands between major population centres.

The challenge to the aircraft manufacturer to meet these market demands is enormous, for the design, development and manufacture of current and future types of aircraft, and will only be met if the critical drivers of cost, performance and environmental constraints are addressed throughout the supply chain. Local economic variations are not expected to affect these forecasts significantly. It is anticipated these will be covered by rescheduling deliveries.

Military

The next European military aircraft programme will address the need for an advanced strike aircraft with an entry into service of about 2015. The programme will develop a future offensive aircraft (FOA) for the UK Royal Air Force and a long-range interdictor for the French Air Force. The military transport market in Europe is dominated by plans for a multi-national future large aircraft (FLA). In a variety of applications, research programmes are investigating the potential for unmanned vehicles. Stealth technology will also play an increasing part in future military aircraft studies. To assure this future growth, continued development of high performance structures and components in advanced materials is critical, with reduced material and production costs. Airframe technology must be supported with advanced metallic and polymer composite materials and related design, inspection and manufacturing technologies.

Selection of materials and manufacturing processes

Key drivers

In the selection of materials and processes for use in the production of aerospace platforms, it is critical that consideration of cost, time to market and platform performance is carried out in a manner which gives the optimum return to both the supply chain and the final customer. The relationship between the key drivers and corresponding technology requirements is shown in figure 1.5. In the current world market scenario, customers are demanding and achieving significant reductions in the direct operating cost of their equipment, whilst ensuring that their customer appeal is enhanced through improved reliability and on-board comfort and facilities. To ensure those demands can be met in the future, whilst maintaining a healthy industrial position, requires a significant reduction in the cost of production through more efficient manufacturing processes, and improved platform performance and maintenance schedules. A detailed breakdown of the different costs is given in figure 1.6. The demands on new materials and processes when considering their introduction into any aerospace platform programme have become much broader, and many apparently promising materials and processes have not delivered the performance required by new projects.

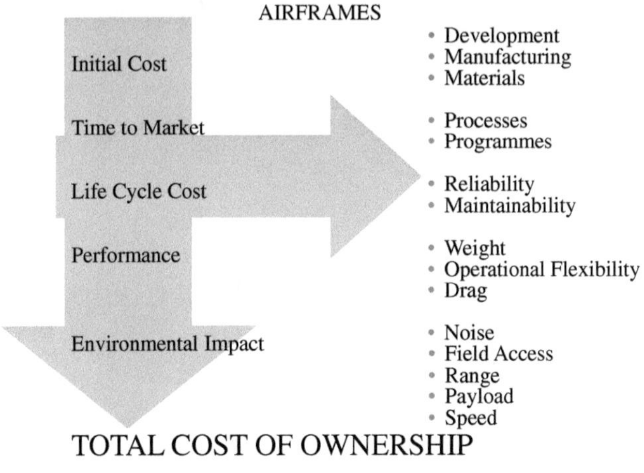

Figure 1.5. Key airframe drivers (left) and technology requirements (right).

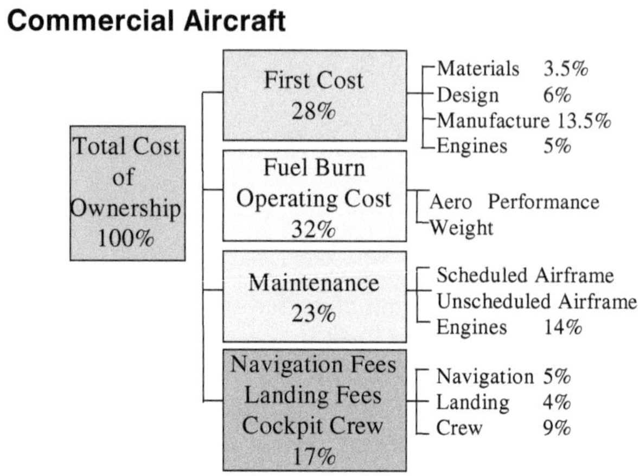

Figure 1.6. Breakdown of aircraft purchase and operating costs.

In the following sections emerging materials and processes are identified and assessed for their likely impact on future aerospace projects. Figure 1.7 shows examples of the breakdown of materials used.

Aluminium alloys

The use of aluminium alloys for aerospace platforms has developed over the past 75 years following initially a high specific strength route based on a high copper alloying content (2000 series) and a high zinc alloying content (7000 series). In parallel, heat treatments were developed to improve and optimize the balance between ultimate and yield strengths, ductility, resistance to the growth of fatigue cracks, toughness and resistance to corrosion. Because of their differing characteristics, the 2000 series have been predominantly

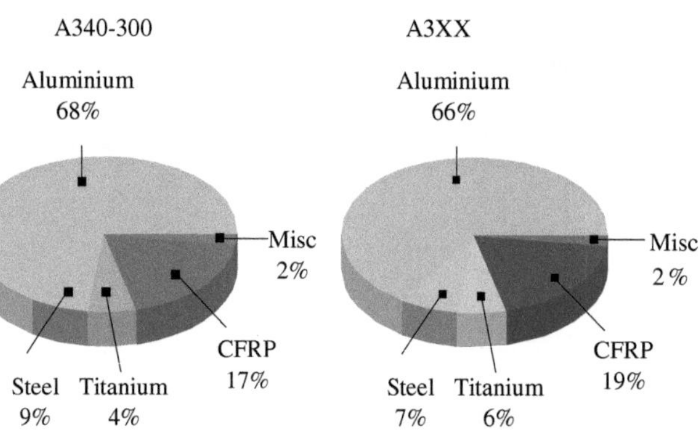

Figure 1.7. Breakdown of materials usage in A340-300 and A3XX aircraft.

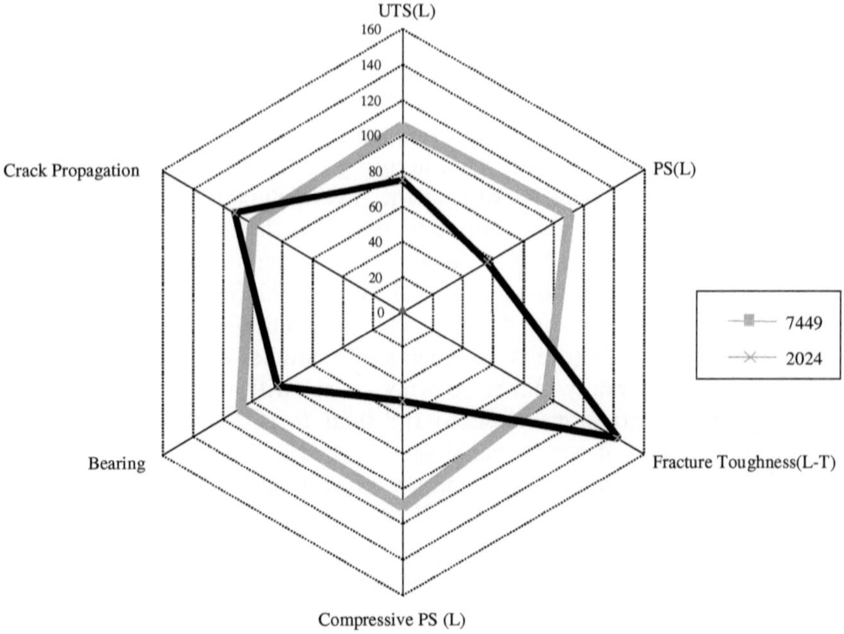

Figure 1.8. Property envelopes for aluminium alloys 7449 and 2024.

used for fuselage skins and lower wing surfaces, whilst the 7000 series have been used for upper wing surfaces, spars and ribs, in sheet, plate and extrusion forms. Both series have been used for forgings. Figure 1.8 shows typical property envelopes.

The emphasis on the improvement of the properties of these generic alloy forms has been influenced strongly by the development of damage tolerance methodology, and the demands of the customer for enhanced maintenance schedules and environmental performance. The pressures of efficiency in the performance of the platform, the production of the platform and the use of more expensive materials has demanded a total reassessment of materials and manufacturing processes to ensure that the drivers of performance, cost and time to market can be met.

The development of aluminium lithium alloys in the past 20 years is one example where the promise of significant benefits has not been realized to date on large commercial aircraft. Improved static and damage tolerance properties with the added bonus of a 10% density reduction was the target, but unfortunately these materials have not yet delivered the required performance for significant primary structure use. In addition they carry a significant cost penalty and it was recognized that the current buy-to-fly ratios would have to be reduced for the product to be attractive to future project directors. Significant effort has been made over the past five years to resolve the outstanding shortfalls and it is now possible to achieve a satisfactory performance for sheet material through improvements in the heat treatment processes. Further work is continuing to extend these properties to plate and extrusion semi-products. Meanwhile new alloys based on the 2000 and 7000 series have been developed to meet the increasing demands for enhanced properties and manufacturing processes, and these are being introduced into new projects such as the A340-500/600.

In parallel, the design of components such as the advanced wings for the A340-500/600 and the A3XX requires the material suppliers to deliver greater lengths of plate, over 30 m, whilst maintaining the current specifications for properties and flatness. The production of thicker plates for spar and rib manufacture is also necessary. Studies are being carried out on the use of integrally machined wing covers with the recent availability of large, wide extrusions close to the final form required with integral stiffeners. Casting of a near-net-shape process is also being developed for significant components, with a pilot study for an A320 bulk cargo door having been successfully completed, and with the result of an industrial application study for the A340-600 passenger door. Both these developments will allow increased material utilization and a reduction in the assembly time and part count.

Manufacturing processes are currently being developed, to meet the demands of production efficiency and platform performance requirements, in conjunction with the future material specifications. These include: forming processes such as shot peening, age-forming and stretching; welding methods such as laser, electron beam and friction stir; treatments including shot peening, protection and cold working; and assembly techniques such as electro impact riveting and welding. These developments must deliver significant savings to the constructor and the airline customer to secure the future of both.

Developments in 2000 and 7000 series aluminium alloys are discussed further in chapters 2 and 9, with novel alloys described in chapter 11. The use of aluminium lithium alloys for helicopter airframes is considered in chapter 4.

Polymer composites

The application of composites to aerospace platforms has been a challenge to manufacturers in overcoming the high costs associated with these materials and associated processing, to enable the benefits to be gained from a demonstrated 10–20% weight reduction. The cost penalty is a combination of the basic material cost, the storage and manufacturing processes and the stringent inspection and test requirements.

Current applications range from military to large civil transport platforms, where significant weight savings have been achieved compared with metallic designs. In large civil platforms the development of primary

Table 1.1. Polymer composite requirements for use in large aircraft wing structures.

Rationalize processes and material specifications	
Reduce basic material costs Resin/Fibre/Pre-preg	
High deposition rates	
Low cure temperature	
Lower tooling cost	
Low cost consumable	
Low part distortion	
General shape/thickness control in manufacture	
Assembly tolerances	
High strain allowables	4000 μe
Less material	~Lower cost
	~Lower weight

structural applications started with the fin for the A310, followed by the fin and dry tailplane for the A320, and the fin and wet tailplane for the A340 and B777. So far there is little expectation for composite fuselage structures on large subsonic commercial aircraft for a considerable time to come.

The next logical step would appear to be an application in the wing box primary structure, and the baseline for the A3XX is defined with a composite outer wing. The development of a composite wing design, however, brings considerable additional problems that can have a significant effect on the weight saving achieved. These can result from the inherent stiffness affecting the design loads of the wing, for example. To achieve a successful design for wing applications, the structural design concepts must be reviewed to match radical manufacturing processes in order to achieve an optimum weight and cost for the component. Table 1.1 lists the requirements.

Significant development work is being undertaken by the material suppliers and the component manufacturers to ensure that the right technologies are in place so that a viable design and manufacturing concept will be achieved. This includes rapid lay-up techniques including non-crimp, drapable fabrics, advanced forms of resin transfer moulding and resin infusion, novel jointing techniques, all-bonded structure, automated assembly methods and non-destructive test methods etc. It is essential to achieve a design that is removed from the 'black-metal' concepts to meet the commercially viable concept that the project teams require.

Other composite applications are being introduced for the A340-500/600 wing, including a thermoplastic J-nose, and inboard lower access panels using glass and carbon fibre with polyphenylene sulphide (PPS) thermoplastic. The use of these thermoplastics facilitates a monolithic, discretely stiffened design. Techniques are also being developed for the forming of large volume structures by Rotomoulding, with current applications being demonstrated for cabin air ducting, landing light drums and slat track cans. These will have a significant effect on both cost and parts count.

The challenge for the composite structure design teams to produce viable, cost effective applications for future projects is being faced with the clear objective of beating the performance of an advanced design and manufacture of metallic components and assemblies.

The use of composites for space structures and propellors is considered in chapters 5 and 6, with new composite systems discussed in chapters 13 and 15.

Titanium alloys

Titanium alloys are used in current airframe applications for medium-sized forgings in high load transfer components such as the engine pylon and undercarriage support fittings on the wing, composite component joints and bolts. Those components which can be produced effectively by superplastic forming/diffusion bonding such as engine pylon secondary structure and lower wing unloaded manhole doors are also in production. Superplastic forming/diffusion bonding has been pursued for use on other components, and demonstrators have been manufactured for spoilers, ailerons, flap beams and fuselage sections. However, to date, none of these applications has proved to be attractive to current project teams when evaluated against the drivers of cost and performance, and the technology may be in waiting for applications on supersonic platforms, where the basic material properties bring greater benefits. Supersonic transport material needs are the subject of chapter 3.

Developments are also currently underway in near-net-shape forging technology, to improve the material utilization, and future application for undercarriages for weight reduction are being pursued. In addition, laser drilled and welded structures are an essential element of the Airbus laminar flow studies. An A320 is undergoing flight trials from Toulouse with a hybrid laminar flow fin. The experimental aircraft is equipped with two titanium fin nose sections, each about 2.5 m long, which have laser-perforated surfaces and an array of sub-surface chambers.

Steels

The use of steel is relatively limited, at about 7% of the airframe structure. Steels are used where the structure requires very high strength and resistance to elevated temperature or where space is restricted. Primarily use of steels is in pylon structure, landing gear and slat track components, as shown in figure 1.9, plus some small standard parts such as bearings, springs, bolts and washers. For example, both high strength alloy steels and precipitation hardened stainless steel forgings are among the steel components used in pylon structures. The materials are selected to provide fatigue and damage

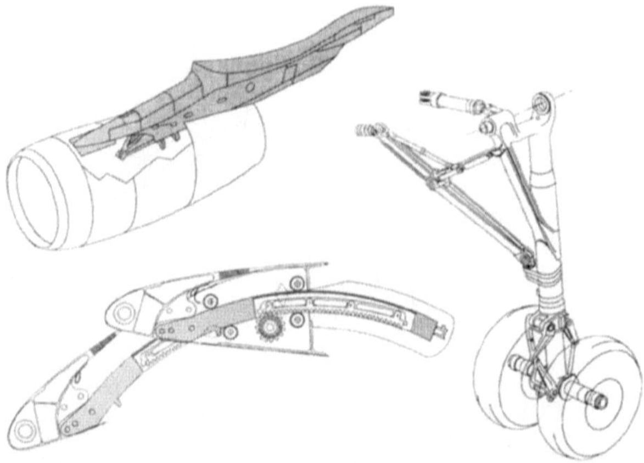

Figure 1.9. Examples of steel components in aircraft.

tolerance, high static strength and stiffness, corrosion resistance and safe use in the range −55 to 250°C. Maraging steel 300M is used extensively on landing gear and slat tracks by virtue of very high strength, good fatigue resistance, simple hardening treatments and ability to maintain close dimensional tolerance.

Summary

Significant cost reductions have already been demonstrated, and further substantial reductions are being pursued, through design, process and tooling optimization, to reduce weight and improve material buy-to-fly ratios. This is required to counteract the increased cost of advanced materials and to ensure that the impact on the direct operating costs of the product is positive. The development of new materials and the associated manufacturing processes is crucial to ensure that the European aerospace industry maintains its competitive position on the world stage by meeting its customer requirements. The overall pattern of developments is shown schematically in figure 1.10. However, new developments can only be acceptable if they bring a win/win result for the manufacturer and operator, and this will require close partnership between the original equipment manufacturer and the material suppliers to ensure optimum solutions are reached for all aerospace platform applications.

Opportunities exist for advanced metallics such as aluminium lithium alloys, and polymer composites in both the thermoset and thermoplastic forms. To cash in the benefits that these new materials promise, innovative manufacturing processes for forming, joining and treating the sub-components of the platform must be pursued with vigour. Some of these

Figure 1.10. Integration of new aircraft design, materials and manufacturing processes.

new processes have been highlighted, such as friction stir welding and advanced composite concepts, which promise to improve significantly both the weight and cost of the product, whilst allowing new aerodynamic and structural concepts to be realized.

Aggressive improvement programmes have been launched by all manufacturers to achieve enhanced levels of operating efficiency. The ability to apply increasingly advanced technologies is key to the future success of the aerospace industry in meeting the customer demands in terms of price performance and reliability, at the same time as ensuring that the investor sees an enhanced return on capital. The potential market growth for aerospace platforms is enormous, and the opportunities for both the original equipment manufacturer and the material supplier are achievable through partnership, innovation and technology to improve the performance of the product to the satisfaction of the customer. There are exciting prospects in developing both metallic and composite primary structural applications; both will undoubtedly find roles for many years to come.

Chapter 2

Advanced materials and process technologies for aerospace structures

Tsugio Imamura

Introduction

Considerable effort is being directed to reducing the weight of aerospace structures and their associated manufacturing costs. Derivatives from current aluminium alloys are being developed to provide materials for low fatigue crack propagation rate and damage tolerance design. For titanium alloys the target is a better balance between strength and fracture toughness and higher fatigue resistance. Process technology is being improved to produce one-piece complex configuration parts, to reduce assembly costs and to reduce weight through fewer parts. For future structures, such as in supersonic transport, further developments in forming technology for lightweight heat resistant materials will be needed.

Materials

Aluminium alloys

Generally the fracture toughness of 2024 is improved by decreasing the volume fraction of constituent particles (Cu_2FeAl_{17}, Mg_2Si etc.), and fatigue crack growth rate at high applied stress intensity ΔK is reduced with improved fracture toughness [1–3]. Investigation shows that: (a) fracture toughness is proportional to the square root of the spacing of the constituent particles; (b) fatigue crack growth rates at low and medium ΔK depend on the size of dispersoids ($Cu_2Mn_3Al_{20}$); and (c) coarse dispersoids reduce the fatigue crack growth rate. Both fracture toughness and fatigue crack growth rate are thus improved by careful control of constituent and dispersoid particles.

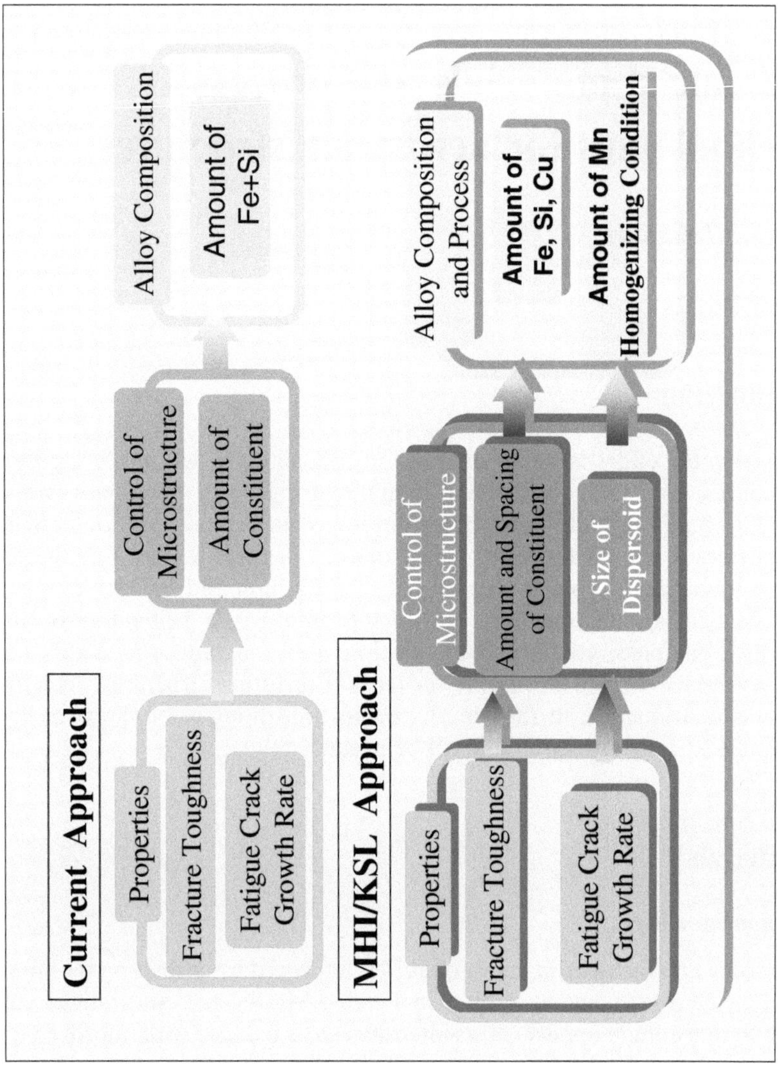

Figure 2.1. New approach to improving fracture toughness and fatigue crack growth rate.

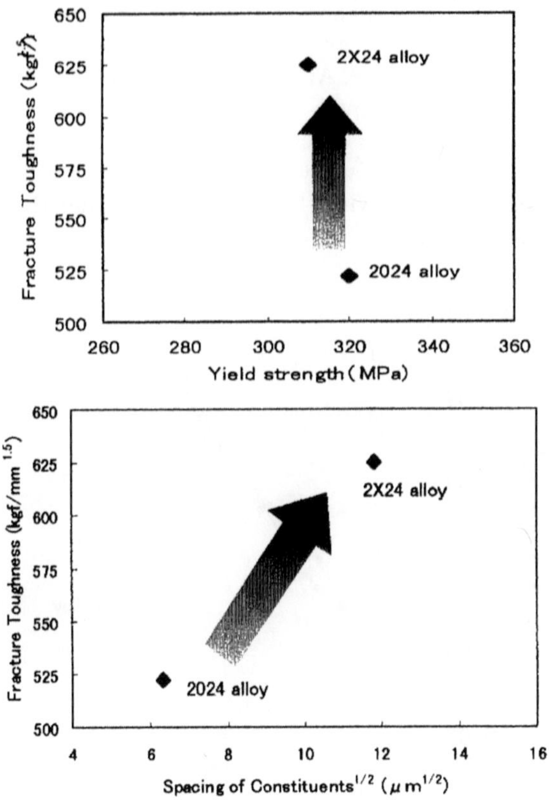

Figure 2.2. Fracture toughness of 2X24 improved by 20% compared with 2024.

This is shown in figure 2.1, giving details of an alloy/process combination developed by Mitsubishi Heavy Industries (MHI) and Kobe Steel.

The fracture toughness and fatigue crack growth rate in the newly designed alloy 2X24 have been measured as shown in figure 2.2 and figure 2.3. 2X24

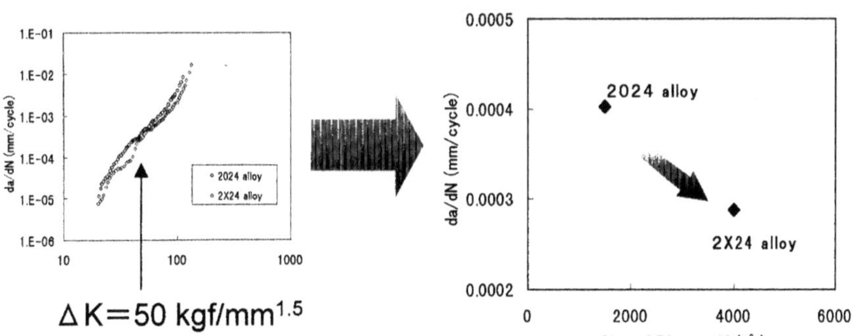

Figure 2.3. Fatigue crack growth rate of 2X24 improved by 50% compared with 2024.

shows approximately 20% higher fracture toughness than 2024 without loss of tensile yield strength, because of an increased spacing of constituent particles in the microstructure. At the same time 2X24 shows approximately 50% lower fatigue crack growth rate over the whole range of ΔK. From scanning electron microscope observations of the fracture surfaces of fatigue test specimens, narrow and wavy striation patterns are observed. Coarse dispersoids are associated with the waviness, showing that they obstruct the crack propagation.

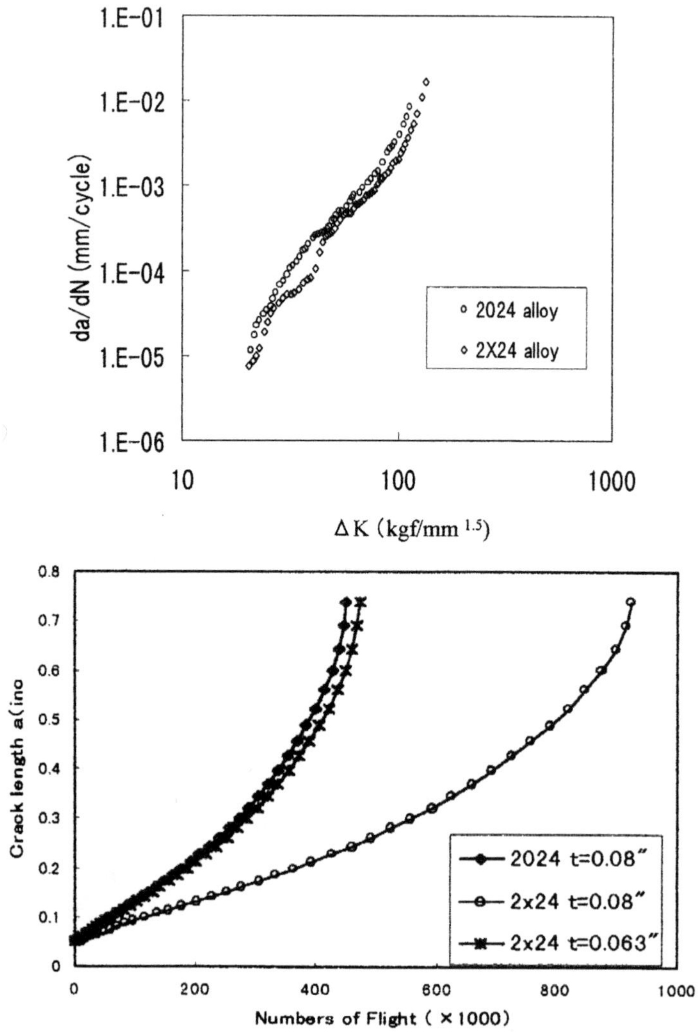

Figure 2.4. Crack growth simulation, showing the possibility of reducing the skin gauge by 21% compared with 2024.

A crack growth simulation and estimation of the margin of safety has been performed on the application of 2X24 alloy for the centre fuselage crown panel of a Global Express business jet, using a modified generalized Willenborg model. The crack growth simulation shows that 2X24 has twice the fatigue life of 2024 in the same thickness gauge, and a 21% thickness reduction can be achieved with the same fatigue life, as shown in figure 2.4.

Titanium alloys

The application of titanium alloys for aerospace is increasing because of good combinations of strength, toughness, corrosion resistance and compatibility with polymer composite materials. For the Boeing 777 aircraft rolled out in 1995, a variety of titanium alloys are used to make up 9% of the total craft weight. For example a large $\alpha + \beta$ Ti-6Al-4V casting is used for the APU duct panel, and a β Ti-10V-2Fe-3Al large forging is used for the landing gear truck beam. Figure 2.5 shows the material distribution of the newest fighter F-22 developed by the US Air Force [4], and the titanium alloy usage is over 30% of the structural weight. The major titanium alloys on the F-22 are Ti-6Al-4V and Ti-6Al-2Sn-2Zr-2Mo-2Cr-Si.

Titanium alloys are also widely used for aerospace structures. The most used alloy is $\alpha + \beta$ Ti-6Al-4V, which has a wide variation of mechanical properties under different heat treatment conditions, as shown in figure 2.6. The alloy has moderate tensile strength and moderate fracture toughness when annealed in the $\alpha + \beta$ region to produce a uniform equi-axed α phase

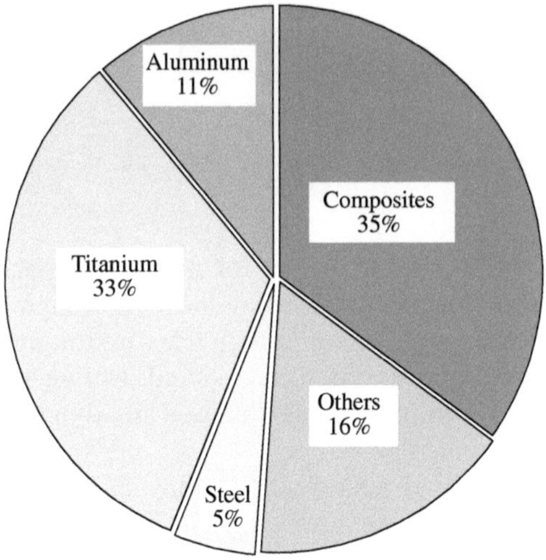

Figure 2.5. Material distribution on the F-22 fighter.

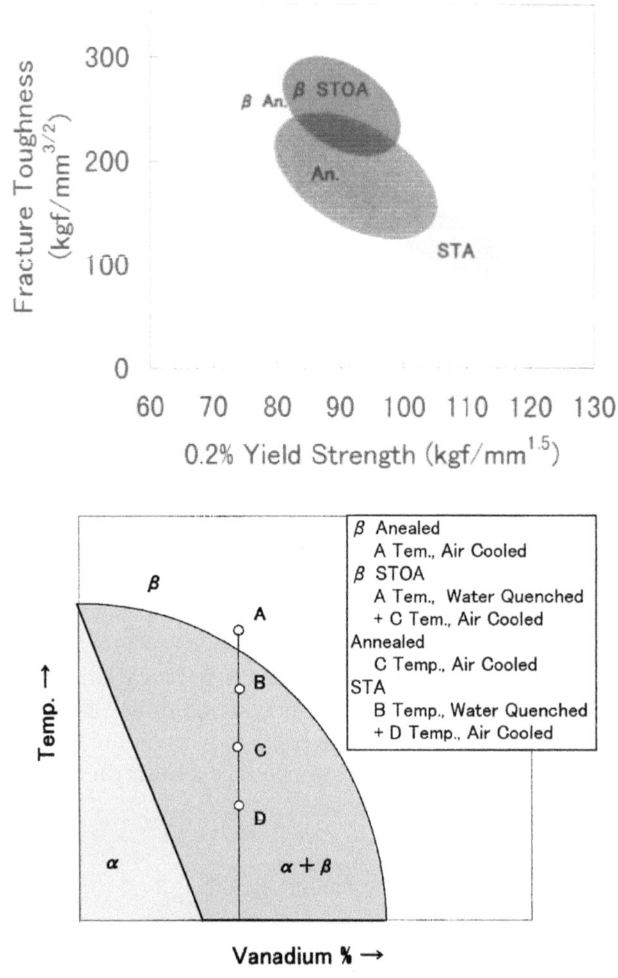

Figure 2.6. Mechanical properties of $\alpha + \beta$ Ti-6Al-4V heat treated under different conditions.

microstructure, and is used generally for many aerospace components. At higher temperatures the β-annealed alloy shows very high fracture toughness but very low strength, and is used in high fracture toughness required parts such as stop fitting. The β solution treated and over aged alloy has superior strength and fracture toughness, and is used for dynamic helicopter parts such as rotors.

Recently an improved heat treatable alloy Ti-10V-2Fe-3Al has become more popular for aerospace applications. The alloy was developed by Timet as a high strength, high fracture toughness alloy. The alloy properties of strength and fracture toughness were re-designed by the airframe

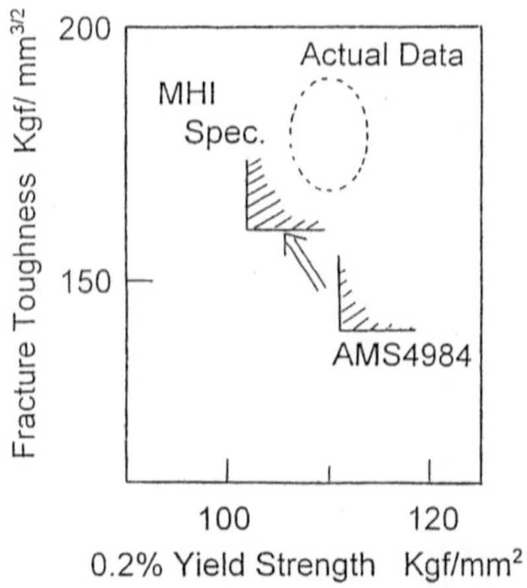

Figure 2.7. Application of Ti-10V-2Fe-3Al alloy for the F-2 fighter.

manufacturer, MHI. At first, Fe segregation was found in the alloy, causing a low β transus temperature locally and resulting in β flecks, i.e. coarse β grain structure areas with low elongation. It was necessary to remove the Fe segregation and ensure uniformity, selecting a higher solution temperature to remove β flecks and ensure high fracture toughness. Figure 2.7 provides an example of a near-net-shape forging using a re-designed Ti-10V-2Fe-3Al alloy for the F-2 fighter.

The new β rich $\alpha + \beta$ alloy SP700 (Ti-4.5Al-3V-2Fe-2Mo) was developed by NKK, and was designed to be more β stable than Ti-6Al-4V, by addition of molybdenum and iron. The SP700 alloy has a fine grain structure and good superplastic formability at low temperature, as shown in figure 2.8 [5]. MHI is developing the application of superplastic formed parts by using SP700 for H-2A rocket components.

Process technology

General

Many unique manufacturing methods have been used to produce aerospace components. Techniques have been developed to optimize difficult-to-work materials and complicated component configurations, resulting from a pursuit of ultimate lightweight structures. On the other hand, manufacturing cost savings are a universal requirement even when there is a limited amount of parts production. Die-less forming has been used extensively to reduce manufacturing costs. Peen forming of complex curvature wing panels, roll forming of stringers and/or frames, and chip forming of the cylindrical

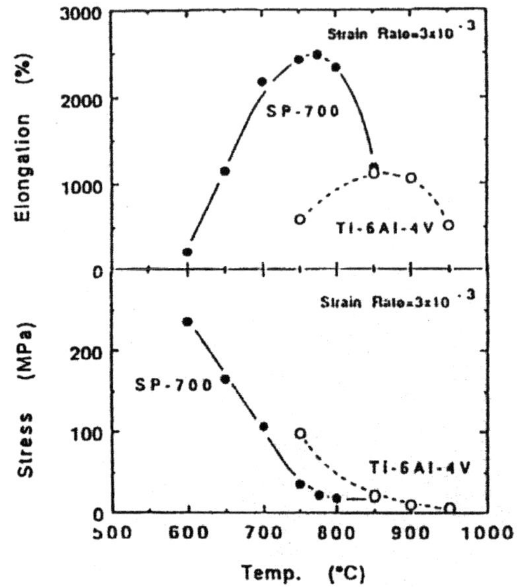

Figure 2.8. Superplastic elongation and flow stress of SP700.

skins of rocket tanks and/or skin panels of airliners are typical examples of representative die-less forming used in the aerospace industries. Aerospace applications also require integrated components, large-size structures and panel thickness control to achieve ultimate weight reduction. Superplastic forming and roll forming are typical examples of technologies combining die-less forming with integrated manufacture and thickness control to achieve substantial cost saving and weight reduction.

Superplastic forming

Superplastic forming was developed as a technology to form integrated single piece structures, which could replace assembled structures with intricate detail and fasteners, by making the most of the exceptional formability of superplastic materials. Superplastic forming can reduce the numbers of parts and fasteners, which leads to considerable cost saving and weight reduction. Figure 2.9 [6] shows a superplastic formed inner skin made of 7475 aluminium alloy, which is to be put together with an outer skin of the same material to form an access-door panel. The conventional door panel consists of 15 to 25 detailed parts assembled with many fasteners. In this particular case, more than 20% weight reduction and 40% cost saving was achieved by using superplastic forming compared with the conventional manufacturing methods. Superplastic formed parts can have as close a tolerance as machined parts, allowing the manufacture of elaborate components like fuel tanks, which require high accuracy and thin wall thickness. Figure 2.10 [7] shows a teardrop

Figure 2.9. Superplastic formed inner skin of an access-door panel.

shaped fuel tank for a satellite fabricated by superplastic forming and electron beam welding. The fuel tank is made of Ti-6Al-4V and has the optimum thickness distribution leading to exceptional weight reduction. The spherical area has a constant thickness of 0.75 mm and the cone area has a thickness distribution varying from 0.93 to 0.56 mm corresponding to the curvature. Since superplastic forming is necessarily accompanied by non-uniform thin out,

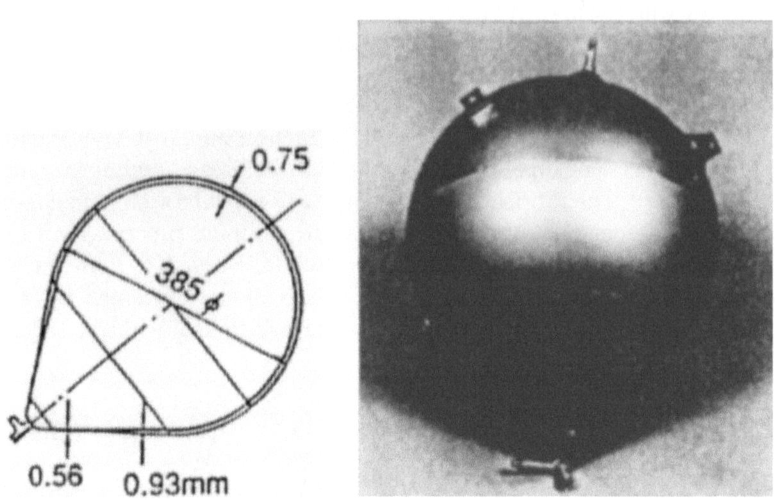

Figure 2.10. Teardrop shaped fuel tank for a satellite with the optimum thickness distribution.

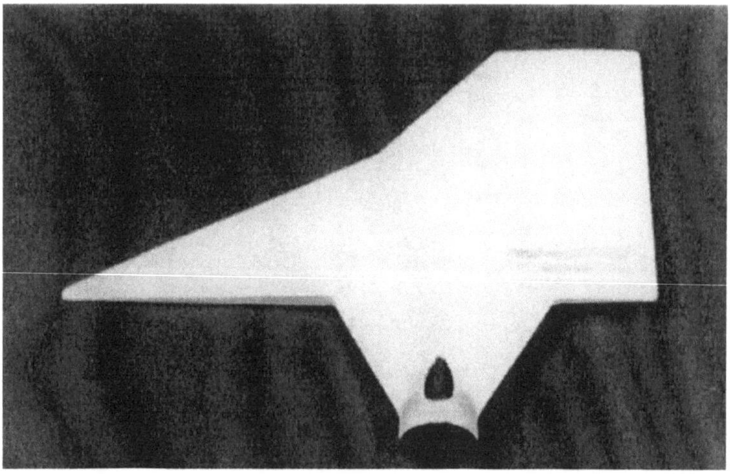

Figure 2.11. Missile fin made of a SiC whisker-reinforced 7075 composite which replaced the conventional fin with weight reduction of more than 50%.

the stock sheets were machined to provide suitable thickness distribution after forming.

Superplastic behaviour has been of great importance as a forming technology for difficult-to-work materials such as metal matrix composites and intermetallic compounds. Superplastic forming has enabled advanced but less workable materials to be plastically formed and therefore become cost competitive, facilitating the practical use of high performance materials and improving the performance of many aerospace components. Figure 2.11 [6] shows a missile fin fabricated by superplastic forming of SiC whisker-reinforced 7075 composites, which replaced the conventional fin with a weight reduction exceeding 50%. The next generation of aerospace components will require further weight reduction and cost saving, and the combined process of superplastic forming and diffusion bonding is a promising technology. The fabrication of various kinds of sandwich panel has been under development, focusing on the way to combine superplastic forming with diffusion bonding and the edge structure needed to join panels to each other. Figure 2.12 shows a hollow fan blade fabricated by 4-sheet superplastic forming/diffusion bonding, and a flat panel with the same core structure.

Roll forming

Stringer and frames, major components of aircraft structure, are manufactured by machining from extrusions or by roll forming from sheet. Figure 2.13 [8] shows a roll-formed stringer having an optimum thickness

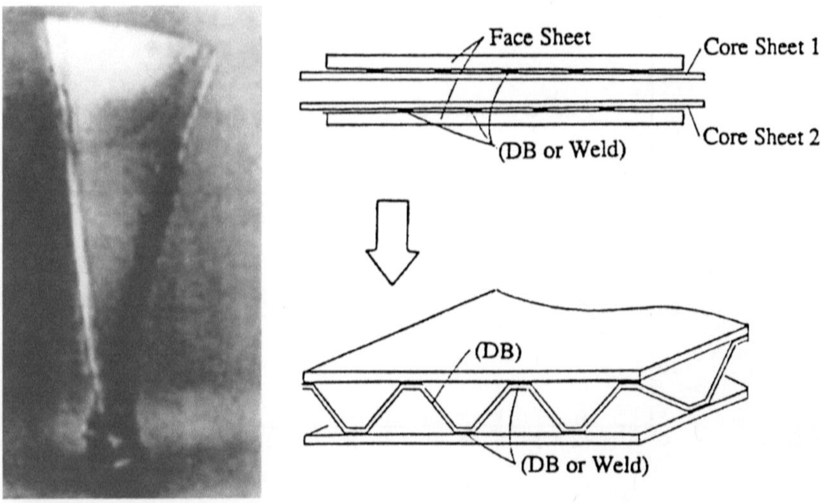

Figure 2.12. Hollow fan blade fabricated by 4-sheet superplastic forming/diffusion bonding and core structure.

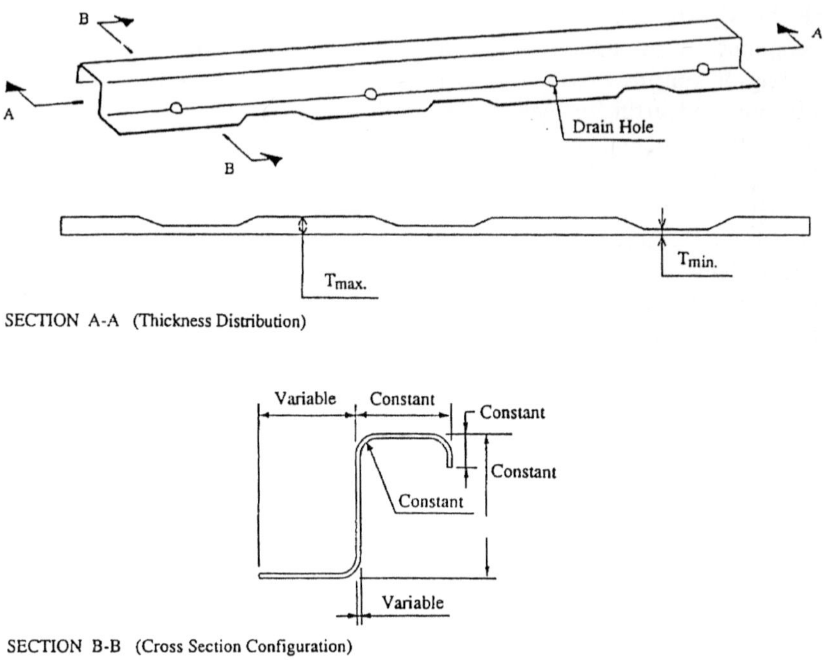

Figure 2.13. Taper-rolled stringer with controlled thickness corresponding to stress distribution.

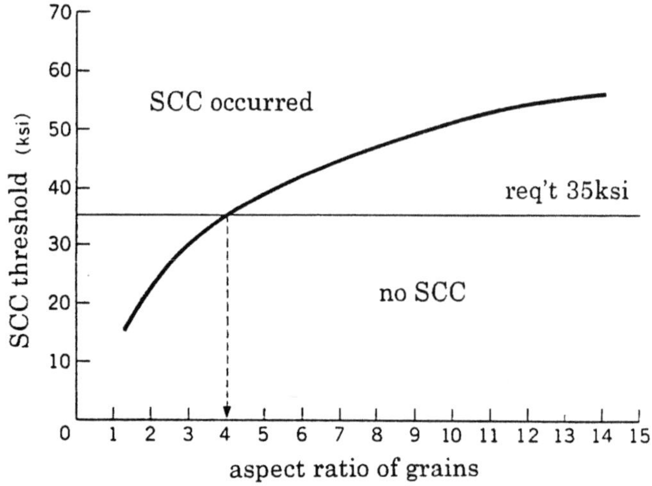

Figure 2.14. Dependence of the threshold stress for stress corrosion cracking on the aspect ratio of grains.

distribution corresponding to variations of stress along the longitudinal direction, which offers 20–30% weight reduction compared with a constant thickness stringer. The optimum thickness distribution is also achieved economically by taper rolling. The overall manufacturing process is taper rolling of annealed strip, solution heat treatment, ten steps of section roll forming and artificial ageing. The roll-formed stringers require a special microstructure, fine enough to endure successive bending operations with a bend radius as small as 1.3 times the thickness. On the other hand, the stringers have to be resistant to stress corrosion cracking, which requires a rather coarse microstructure in directions perpendicular to the applied stress. Thus the stringer strips need a grain structure with a large aspect ratio, fine in the transverse direction to maintain formability, and coarse in the transverse direction to provide good stress corrosion cracking resistance. Figure 2.14 [8] shows the dependence of threshold stress for stress corrosion cracking as a function of the aspect ratio of the grains, together with the flow stress during stringer manufacture. In order to maintain the large aspect ratio grain structure in strip rolled to various degrees at different positions, it was necessary to adopt special recovery heat treatments between the taper rolling and solution heat treatment steps.

Summary

Recently, new metallic materials technologies have been developed to enhance lightweight structure and reduce costs in aerospace applications.

Strength, fracture toughness and corrosion resistance must be optimized at the same time as reducing manufacturing costs, through the development of near-net-shape forming, and by using integrated components with reduced part numbers, for example by superplastic forming. These development activities show steady progress and achievement.

References

[1] Speidel M O 1975 *Sixth International Light Metals Conference* Lepven/Vienna p 67

[2] Rice J R and Johnson M A 1970 *Inelastic Behavior of Solids* p 641

[3] Staley J T *et al* 1970 *Inelastic Behavior of Solids* p 641

[4] JMIA 1992 Feb.

[5] Ogawa A, Fukai H, Minakawa and Ouchi C *Beta Titanium Alloys in the 1990s* p 513

[6] Tsuzuku T, Takahashi A and Sakamoto A 1991 *Superplasticity in Advanced Materials* ed S Dori, M Tokizame and N Furushiro (The Japan Society for Research on Superplasticity) p 611

[7] Takahashi A , Shimizu S and Tsuzuku T 1999 *J. Japan. Society Res. Superplasticity* **31**(356) 1128

[8] Hirota K, Ibaragi M *et al* 1996–5 *Mitsubishi Heavy Industries Technical Review* **33**(3)

Chapter 3

Materials for supersonic civil transport aircraft

Yann Barbaux and Jacques Cinquin

Introduction

The consequences of the drastic economical and technical requirements for future supersonic civil transport [1] on the materials selection for the different parts of the aircraft structure have been detailed and discussed elsewhere [2]. As a result, Aerospatiale, BAe and DASA decided to increase their effort on materials studies and to launch specific research programmes on aluminium alloys and carbon fibre reinforced polymers (CFRPs). Major research programmes were initiated on aluminium alloys in 1992 [3] and on organic matrix composites in 1994 [4].

Aluminium alloys

The work programme of recent research was divided into two main tasks, corresponding to the study of the two factors assumed to influence directly the creep resistance and the thermal stability of metals:

- Task 1: selection of the main precipitation system
- Task 2: optimization of the chemical composition and of the process parameters.

The critical analysis of existing data resulted in the selection of 33 chemical compositions, from the four alloy systems given in table 3.1. These alloys were direct chill (DC) cast and rolled down to 14 mm thick plates and 1.6 mm thick sheets on laboratory equipment at DERA, British Aluminium and Pechiney. They were then tested for creep, thermal stability and corrosion. Based on the results obtained on these alloys, a selection of 14 different

Table 3.1. Selected alloy systems.

Alloy system	Main precipitation system
Al-Cu (2001 type)	θ' (Al_2Cu)
Al-Cu-Mg (2024 or 2618 type)	S' (Al_2CuMg)
Al-Mg-Si-Cu (6013 type)	λ' ($Al_5Cu_2Mg_8Si_7$)
	β' (Mg_2Si)
Al-Li-Cu-Mg	δ' (Al_3Li)
	T1 (Al_2CuLi)

compositions from the S' and $\alpha' + \beta'$ precipitation systems was made on which the effect of minor alloying element and thermo-mechanical process variations was studied.

Results obtained in this project [5] were very satisfactory: all the alloys tested presented a creep behaviour and a fracture toughness much improved as compared with CM003 alloy (enhanced 2618), which was, at that time, the best reference in terms of creep/damage tolerance compromise. This is illustrated by figure 3.1 and table 3.2, which present respectively creep results in accelerated conditions and fracture toughnesses on compact tension specimens of three of the alloys (labelled A1, C1 and D6), in comparison with creep results from CM003 and fracture toughness results from 2024. Creep life times were extended by a factor of up to seven under different

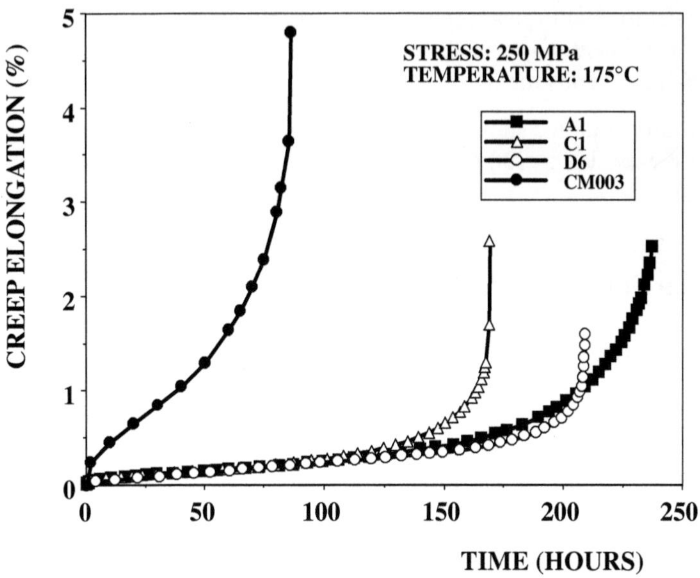

Figure 3.1. Creep curves at 175°C/250 MPa.

Table 3.2. Fracture toughness from R-curves on 400 mm wide compact tension specimens.

Alloy	Kc (MPa/m^{-2})	Kc0 (MPa/m^{-2})
A1 T6 (mod. 2650)	125	90
C1 T6 (mod. 6056)	160	110
D6 T6 (Al-Cu-Mg-Ag)	158	110
2024 T3	130	90

creep test conditions (including 150°C/250 MPa) compared with CM003, and fracture toughness values from R-curves were equal to or better than the damage tolerant 2024 T3 reference. Industrial sheets and plates from the two most promising compositions, a modified version of 2650 Al-2%Cu-Mg alloy and an optimized version of 6056 alloy, entered an exhaustive evaluation programme in 1997, and the results confirm the improvement in creep behaviour over CM003, although the benefit is reduced compared with the laboratory tests.

In parallel with the development of improved alloys, Aerospatiale has also started studies on the interactions between creep and fatigue on notched coupon specimens and on specimens representative of technological details such as pocket recess or assemblies. These studies are based on the development of two parallel methods:

- a modelling approach combining thermo-elasto-plastic finite elements and physical/metallurgical prediction of creep damage
- an experimental approach with the development of specific test equipment capable of reproducing close to real exposures on technological specimens.

The results obtained show a slight detrimental effect of 5000 and 10 000 hours of creep exposure at 130°C on the fatigue behaviour of notched specimens in 2650 alloy.

Carbon fibre reinforced polymers

It has already been published [2] that composite materials with carbon fibres and polymeric matrices are candidates to achieve the required weight savings on future supersonic civil transport. The main requirements are acceptable properties regarding subsonic flight specifications (i.e. damage tolerance), and thermal stability in supersonic flight conditions. Different types of matrices are under investigation for Mach 2.05 applications with IM or HR fibres, including second generation epoxy, cyanate-based, thermoplastic and bismaleimides, as shown in table 3.3. The work programme was divided into two main research areas:

Table 3.3. Candidate polymer matrices for carbon fibre reinforced polymers.

	State of knowledge
Second generation epoxy	Well-known process Expected service temperature: 120°C
Cyanate based systems	Process similar to epoxy system New products on the market Expected service temperature: 150°C
Thermo-plastic	Potential hot forming process Expected service temperature: 180°C
Bismaleimides	Processing generally with post-curing Expected service temperature: 180°C Low damage tolerance compared with 2nd generation epoxy

- the influence of long-term thermal ageing on carbon fibre reinforced polymer physical and mechanical properties
- the long-term behaviour of carbon fibre reinforced polymers under complex thermo-mechanical loading.

The influence of thermal ageing has been studied by isothermal ageing up to 4000 hours at different temperatures from 100°C to 180°C, and by thermal cycling over the ranges −50°C to 120°C or 180°C for up to 1000 cycles. The cumulative time at the maximum temperature for 1000 thermal cycles is equivalent to 4000 hours under isothermal conditions. Different properties have been investigated after these thermal ageing exposures, such as filled hole compression, compression after impact, glass transition temperature and microstructure. The influence of thermal ageing on the mechanical properties can be related to the degree of curing of the matrix, and also to the chemical type of the matrix. Figures 3.2 and 3.3 clearly show a post-cure effect on the second generation epoxy, not fully transformed during the initial curing. For the cyanate system, there is a real mechanical property degradation when isothermal ageing is performed above 160°C. Figures 3.3 and 3.4 show that the mechanical property degradation, or the post-cure effect, appears during the first 1000 hours of ageing. These first tests were performed at a higher temperature than the service temperature, corresponding to Mach 2.05, in order to obtain in a short time the first indications of thermal ageing response for the different families of matrix.

Another important point is the influence of thermal ageing on the damage tolerance properties of the carbon fibre reinforced polymers. The results, shown in figure 3.5, have been obtained with compression after impact (CAI) tests performed on bismaleimide composites. Different parameters have been investigated such as the duration of ageing up to 4000

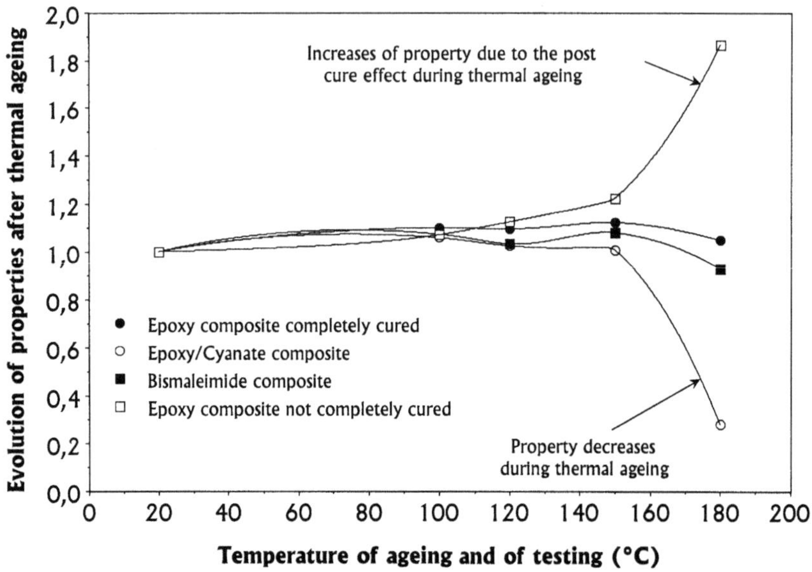

Figure 3.2. Filled hole compression after 2000 hours of isothermal ageing.

hours, the thermal cycling effect, the position of the impact (before or after thermal ageing), and the temperature of ageing, 120°C or 180°C. These results indicate that the position of the impact before or after the thermal ageing is an important parameter. If the maximum temperature is 120°C,

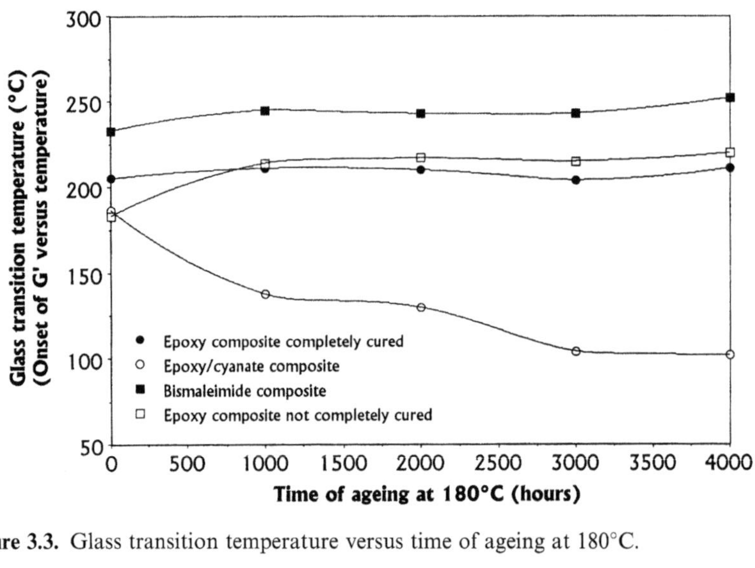

Figure 3.3. Glass transition temperature versus time of ageing at 180°C.

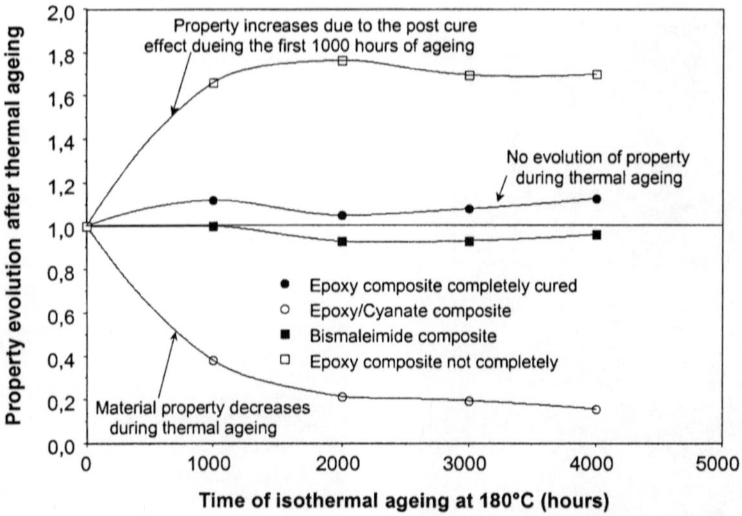

Figure 3.4. Filled hole compression at 180°C after isothermal ageing at 180°C.

no degradation of properties is observed up to 4000 hours of ageing or 1000 cycles. In these conditions, no oxidation or microcracks are observed in the composite materials. If the temperature of ageing is 180°C, under isothermal conditions we do observe oxidation on the exposed edges of the samples, as shown in figure 3.6(*a*). If the ageing is done under thermal cycling conditions

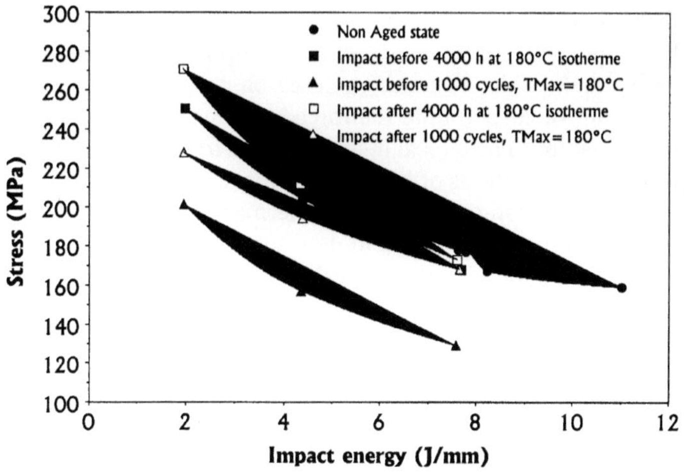

Figure 3.5. Compression after impact test performed after thermal ageing on bismaleimide composite.

(a) (b)

Figure 3.6. Micrographic observation of a bismaleimide composite. (a) 4000 hours at 180°C under isothermal conditions. (b) 1000 cycles with $T_{max} = 180°C$ and $T_{min} = -50°C$.

with $T_{max} = 180°C$ and $T_{min} = -50°C$, we do observe microcracks inside the composite material, as shown in figure 3.6(*b*).

Additionally to the study of the effect of thermal ageing, Aerospatiale developed specific creep test procedures on $[\pm 45°]_{4S}$ specimens to test the creep behaviour of various composite materials. Figure 3.7 presents the creep behaviour of the different candidate composites for supersonic aircraft.

Aerospatiale is also beginning a research programme to assess the long term behaviour of carbon fibre reinforced polymers under complex thermo-mechanical loading. The typical flight spectrum for the future supersonic civil transport in the hypothesis of Mach 2.05 is presented in figure 3.8. This flight spectrum induces strong creep–fatigue–thermal cycling interactions. In order to test the degrading effects of 'cold–hot' thermal cycling and low frequency fatigue cycling compared with classical creep testing, on the residual properties of the composites after exposure, three specific accelerated thermo-mechanical cycling conditions were defined, as shown in figure 3.9. Each cycling type corresponds to 10 000 hours at 120°C. Compared with typical flight conditions, the maximum temperature has been increased by 20°C and the maximum stress has been doubled for test acceleration.

A specific testing apparatus has been developed to perform the three cycling spectra (figure 3.9). These cycling spectra have been applied to three composite systems with the same fibre: one bismaleimide, one cyanate

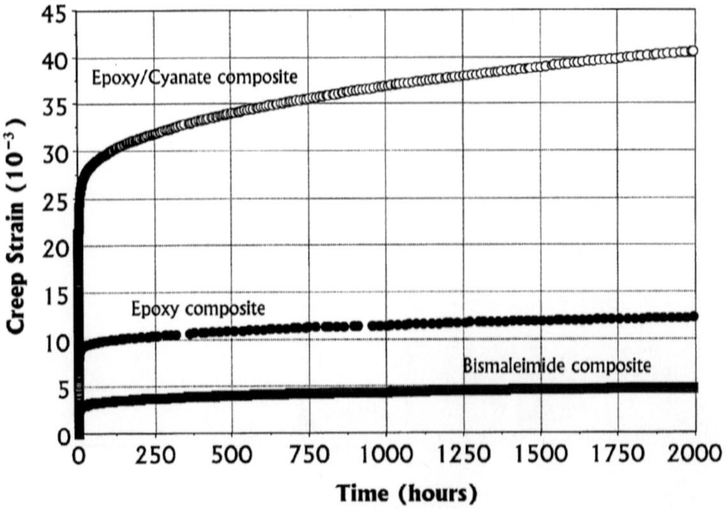

Figure 3.7. Creep testing on $[\pm 45°]_{4S}$ laminates at 120°C, 70 MPa.

and one epoxy. First results are available on the cyanate matrix composite. Table 3.4 presents residual properties after cycling exposure on quasi-isotropic open hole tension (OHT) and filled hole compression (FHC) specimens. These results tend to show that 'cold–hot' thermal cycling determines the composite compression properties. The duration of exposure is limited compared with what has to be justified (at least 60 000 hours). This means that long-term tests have to be carried out and special care has to be paid to the development of reliable models, able to predict long-term behaviour from short-term accelerated tests.

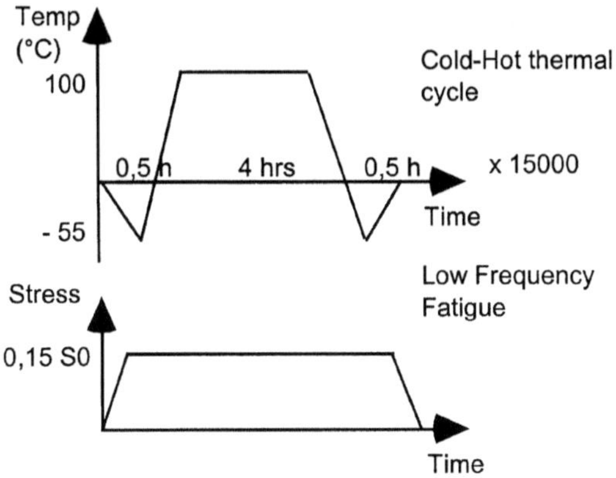

S0 : critical static strength at 100°C

Figure 3.8. Typical flight spectrum of supersonic aircraft.

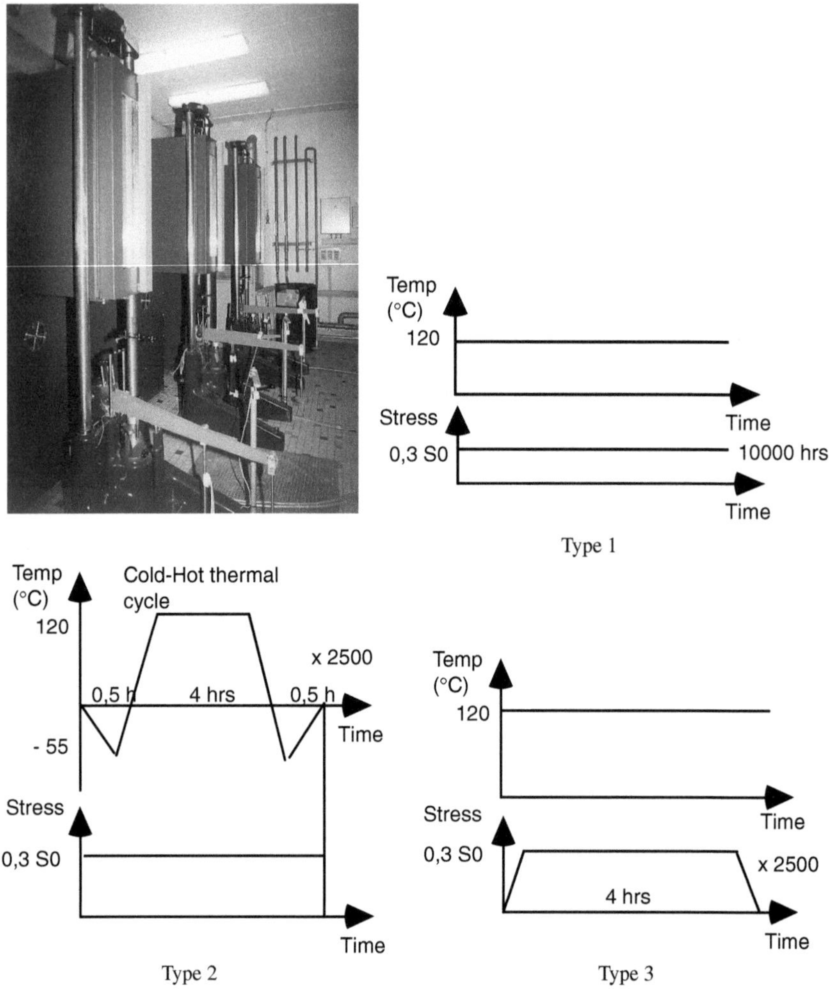

Figure 3.9. Creep facilities with thermal cycling chamber (−55°C + 200°C).

Table 3.4. Effect of cycling on the residual properties of cyanate matrix composites.

	Loss of stress (% of initial stress)	
	FHC	OHT
Type 1 (creep)	−3.9	+2
Type 2 (thermal cycle)	−10.4	+2.9
Type 3 (fatigue)	−5.7	+1.9

Summary

The pre-design studies conducted at Aerospatiale, BAe and DASA indicate that, because of the drastic economical and technical requirements defined for future supersonic civil transport aircraft, an important share of the structure of this aircraft will have to be made out of polymeric matrix composites and advanced lightweight aluminium alloys.

For the aluminium alloys, studies were oriented in two directions:

- the development of improved alloys
- the analysis of the behaviour of aluminium components in creep–fatigue interaction conditions.

Concerning the first topic, recent research has resulted in the development of new low density alloys, derived from 2650 and 6056 families, with much better creep resistance than the 2618A Concorde alloy, combined with a fracture toughness better than the reference 2024 T3 'subsonic' alloy. For the carbon fibre reinforced polymers, the test procedure for evaluating the damaging of these materials in real cycling conditions has been established, and pilot test equipment has been designed and built. Key points that will need to be studied in more detail have been identified, such as the influence of thermal ageing on the damage tolerance properties of the composite materials and the creep–fatigue–thermal ageing interactions.

References

[1] Swadling S J 1993 *J. de Physique IV* **3** 11–30

[2] Barbaux Y *et al* 1994 *Proceedings of EAC '94* Toulouse 25–27 October, pp 433–439

[3] Patri G and Frison G 1994 *Revue Aerospatiale* **108**

[4] Barbaux Y and Polmear I J 1995 *Proceedings of PICAST 2-AAC 6* Melbourne 20–23 March **2** pp 515–520

[5] Barbaux Y *et al* 1995 *Proceedings of EUROMAT '95* Padova, September

Chapter 4

Aluminium–lithium alloys in helicopter airframes

Alan Smith

Introduction

Lithium has always been of interest to research metallurgists as it is one of only eight elements whose solid solubilities in aluminium exceed 1 at% and can promote precipitation strengthening effects, thereby offering the possibility that useful engineering alloys could be developed. It was realized in the early 1900s that binary aluminium–lithium alloys were of little industrial use due to low strength, low ductility and high brittleness, but that lithium additions could enhance the properties of aluminium alloys containing other soluble elements such as copper and zinc. In particular, with lithium being the lightest metallic element, its incorporation at appropriate levels into aluminium based alloys could lead to useful density reductions. The first such alloy to be fully commercialized was the 2020 alloy registered in the late 1950s by Alcoa, of nominal composition Al-4.5Cu-1.1Li-0.5Mn-0.2Cd. Due primarily to its high compressive strength, increased modulus (arising from lithium additions in excess of 1 wt%) and enhanced elevated temperature strength and stability, with reduced density a secondary advantage, 2020 was widely incorporated in the wings and horizontal stabilizer of the RA-5C Vigilante Mach 2 attack and reconnaissance aircraft. A total of 117 Vigilantes was built from 1958 onwards incorporating 2020, giving over 25 years of active and operational service with the US Navy without major problems, but special design and manufacturing procedures had to be instigated to accommodate low toughness, low ductility and relatively high notch sensitivity. These deficiencies led to difficulties in meeting the increasing damage tolerance requirements of the mid to late 1960s and, consequently, there were no further applications for 2020 after the last Vigilante and spares were made.

Aluminium–lithium (Al-Li) activities waned for several years until the early 1970s when oil price increases prompted a renewal of interest from aircraft manufacturers and operators. However, this time, the potential for structural weight savings from reduced density was of paramount importance, and coincided with the aluminium producers' enthusiasm for new product developments to counter the serious inroads being made into their markets by the increasing popularity and use of lightweight organic based composite materials. A widely held belief in the 1970s was that many of the problems associated with 2020, and Al-Li alloys in general, were attributable primarily to coarse microconstituents arising from the ingot metallurgy route hitherto routinely used for billet manufacture. Many of the renewed Al-Li developments consequently employed rapid solidification powder metallurgy techniques, particularly in the USA, even though scaling up to the billet sizes needed for aircraft production would be extremely difficult and expensive. Meanwhile, particularly in the UK and France, advances in the fundamental understanding of Al-Li submicroscopic strengthening mechanisms led to the development of new ingot metallurgy alloys designated 8090 (UK and France), 8091 (UK), 2091 (France) and CP 276 (France), while alloy 2090 emerged from the USA. Today, of all the above powder and ingot metallurgy developments, only 8090 survives as a fully commercialized alloy with actual commercial applications.

Current commercial aluminium–lithium alloys

Aluminium–lithium alloy development

Aluminium–lithium alloy 8090 exhibits a density reduction and a stiffness increase, each of approximately 10% compared with other aluminium aerospace alloys. As such, 8090 may be described primarily as a reduced density alloy, as were alloys 8091, 2090, 2091 and CP 276. The inclusion of lithium in certain aluminium alloy compositions can also enhance a number of other properties and has led to the development and commercialization of alloys 2195, 2097 and 2197, in which density reductions are less than those of 8090 and are not the sole objective. Alloy compositions are given in table 4.1.

To date, the greatest use of aluminium lithium is that of alloy 2195 in the construction of the US Space Shuttle external fuel tank, where weldability, ultra high strength, enhanced cryogenic fracture toughness and reduced density are exploited to give an approximate 3600 kg weight saving compared with 2219 plate. The second widest use of Al-Li is in Europe, where 8090 is used extensively to reduce structural weight on the Anglo-Italian EH101 helicopter and this will be discussed further in this chapter. Alloy 8090 sheet is also fabricated in Japan by Mitsubishi Heavy Industries for a non-structural application on the Boeing 777 aircraft. Alloys 2097 and 2197 were developed

Table 4.1. Composition ranges of current commercially available Al-Li alloys.

	Chemical composition (weight per cent)										
Alloy	Si	Fe	Cu	Mn	Mg	Cr	Zn	Li	Ag	Zr	Ti
AA8090	0–0.20	0–0.30	**1.0–1.6**	0–0.10	**0.6–1.3**	0–0.10	0–0.25	**2.2–2.7**	0–0.05	**0.04–0.16**	**0–0.10**

Ingot manufacturer:[a] British Aluminium (UK)
Product forms available:[b] sheet, plate, extrusions, forgings
Nominal density: 2.55 g cm^{-3}
Use: reduced density 'medium strength' and 'damage tolerant' applications

	Chemical composition (weight per cent)										
Alloy	Si	Fe	Cu	Mn	Mg	Cr	Zn	Li	Ag	Zr	Ti
AA2195	0–0.12	0–0.15	**3.7–4.3**	0–0.25	**0.25–0.8**	0–0.05	0–0.25	**0.8–1.2**	**0.25–0.6**	**0.8–0.16**	**0–0.10**

Ingot manufacturer:[a] McCook Metals[c] (USA)
Product forms available:[b] plate
Nominal density: 2.71 g cm^{-3}
Use: applications requiring enhanced cryogenic properties, ultra high strength and weldable

	Chemical composition (weight per cent)										
Alloy	Si	Fe	Cu	Mn	Mg	Cr	Zn	Li	Ag	Zr	Ti
AA2097	0–0.12	0–0.15	**2.5–3.1**	**0.10–0.6**	0–0.35	0–0.05	0–0.35	**1.2–1.8**	0–0.05	**0.08–0.16**	**0–0.15**
AA2197	0–0.10	0–0.10	**2.5–3.1**	**0.10–0.5**	0–0.25	0–0.05	0–0.05	**1.3–1.7**	0–0.05	**0.08–0.15**	**0–0.12**

Ingot manufacturer:[a] AA2097, ALCOA (USA); AA2197, McCook Metals[c] (USA)
Product form available:[b] plate
Nominal density: 2.64 g cm^{-3}
Use: applications requiring enhanced fatigue resistance

N.B.: Heavy type denotes deliberate alloying additions.
[a] Fabrication into wrought products may be carried out by companies other than the ingot manufacturer. [b] All plate and sheet is unclad. [c] Formed mid 1998 from reorganization and partial sale of Reynolds Metals.

primarily to exploit the significantly slower fatigue crack growth rate which results from lithium additions to aluminium–copper alloys, and they are replacing 2124-T8511 plate for the bulkheads in the US Air Force F-16 high performance aircraft.

Use of 8090 on the EH101 helicopter

The EH101 is a large, three engined helicopter, designed, marketed and built jointly by Westland of England and Agusta of Italy. The primary role is for maritime operation in antisubmarine warfare, ship surveillance, and search and rescue roles. Additionally, civil and utility variants are available, capable of carrying up to 30 passengers with three crew or, when fitted with a rear loading ramp, providing troop/vehicle transportation. The EH101 initial design commenced in the late 1970s and, since this pre-dated full commercialization of the new generation of Al-Li alloys, was based upon conventional 2000, 6000 and 7000 series aluminium alloys. Although the prototypes were constructed primarily in these alloys, the decision was made to change extensively to 8090 on production aircraft in the late 1980s. Full EH101 production commenced in 1995 when the first order for 44 helicopters was received from the Royal Navy and it is on these and future EH101s that full use of alloy 8090 is and will be made. This accounts for over 90% of all aluminium alloys used in construction of the airframe and results in a structural weight saving of approximately 180 kg per aircraft. Table 4.2 summarizes the 8090 product forms, tempers and areas of application.

The optimized vibration characteristics of the original design were based upon the stiffness characteristics of conventional aluminium alloys. In order to maintain identical vibration levels in specific components made from 8090, the higher inherent stiffness has necessitated reduction in section size, giving a 2–3% structural weight saving in addition to that of 9–10% arising directly from the lower alloy density.

Advantages of 8090

In addition to the reduced density and increased elastic modulus, the use of alloy 8090 has the following advantages compared with conventional aluminium alloys:

1. Significantly lower fatigue crack growth rate, hence improved fatigue life.
2. Due to the formation of submicroscopic strengthening precipitates during natural ageing, conventional aerospace aluminium alloy sheet such as 2014A and 2024 requires re-solution heat treatment just prior to forming if adequate cold formability is to be achieved. In contrast, natural ageing effects are essentially absent in solution heat treated and stretched 8090 sheet, thereby facilitating cold forming in the T3 temper, with cost, production planning and logistic advantages.

Table 4.2. AA8090 products forms and application on the EH101 helicopter.

Product form	End use condition	Substitute for	Application
Sheet T81	Solution heat treated, quenched, controlled stretched and precipitation heat treated to an underaged condition	AA6068-T6 (BS L113) AA2024-T3 (BS L109) AA2024-T4 (BS L110)	– Forward fuselage lower structure, cabin side, outer skins – Rear fuselage skinning and stringers – Instrument panels, consoles, avionics cabinets – Gearbox fairing substructure – Sundry sheet metal parts, including superplastically formed sheet
Sheet T621	Re-solution heat treated by the user (for enhanced cold formability), quenched and precipitation heat treated to an underaged condition (same ageing parameters as for T81 temper)	AA6082-T6 (BS L113) AA2024-T4 (BS L110)	
Sheet T8	Solution heat treated, quenched, controlled stretched and precipitation heat treated to near peak aged condition	AA2014A-T6 (BS L157, L159, L165, L167)	
Sheet T6	Superplastically formed components	AA2024 (Assemblies) AA2007 (SPF parts)	
Extruded sections and tube T8511	Solution heat treated, quenched, controlled stretched and precipitation heat treated to a near peak aged condition	AA7075-T74511 (BS L160)	– Numerous standard profiles, e.g. T sections, C sections, F sections, cruciform sections – Hollow seat tracks and floor beams – Square cross-section tubes for instrument racking, conduits for cables and ladders – Over 40 different profiles used, mostly as 7000 mm lengths which are incorporated lengthwise in the main cabin
Die forgings T852	Solution heat treated, quenched, controlled stretched and precipitation heat treated to a near peak aged condition	AA7010-T7451 (plate to DTD 5130A)	– All structural frames in main cabin machines from AA8090 cold compressed die forgings. These include side frames, roof frames, intercostals and sides of main undercarriage box. A total of 16 different forging configurations are made from which 38 different components are machined per aircraft

3. Although technically possible, the superior corrosion resistance of 8090 makes cladding of sheet unnecessary.
4. Using most established techniques, the weldability of 8090 is comparable with that of conventional weldable 6000 series aluminium alloys, but with higher strength levels.
5. Attributable partly to the higher processing temperatures which 8090 can withstand, extrudability is significantly enhanced compared with conventional high strength alloys such as 7075. A number of complex one-piece extrusions such as hollow seat tracks and floor beams are used on the EH101, and weight savings are maximized by eliminating the need to bond adhesively several constituent sections which would have been the case with 7075.
6. Special processing of 8090 sheet can optimize superplastic forming capability, thereby offering the possibility of manufacture of complex shapes.
7. The electrical conductivity of alloy 8090 is approximately half that of conventional aluminium alloys. Although this should be considered when assessing lightning strike issues, it does provide a convenient means of differentiation if alloys become inadvertently mixed.

Disadvantages of alloy 8090

Whilst the positive aspects of alloy 8090 are such that extensive use is made on the EH101, there are nevertheless some disadvantages which should be noted:

1. Attainment of medium and high strength levels is critically dependent upon the application of post-solution heat treatment cold work. This is due to the need for co-precipitation of both δ'-Al_3Li and S'-Al_2CuMg intermediate phases during artificial ageing, and, whilst homogeneous nucleation of the former readily occurs, the copper levels present in 8090 allow only heterogeneous nucleation of S'-Al_2CuMg. Cold work-induced dislocation networks are suitable nucleation sites and may be introduced into sheet, plate and extrusions by post-solution heat treatment stretching, usually only by the metal manufacturer. Although not usually used in the as-supplied, fully heat treated condition, it does have the implication that if re-solution heat treatment is necessitated for the forming of complex parts in sheet or severe bending of extruded sections, then it will usually be impossible to apply subsequently the requisite cold stretching operation. Subsequent medium strength levels associated with the T8 sheet and T8511 extrusion tempers will therefore generally not be achievable. This drawback is partially countered by the minimal or absent natural ageing effects in 8090 such that all but the most complex of components can generally be made by cold forming sheet in the

stretched T3 condition. Minor manipulation such as joggling is also possible in relatively thin walled extruded sections in the T3510/T3511 tempers.

In the case of forgings, post-solution heat treatment cold work can be introduced by cold compression, but will only be effective if sufficient plastic deformation occurs in order to shear the metal relatively freely and without undue constraint from the adjacent regions in the components. These requirements place limits on the configuration and section size of die forgings which can be effectively cold compressed (triaxial compression techniques such as HIPping would not achieve the desired effect since metal shearing does not occur). Furthermore, since the property values are critically dependent upon the degree of cold work in 8090, it is essential that the cold compression levels are uniformly distributed throughout the forging. This is achieved to a high degree in the 8090 die forgings for the EH101 main cabin frame by the use of two sets of dies for each component, one set for the initial hot forging and a second set specifically for cold compression after solution heat treatment. Sufficient metal flow would not be achieved in forgings of a more bulky nature such as undercarriage cylinders and such components therefore remain in the heavier conventional alloy 7010-T74.

2. Metallic lithium is an inherently costly material and inevitably the price of the final Al-Li alloy is significantly higher than that of conventional aluminium alloys. However, although usually difficult to quantify, this can be partially offset against both the design and manufacturing advantages of the alloy. It should be remembered that material cost is frequently only a fraction of that of the completed component.
3. It is desirable to segregate Al-Li scrap from that of conventional aluminium alloys. Mixed aluminium alloy scrap is usually reprocessed by a secondary aluminium smelter to produce relatively pure aluminium ingots for use in the non-aerospace foundry industry. Most alloying additions and impurities are removed during melting and filtration, with the residual low levels generally having no discernible effect upon properties. However, even very low levels of residual lithium can lead to significant deterioration in the castability of the aluminium–silicon–magnesium alloys widely used for general engineering applications.

Future aluminium–lithium developments

Although the current commercially available Al-Li alloys appear to have been successful in overcoming many of the problems of their predecessors, the new Al-Li technology is still in its infancy. A major impediment to wider use of Al-Li alloys may be the change in emphasis within the aerospace industry from performance at any price to a drive to minimize the cost of manufacture, ownership and operation. Although the current uses of Al-Li alloys are niche applications, Airbus and BA are involved in extensive evaluation programmes to assess the suitability of these materials for future commercial fixed wing aircraft applications. These give rise to optimism that we are on the verge of significant expansion in Al-Li use and that it will become one of the most important aerospace materials of the early twenty-first century.

Building upon the experience gained through use of Al-Li on the EH101, some further topics which appear to be worthy of research and development for exploitation in future aerospace applications are as follows:

1. *Development of an Al-Li based shape casting alloy.* Due primarily to concerns over corrosion of magnesium alloys, all shape castings on the EH101 consist of the aluminium–silicon–magnesium alloy A356. Weight savings could be achieved if use could be made of reduced density Al-Li based castings which have received little research interest and are not available commercially. Whilst a great deal of both academic and industrial research has been directed at compositions for *wrought* products, there is a wide range of possibilities for casting alloy compositions awaiting investigation. As far as is known, the effects of lithium additions to aluminium–silicon alloys have hardly been researched and it is possible that new and complex intermetallic phases have yet to be discovered. In addition to alloy research, the actual manufacturing technology associated with aluminium shape castings would also need development. Particular problems to be resolved would be the propensity of molten Al-Li to attack aggressively all conventional sand/investment mould materials, as well as excessive hydrogen pick-up from the atmosphere resulting in gross porosity. The direct chill (DC) casting route for making ingots for fabrication to wrought products addresses these issues effectively, but a different system would be needed for pouring into sand/investment moulds.
2. *Development of new wrought Al-Li alloys which do not require post-solution treatment cold work.* As previously mentioned, the need for post-solution heat treatment cold work is problematic for die forgings of certain configurations, as well as for medium strength sheet to be formed into complex shapes. Even with cold work, 8090 cannot always match the high strength and high toughness of some of the more advanced conventional aluminium alloys. Development of a low density, high strength Al-Li alloy with improved toughness and not dependent upon cold working would significantly extend the application of Al-Li, with beneficial effects upon costs. Some progress has been made in the UK. Two experimental heat treatable alloys have been developed, the initial ingots of which are produced by DC casting. A third, non-heat treatable, dispersion strengthened alloy has emerged, ingots being made by mechanical alloying. All three of these alloys have yet to undergo extensive evaluation

and possibly further research and development efforts to bring them to full commercialization. However, mechanical alloying may hold the greatest promise for the future, with the potential to overcome many of the problems of microsegregation and limits to solid solubility experienced in DC cast compositions with high levels of alloying additions, especially lithium.

3. *Development of Al-Li rivets.* All rivets used on the EH101 are those commercially available in conventional aluminium alloys 2017 and 7075. There is scope for further weight savings if lightweight Al-Li rivets were commercially available. The superior fatigue characteristics of Al-Li alloys could be particularly advantageous, as the integrity of many assemblies is often primarily dependent upon the fatigue resistance of the riveted joint. Some work was carried out in this respect in the early days of 8090 development, but it was concluded that this particular alloy was not suitable for rivet manufacture, due to lack of cold formability. However, this may not necessarily now be the case due to the significant advances in metal cleanliness and compositional optimization which have occurred in the intervening years. Additionally and/or alternatively there is scope for alloy development to produce compositions specifically for use as rivets. Solid rivets would probably be the initial target, while blind rivets could present a greater challenge. The ingots for production of the rivet wire could be made by either DC casting or mechanical alloying, depending upon the alloy composition.
4. *Ways to make Al-Li alloys cheaper.* Using established DC casting, manufacture of Al-Li ingots currently entails use of the highest purity (and most expensive) grades of elemental lithium, as the standard sodium and potassium impurities in this metal are particularly deleterious to fracture toughness and ductility of the subsequent alloy. There may be possibilities to reduce overall costs by, for example, using less expensive and lower grade lithium at the outset, but exploiting the vacuum refining routes used by the steel and superalloy industries to remove the impurities from the resultant Al-Li alloy.

Summary

A new generation of lithium-containing aluminium alloys has achieved full commercial status and is being used successfully in a limited number of aerospace applications. There nevertheless exists a number of areas requiring further research and development which are suitable for close academic and industrial collaboration, exploitation of which could lead to wider use of this increasingly important class of materials.

Chapter 5

High performance polymers and advanced composites for space application

Rikio Yokota

Introduction

The AIAA reported on the state of space technology in SPACE 2000–2020. New and more advanced polymer composites having lighter, stronger, more dimensionally stable, and stiffer properties were predicted as being needed to develop space platforms to exploit the solar system [1]. Although international space station programmes have been delayed due to worldwide economic problems, there have been many technically notable successes, such as on board satellite repair technology, planetary exploration using space vehicles such as the US Mars pathfinder, deployment of large space structures, and the development of application satellites. During this period, graphite/epoxy carbon fibre reinforced polymer composite (CFRP), which is an advanced polymer composite, has been developed actively and used in spacecraft primary structures. It is still almost the only suitable material for these primary structures. High temperature matrix resins such as addition-type polyimides and high temperature thermoplastic resins are under development (see chapters 13 and 15), but are not yet familiar in space [2].

Aromatic polyimides are well known to have excellent thermal, mechanical, and electrical stability based on their hetero-aromatic structures, and various high performance polyimide films have been developed and nominated as suitable materials for the membranes of flexible large structures, such as extendable solar arrays [3]. This chapter discusses the technological development of high performance polymers and advanced polymeric composites for space applications in Japan, covering the following topics:

1. advanced composites for spacecraft primary structures;
2. advanced composites for the newly developed three-stage M-V rocket;

3. high performance polymeric materials and composites for flexible and/or rigid extendable structures; and
4. heat resistant matrix resins.

Advanced composites for spacecraft primary structures

In Japan, space activities that are conducted in various government ministries and agencies are coordinated by the Space Activities Commission. The Institute of Space and Astronautical Science (ISAS) is the central institute for scientific research in space, and the National Space Development Agency (NASDA) is in charge of the development of application satellites and their launchers. ISAS was founded in 1981 by reorganization of the Institute of Space and Astronautical Science, University of Tokyo. The present ISAS is a national research institute conducting inter-university research in cooperation with researchers from universities in Japan and other countries.

Table 5.1. Summary of ISAS scientific satellites

Date	Name	Weight (kg)	Application
1970.2	OHSUMI	24	test satellite
1971.9	SHINSEI	66	cosmic ray *et al*
1972.8	DENPA	75	plasma *et al*
1974.2	TANSEI-2	56	test satellite
1975.2	TAIYO	86	solar x-rays *et al*
1977.2	TANSEI-3	129	test satellite
1978.2	KYOKKO	126	auroral image *et al*
1978.9	JIKIKEN	90	plasma *et al*
1979.2	HAKUCHO	96	x-ray stars *et al*
1980.2	TANSEI-4	185	test satellite
1981.2	HINOTORI	188	solar flares *et al*
1982.2	TENMA	216	x-ray galaxies *et al*
1984.2	OHZORA	216	upper atmosphere
1985.1	SAKIGAKE	138	test spacecraft
1985.8	SUISEI	140	Halley's comet
1987.2	GINGA	420	x-ray sources *et al*
1989.2	AKEBONO	295	auroral and plasma
1990.1	HITEN	140	lunar swingby
1991.8	YOHKOH	420	solar flares
1992.7	GEOTAIL	1008	ISAS and NASA
1993.2	ASCA	420	x-ray sources *et al*
1995.3	S.F.U.	4000	space experiments
1997.2	HALCA	830	space VLBI antenna
1998.7	NOZOMI	540	Mars exploration

ISAS succeeded in launching the first Japanese satellite OHSUMI into orbit in 1970. Since then, 24 scientific and test satellites have been launched, including SUISEI and SAKIGAKE to explore Halley's comet in 1986. To improve performance and increase payload capability, launchers and satellites require the development of primary and secondary structures that are much lighter than those produced from conventional metals. Since the main structural configuration of SUISEI was made of carbon fibre reinforced epoxy, ISAS used advanced composites for the primary structures of all the satellites.

Table 5.1 shows the list of all the satellites and spacecraft launched by ISAS. ASCA, launched in 1993, is a structurally advanced spacecraft to investigate solar sources. Because the focal length of ASCA's ray mirror is so long, a high-precision Extendable Optical Bench (EOB) construction, made from carbon fibre reinforced polymer, was developed for the primary structure as shown in figure 5.1 [4]. All the tubes of the EOB are made of carbon fibre composite, using high modulus type fibres. The tubes are laminated so that the longitudinal thermal expansion coefficient of the

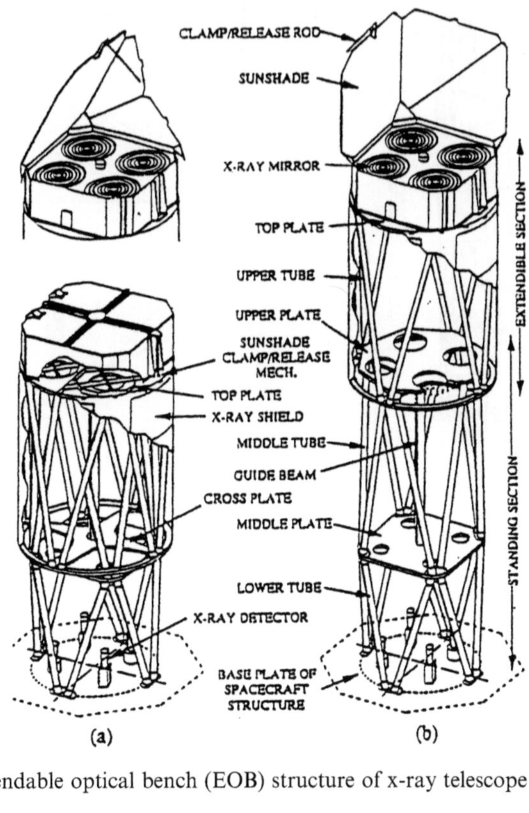

Figure 5.1. Extendable optical bench (EOB) structure of x-ray telescope satellite ASCA.

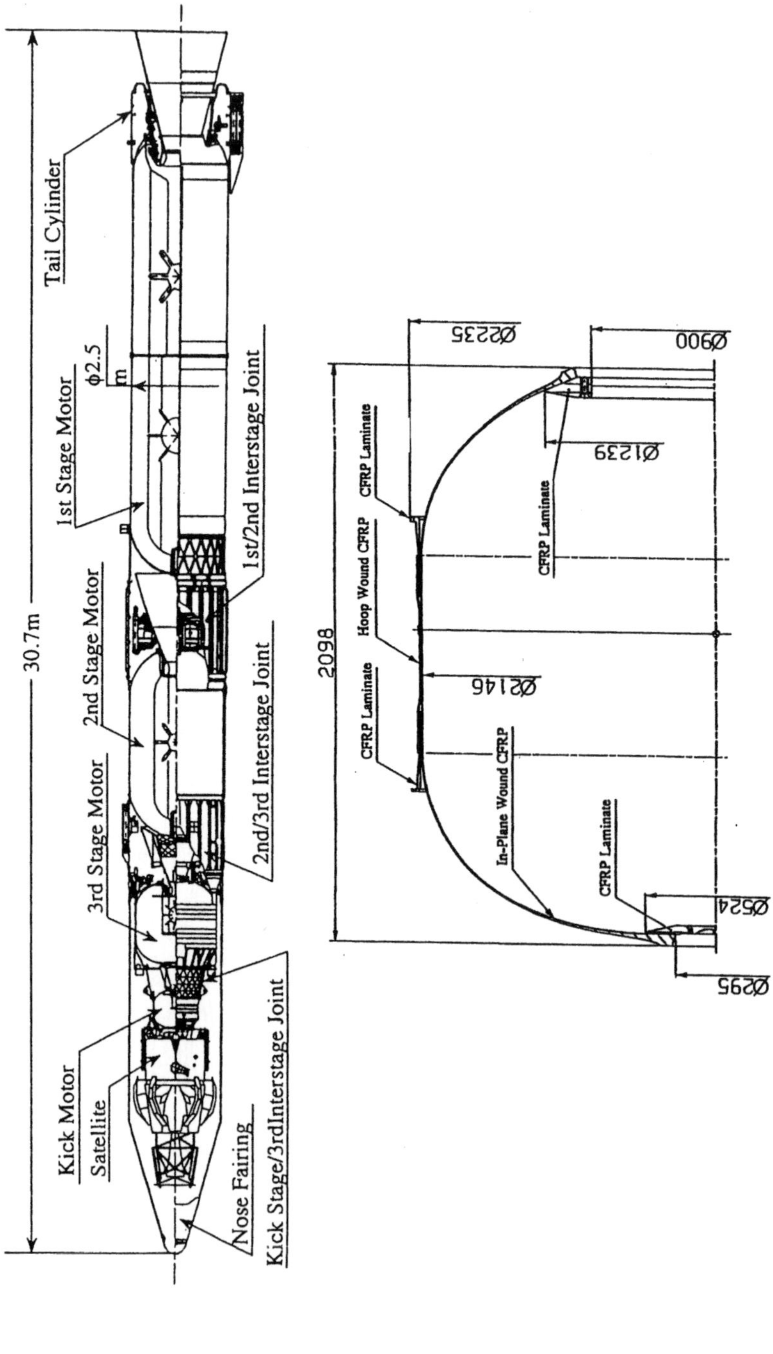

Figure 5.2. Outline of M-V rocket and M-34 filament wound carbon fibre reinforced polymer motor casing.

tubes can be made zero or even negative. The lamination configuration is designed such that the negative expansion of the tubes compensates the positive expansion of the metal top plates.

Advanced composites for the M-V rocket

M-V is the new generation satellite launcher of ISAS [5]. The first flight of M-V-1 was successfully completed in February 1997. M-V is a 30 m long, 2.5 m diameter, 130 ton, three-stage solid propellant rocket, with a 2 ton launch capability to low earth orbit (LEO). As shown in figure 5.2, the third stage M-34 rocket motor casing is made of filament wound carbon fibre reinforced epoxy. A filament wound motor casing was selected not only because of its potential high performance but also because of its cost advantage over a titanium alloy casing. To accommodate more than 10 tons of M-34 rocket motor solid propellant, the interior volume of the casing is $6.0\,m^3$. The maximum operating pressure is more than 6.0 MPa. The large nose faring is made of an aluminium honeycomb sandwich shell with carbon fibre composite face sheets. The M-34 nozzle introduced a deployment system of tapered double reverse helical spring extensors, made of glass fibre reinforced composites (GFRP).

High performance composites for flexible structures

A flexible solar array is an attractive example of applying a flexible and extendable advanced composite with aromatic polyimide film. Japan's spacecraft SFU retrieved by the Space Shuttle in January 1995, deployed two large flexible solar arrays in low earth orbit. Because of a high glass transition temperature T_g and outstanding mechanical properties even at very low temperatures, aromatic polyimide is the most successful, widely used polymeric material in space [6]. Until ten years ago, spacecraft were equipped with rigid power generators of the solar paddle type. As spacecraft became larger they required much more electric power, and flexible solar arrays are the most attractive way for power generation. Figure 5.3 illustrates the SFU spacecraft and the solar array configuration with extendable mast. The deployed wing is 2.4 m wide and 9.7 m long. The solar array is composed of two boards and the mast canister. The extendable/retractable mast is continually coilable, consisting of three glass fibre reinforced polymer spring rods (longerons) and radial spacers. The main source of the spring force is generated by the bending strain energy of the glass fibre reinforced polymer longerons. The radial spacers were made of moulded UPILEX-R. No mechanical backlash exists, because there are no pin-joint hinges, resulting in high dimensional stability. Each array blanket consists of 48 hinged

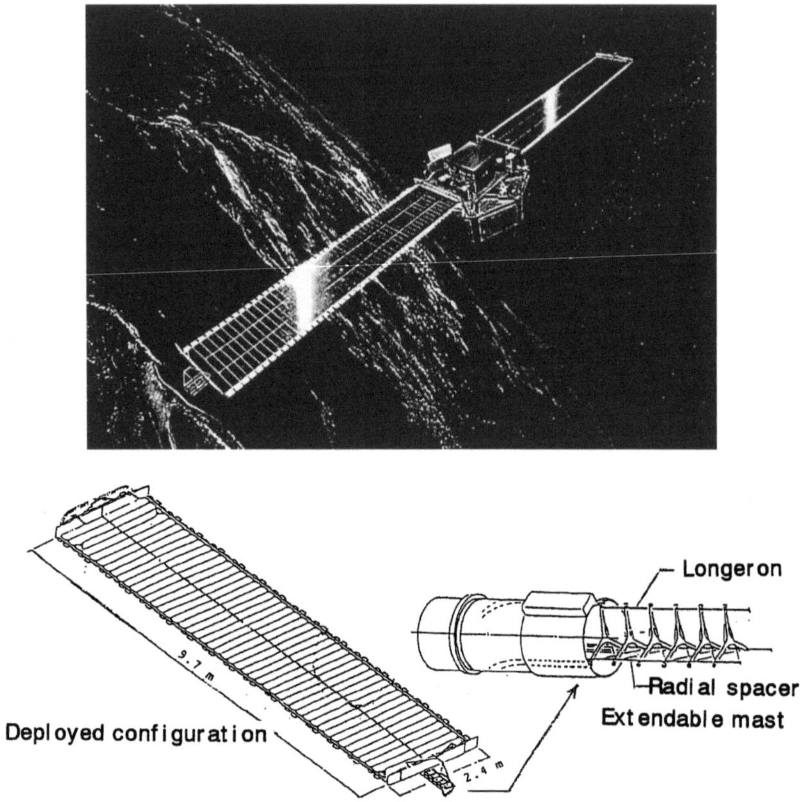

Figure 5.3. Illustrated SFU spacecraft and deployed configuration of its solar array with the extendable mast.

polyimide panels (films), 202 mm wide and 2400 mm long. About 27 000 solar cells are mounted on the two array blankets and generate 3.0 kW power. The silicon cells are 100 μm thick with 100 μm cover glass bonded by S-691-RTV silicon type adhesive on to the polyimide panels [3].

Development of a large deployable antenna in Muses-B is another advanced technology in space [7]. Figure 5.4 illustrates the Muses-B antenna launched in 1997 to be used aboard the satellite for Space-VLBI (Very Long Baseline Interferometry). A 10 m diameter parabolic antenna with mesh surface was successfully deployed with steps extending the six extendable masts in low earth orbit. Figure 5.5 shows a deployment test of the MUSES-B flight model antenna. This incredibly complicated system consists of 6000 fine cables of high modulus Kevlar 149 aramid covered by a CONEX aramid net. Because of the requirement for high surface accuracy, each cable must keep precisely its length without creep under tension in space. It is known that a high strength Kevlar

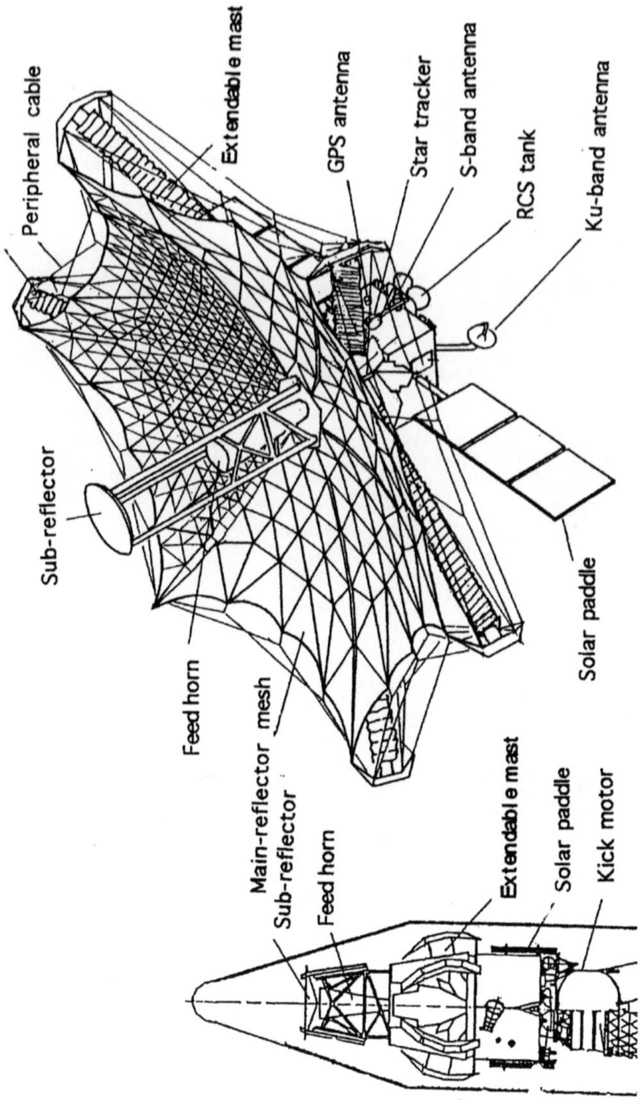

Figure 5.4. Configuration of MUSES-B antenna: (1) stowed in the nose faring, (2) deployed.

(*a*)

(*b*)

(*c*)

(*d*)

Figure 5.5. Sequential deployment test of MUSES-B flight antenna: (*a*) stowed; (*d*) after deployment.

149 cable exhibits very little elongation as stressed and has negative expansion over a wide range of temperatures. The tension of each cable and each extendable mast was controlled strictly, using tensioners on the top of each mast. This is the first application of high performance organic fibres for a large deployable parabolic antenna surface in space.

Heat resistant matrix resins

Aromatic polyimides are used widely in industry, because they possess high thermal stability, good mechanical properties, excellent electrical properties and excellent environmental stability [8]. Figure 5.6 shows the chemical structures of commercially available, space application, polyimide films. Metallized polyimide films, called flexible multi-layer thermal insulations (MLI), are now indispensable for passive thermal control systems of spacecraft as well as for flexible solar arrays. However, polyimides with these outstanding properties often give poor processability even in the addition-type oligoimide such as BMI and PMR-15. Because of their intermolecular ordered structure, aromatic polyimides have poor molecular mobility beyond the glass

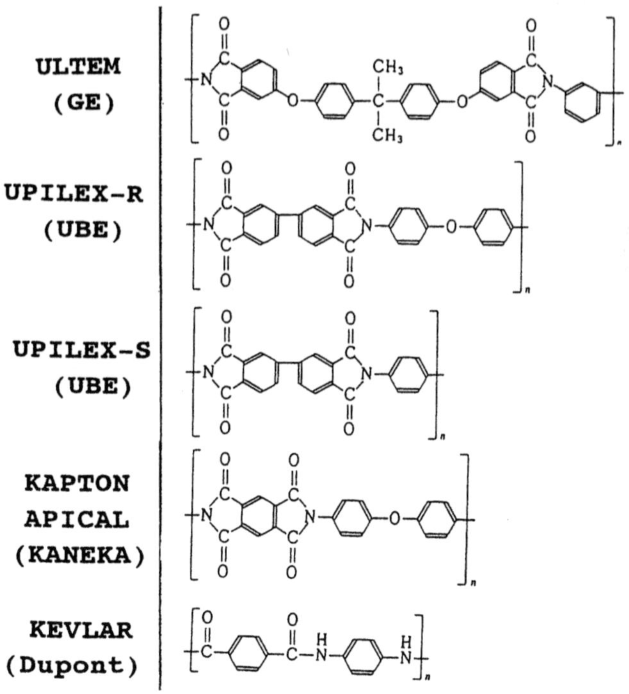

Figure 5.6. Chemical structures of commercially available, space application polyimide films.

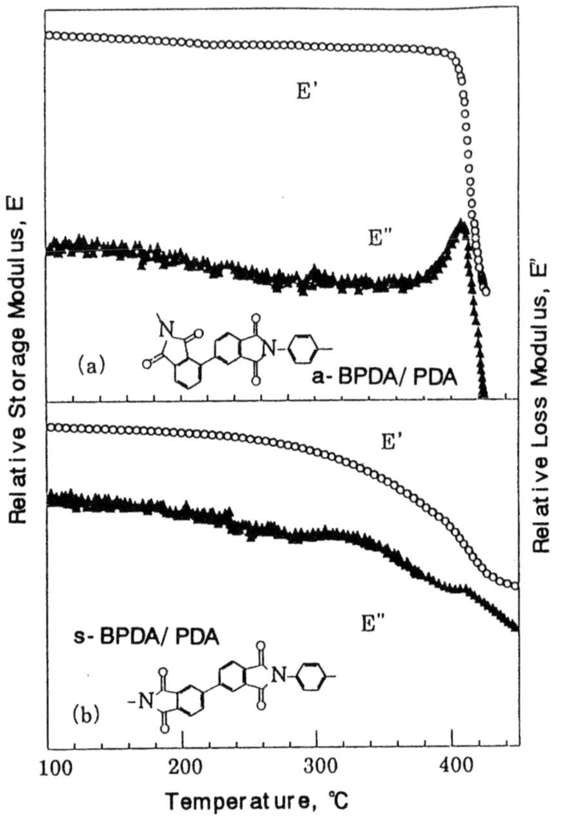

Figure 5.7. Temperature dependence of storage and loss moduli for isomeric biphenyl-type polyimide films.

transition temperature T_g, resulting in extremely severe processing conditions for structural applications [9]. Figure 5.7 shows the temperature dependence of storage and loss moduli for isomeric biphenyl polyimide films.

The asymmetric biphenyl polyimide, a-BPDA/PDA has a very inflected chain structure compared with the semi-rod s-BPDA/PDA polyimide, as shown in figure 5.7. The a-BPDA/PDA annealed at 400°C shows a higher T_g than the s-BPDA/PDA treated under the same conditions [10]. The difference in the extent of the storage modulus decrease at T_g for the two polyimides in figure 5.7 is attributed to the difference in the intensity of intermolecular interactions, suggesting an improvement in the processability of a-BPDA derived polyimide resins.

Figure 5.8 shows thermogravimetric results under nitrogen for the novel poly{(phenylsilylene) ethynylene1,3-phenyleneethynylene} (MSP-1), polyimide PI(BPDA/PDA), and their 1/1 blended films. Inorganic/organic

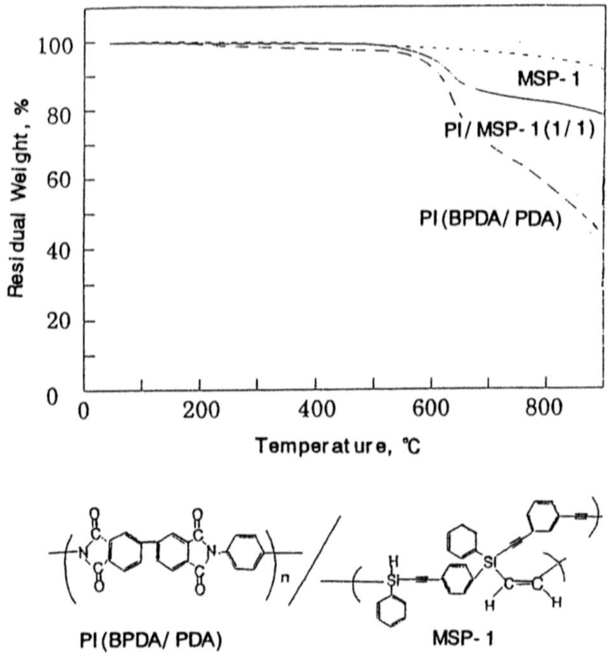

Figure 5.8. Thermogravimetric data in N_2 flow of the novel poly {(phenylsilylene) ethynylene-1,3-phenyleneethynylene} (MSP-1), PI(BPDA/PDA), and their 1/1 blend films.

polymer blends of this type can be expected to form a class of new thermally stable, toughened matrix resins [11].

Summary

Advanced polymer matrix composites are important in a range of space applications, notably in primary satellite structures, rocket casings, flexible/extendable structures such as antennae and as thermal insulators.

References

[1] Brodsky R F and Morais B G 1982 *Aeronautics & Astronautics* May pp 54–65

[2] Yokota R 1995 *Proceedings of 1st China–Japan Seminar on Advanced Engineering of Plastics, Polymer Alloys and Composites* (Society of Polymer Science, Japan) p 100

[3] Shibayama Y *et al* 1991 *Proceedings of the European Space Power Conference, Florence, Italy* p 735

[4] Onoda J *et al* 1994 IAF-94-1.1.174, Israel

[5] Onoda J *et al* 1994 ISTS-94-b-19, Japan

[6] Yokota R *et al* 1997 *Proceedings of 7th Symposium on Materials in a Space Environment, Toulouse, France* (ESA/ONERA)
[7] Takano T *et al* 1996 ISTS-96-e-18, Japan
[8] Yokota R 1996 *Structure and Design of Photosensitive Polyimides* ed K Hone and T Yamashita (TECHNOMIC) chapter 3
[9] Serafini T T 1984 *Polyimides: Synthesis, Characterization and Applications* ed K L Mittal (New York: Plenum) vol 2 p 957
[10] Hasegawa M, Yokota R, Sensui N and Shindo Y *Macromolecules* in press
[11] Yokota R, Ikeda A and Itho M 1998 *Polymer Prep. Japan* **47** 643

Chapter 6

Advanced polymer composite propeller blades

Mike Burden, Roy McCarthy and Brian Wiggins

Introduction

The major advantages of carbon fibre composites are their high specific stiffness, strength and excellent fatigue resistance when compared with metallic alloys [2, 3]. Comparative data for materials are shown in table 6.1. These properties give rise to the principal drivers for the use of composites in aircraft components, which are (a) improved performance resulting from the ability to optimize component shape, form and mechanical properties and (b) reduced weight which improves the efficiency of the component and aircraft and allows an increased payload or range for the aircraft.

For some applications composites can also give cost reduction both in initial cost and cost of ownership (e.g. when replacing expensive alloys such as titanium or when the component is structurally complex or unidirectionally loaded). At present, carbon fibre composite components are generally more expensive than their metallic counterparts because of the high cost of the carbon fibre. The majority of aerospace carbon fibre composite components are manufactured using a pre-preg process in which the fibre, already pre-coated with resin, is formed in tooling under heat and pressure. This manufacturing route uses relatively expensive raw material as shown in table 6.1, which indicates the price of pre-preg fibre, compared with aluminium and titanium alloy forgings. Whilst the military aerospace market will often accept a higher component price to obtain improved performance, at the present time the major driver in the civil aerospace market is cost reduction. The drive to improve material properties has tended to produce more expensive fibres and resin. The high modulus or high strength carbon fibres are three to five times more expensive than the normal aerospace fibres. The new single component resins RTM6 and PR500 are four to five

Table 6.1. Comparative materials properties and costs for a range of potential propellor blade materials.

	Density (g/cm^3)	Young's modulus (GPa)	Tensile strength (MPa)	Specific modulus (GPa)	Specific tensile strength (MPa)	Material cost (£/kg)
Carbon fibre epoxy:						
UD Pre-preg[c]	1.5	130	1750	87	1167	£50–90[a]
UD RTM[d]	1.5	130	1650	87	1100	£15–40[a]
Aluminium alloy	2.8	74	470	26	168	£10–15[b]
Titanium alloy 6AL 4V	4.4	106	1000	24	227	£25–40[b]

[a] Fibre and resin cost-no processing.
[b] Forging.
[c] 60% volume fraction carbon fibre.
[d] 57% volume fraction carbon fibre.

times more expensive than the established two-part epoxy resins. New fibre and resin systems are discussed further in chapters 13, 14 and 15.

Due to these higher costs, composites are finding it difficult to replace metals on civil aircraft and the focus on technical development is changing to find cheaper methods of manufacture with less emphasis on improving properties. There has, therefore, been a large amount of interest recently in resin transfer moulding (RTM) and similar processes because of the lower raw material costs associated with these manufacturing routes, as shown in table 6.1.

Background

The propeller blade is in many ways a very suitable application for carbon fibre composites. The first generation of propeller driven aircraft used blades made from the natural composite, wood. After the Second World War, this construction was mostly replaced by aluminium alloys as higher thrusts and forces were required. In the 1960s, Dowty began designing and manufacturing glass and carbon fibre polymer composite propeller blades, initially for hovercraft applications and then aircraft. Today, nearly all the large propellers for new aircraft have carbon fibre blades, as shown in figures 6.1 and 6.2.

Ideally, the blade should be as light as possible to minimize weight and reduce centrifugal loading on other parts of the propeller such as the hub and blade bearings, which can then also be lighter. Carbon fibre composites are ideally suited to blade manufacture since blades must have high specific strength and stiffness. The centrifugal force and aerodynamic loading on a

Figure 6.1. Lockheed Hercules C-130J with composite blades.

blade produces mainly unidirectional stresses parallel to its axis, which again favours a fibre reinforced composite construction. Significant vibratory aerodynamic bending moments are also present, caused by the propeller disc acting at an angle to the airflow. A propeller is in effect a very good rotating fatigue machine. The blade material must therefore have good specific fatigue strength. Carbon fibre composites have excellent fatigue properties, although these do vary depending on the fibre lay-up and loading patterns. As a result of the vibratory excitation, the blade has to be designed in such a way as to ensure that no natural frequencies are excited in the propeller

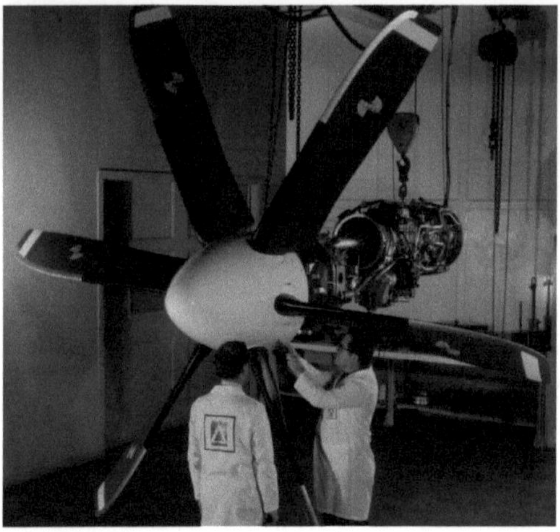

Figure 6.2. Composite blades.

running range. Composites allow the lay-up and fibres to be tailored to avoid harmful resonance conditions without significantly impacting the strength or weight of the blades. In addition to the structural requirements of the blade there is an environmental requirement to resist erosion from stones and grit. Aluminium alloy blades have to be dressed frequently to remove notches which can significantly reduce fatigue life. Composite blades, however, when suitably treated, require little maintenance, are easy to repair and last significantly longer than metal blades. For example, aluminium alloy aircraft blades generally are significantly worn after only 10 000 flight hours whilst composite blades can last more than three times longer. The cost of ownership of the composite blades is therefore lower and life cycle costs of running turboprops can be significantly reduced.

Blade design

The outside profile of a propeller blade is defined primarily by aerodynamic and acoustic requirements with the need to optimize cruise efficiency, take-off thrust and noise. Whilst metal propeller blades are solid, composite manufacturing techniques allow sandwich structures or hollow box structures to be produced with little extra difficulty. These constructions can reduce the weight of a composite blade compared with a solid structure and optimize the benefit of the composite material and manufacturing techniques. For a given loading the blade structure is designed to optimize the weight whilst

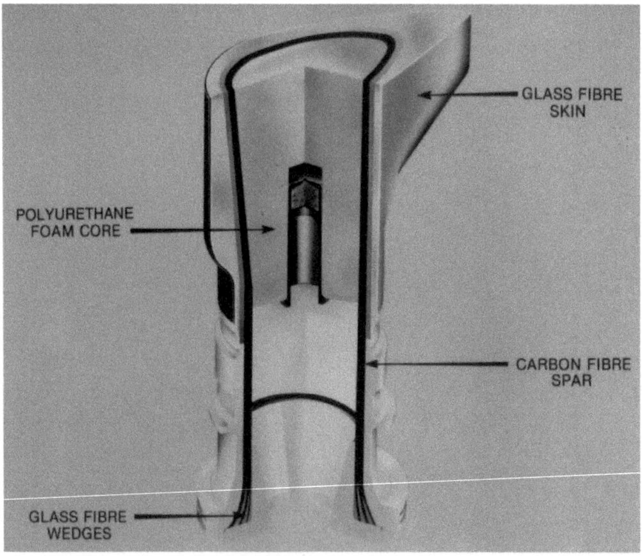

Figure 6.3. Composite blade construction.

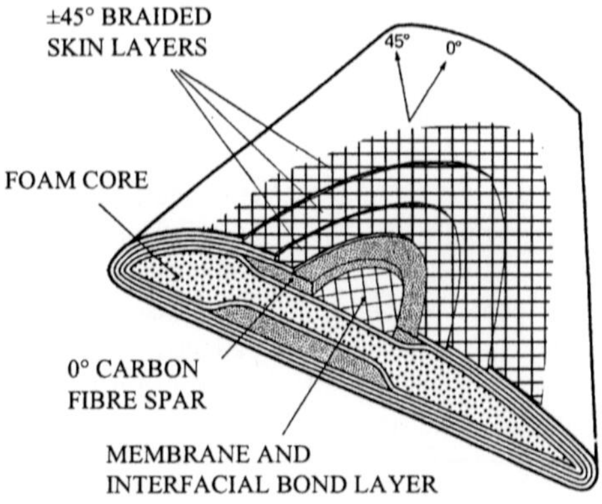

Figure 6.4. Blade root construction.

maintaining an acceptable level of strain in the composite. The composite construction used by Dowty for propeller blades has developed over 30 years and is shown in figure 6.3.

The main load-carrying members are spars consisting of layers of mainly unidirectional carbon fibre material designed to resist the centrifugal and bending loads on the blade. These spars extend from the root to the tip of the blade, reducing in thickness towards the tip where the loads are reduced. The outer shell consists of fibres oriented at ±45° to the blade axis. This provides the torsional strength and stiffness to the blade and also gives improved impact resistance. The structural foam core provides a method of transferring shear stress between the two composite spar beams. The arrangement of the composite to metal root is illustrated in figure 6.4. The carbon spars, which are basically flat planks, in the blade aerofoil go through a transition to join together to form a cylinder. This cylindrical spar is connected to the metal outer sleeve through a mechanical wedge system by inserting glass fibre wedges between the layers of carbon fibre. The introduction of an inner sleeve totally encloses the resulting annular wedge. This design of blade retention has an outstanding safety record with no blade losses over 20 years of service and 75 million flying hours. An important feature of the composite construction is an aluminium lightning braid which runs both on the pitch face and camber face from the blade tip to the metal outer sleeve giving lightning protection.

Blade manufacture by resin transfer moulding

When Dowty initially became interested in producing composite propeller blades, it investigated and experimented with both pre-preg and resin

Table 6.2. Advantages of resin transfer moulding (RTM) compared with pre-peg manufacturing routes.

RTM	Pre-preg
Low material cost	High material cost (2 × RTM)
Minimal material storage problem	Material has short life at room temperature and must be stored in a freezer
Close control on part thickness and fibre content	For parts with accurate thickness control matched metal moulds and computer controlled presses are required
Complex parts can be moulded in one shot	Difficult to mould complex parts in one shot

transfer moulding manufacturing processes. Composite blades were manufactured by both methods so that the relative advantages and disadvantages could be assessed. Resin transfer moulding was selected as being the best route for these complex constructions. This process has been continually developed over many years since 1967 into a very effective production method for structurally loaded carbon fibre composite components. The advantages over the competing pre-preg manufacturing route are shown in table 6.2.

In the resin transfer moulding process dry fibre preforms are injected with resin. Considerable effort has been put into methods of dry fibre reinforcement preform assembly to ensure fibres are placed and stay in the orientation and position specified. Fibres and fabric layers are held together with powder binder. A number of the carbon fibre layers are built-up in a shaped former, and then consolidated under heat and pressure to produce a rigid preform which can be handled easily, as shown in figure 6.5. These preforms are inserted into the blade mould and a core of low density polyurethane foam is formed *in-situ*, as shown in figure 6.6. The remaining outer ±45° fibre layers in the blade skin are produced using a braiding machine, as shown in figure 6.7. The braiding process [4] has a number of advantages compared with normal fabric assembly or alternative methods. It uses one of the cheapest forms of carbon fibre available and is able to lay down +45° oriented fibres at a very rapid rate producing a semi-woven structure which locks on to the blade shape. The required properties can be tailored by using mixtures of fibres, e.g. carbon and glass. The woven structure has a high impact resistance and the quality is more consistent than that produced by hand assembly. Outer metal root sleeve and glass cloth wedges are assembled together with edge reinforcement layers and aluminium braid lightning protection. The complete blade assembly is then placed in the blade mould for resin transfer moulding, as shown in figure 6.8.

The resin transfer moulding process is carried out with the blade mould mounted in a vertical position with the blade root uppermost. The mould is

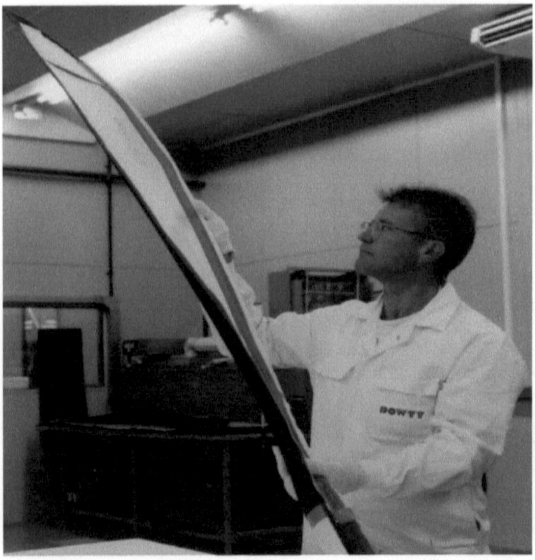

Figure 6.5. Blade preform.

heated to reduce resin viscosity and provide a satisfactory curing period. It is sealed to enable a vacuum to be applied to the preform at the top. The resin feed is at the base of the mould. The epoxy resin used by Dowty has been selected because of its low cost, excellent mechanical properties, low viscosity

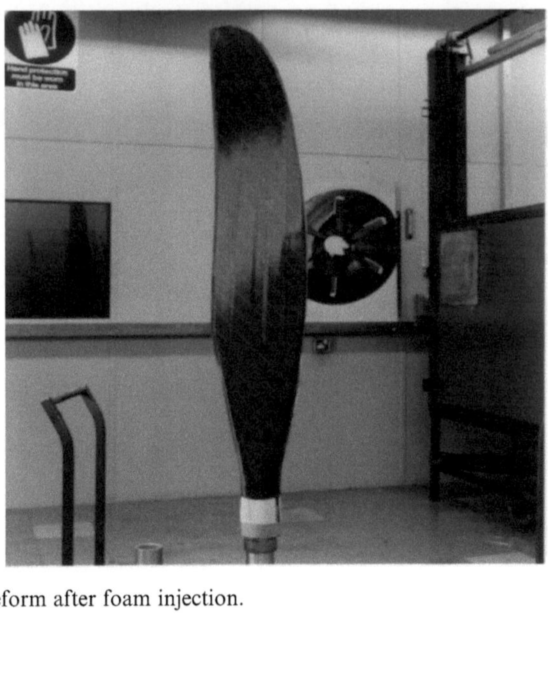

Figure 6.6. Preform after foam injection.

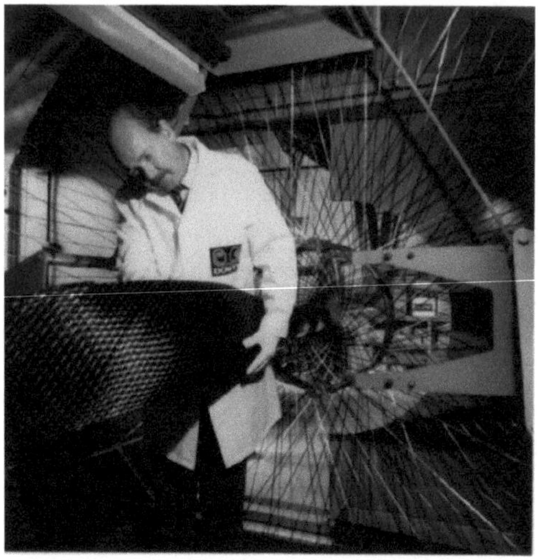

Figure 6.7. Braiding process.

and satisfactory life at elevated process temperatures. It is pumped at a controlled rate into the mould. After curing for a suitable period the blade is removed from the mould and deflashed, and the root assembly is completed. Satisfactory wetting of the fibres and minimum porosity are dependent on the quality of the resin in the preform, correct feed method and tooling, and consistent vacuum and feed rate. Control of these parameters at Dowty enables resin transfer moulding composites to be manufactured with high

Figure 6.8. Preform and resin transfer mould.

fibre volume fractions (up to 65%) and void contents of less than 1%. Because of the critical application, each blade produced is examined using ultrasonics and radiography to ensure that the major types of defects in composite structures, such as porosity, disbond and localized fibre waviness (wrinkles), are minimized to meet the required standards.

To provide wear and erosion resistance the blade is spray coated all over with polyurethane elastomer. Replaceable nickel electroform leading edge guards and de-icing boots are bonded in position to complete the structure.

Blade testing and fatigue issues

Carbon fibre composite structures are currently designed with very conservative margins, another factor reducing their competitive position relative to metals. Whilst a propeller blade is an excellent example of a suitable application for carbon fibre composites it also illustrates that there is still further potential if design rules and basic understanding can be improved. A propeller blade has to satisfy a number of design criteria which then have to be validated by test for certification. These include impact, lightning strike, environmental resistance, static strength and fatigue. The fundamental design criterion for a propeller blade is fatigue life. The initial design predictions of operating stresses and strains within a blade have to be confirmed by extensive flight testing with a strain-gauged propeller. From these results a worst case fatigue spectrum can be generated for the

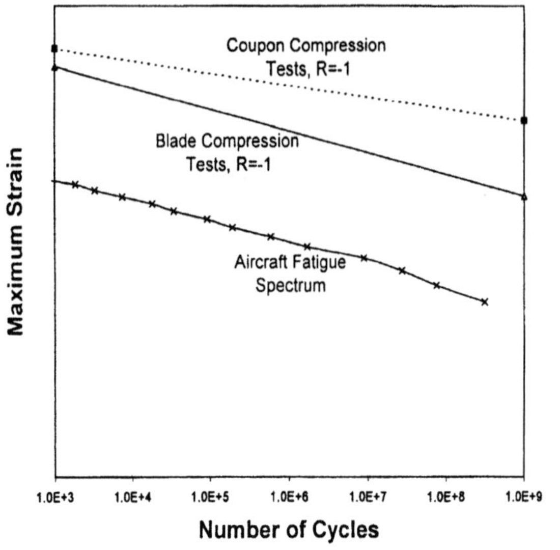

Figure 6.9. Fatigue data.

strains on a propeller blade for the life of an aircraft. Extensive fatigue testing is carried out on coupons, sections of blade, and on full blades, from which S/N curves and Goodman diagrams can be determined. A fatigue life assessment can then be made for the blade. Unfortunately, design tools for fatigue life assessment of composites are not entirely satisfactory [5, 6]. As a result significant safety factors are applied when determining life. Figure 6.9 illustrates this point, showing that the failure S/N curve for the blade is significantly above the worst case fatigue spectrum for the aircraft.

Figure 6.9 also shows another feature of composites which complicates the design process. The S/N curves generated for coupons are different from those of full size blades. This scale effect for composites has been investigated previously [7], as shown in figure 6.10, but is not fully understood. As a result, at the present time, extensive full scale testing on components is necessary. To reduce development time and cost, better design tools are required to determine the effect of lay-up on compressive strength and compressive fatigue. Very rarely is it possible to use a completely unidirectional structure, and a proportion of layers with other orientations will be required. Predicting the effect of these layers on compressive strength and fatigue is complex, and testing is invariably required. Whilst still very good compared with metals, the compressive strength and compressive fatigue properties of carbon fibre composites are well below the equivalent tensile properties. The development of hollow carbon fibres of larger diameter should lead to improved compressive properties, but probably at increased cost, which would be against current market drivers.

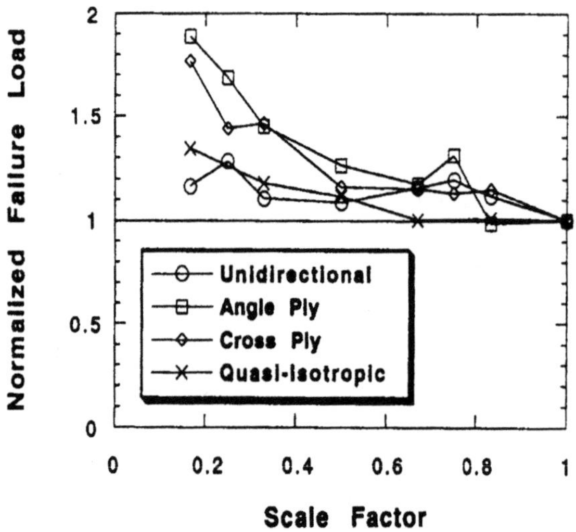

Figure 6.10. Scaling effect under flexural loading – taken from reference [7].

To improve the competitiveness of composite components in the future, accurate predictive design tools are necessary to reduce design and development costs relative to metals.

Future developments in polymer composite processing

In the UK, both component manufacturers and material suppliers are working on collaborative programmes to assess the economic and structural benefits that may be obtained using the resin transfer moulding process and its variants, as compared with the autoclave/press moulding of pre-pregs and metallic manufacture. The resin transfer moulding method of processing uses fibre and resin materials in the lowest cost form which should give it the potential to be the most economic route for many applications. As well as resin transfer moulding there are a number of similar dry fabric processes being developed which show potential for the future. These are SCRIMP, RIFT and RFI.

SCRIMP is Seemann Composites resin infusion moulding process, developed in the United States by Seemann Composites. This technique uses a one sided tool on to which dry reinforcement is laid and then covered with a special resin transfer medium before being sealed in a vacuum bag. Vacuum is applied and resin is distributed all over the component. This is an excellent method for large-area components where a one sided finish is acceptable and some variations in thickness can be tolerated.

RIFT [8] is resin infusion under flexible tooling. This is a variation on SCRIMP using a different resin transfer medium, with very similar results.

RFI is resin film infusion, a combination of resin transfer moulding and pre-preg autoclave techniques using a one sided tool on to which layers of dry reinforcement are laid and interleaved with resin film in a part cured condition. Alternatively, a thick resin film may be applied on top of the complete dry fibre reinforcement stack. A vacuum bag is fitted to which vacuum and heat are applied to force the resin into the reinforcement. Autoclave pressure may also be applied to achieve high fibre volume fraction. This process ensures resin is available in difficult areas of the moulding. Resin films are costly, which limits the economics of the process.

The key to manufacturing cost-effective structural parts by resin transfer moulding or one of the other similar variants of this process described above is the preforming of the reinforcing fibres into optimum directions for component loading and holding them in these positions during resin injection. Collaborative programmes are under way with universities and major aerospace companies to develop methods of automated fibre placement and triaxial braiding to produce net shape preforms at minimum cost. Resin and fabric suppliers are also involved to optimize methods for binding the reinforcement fabrics during cutting and assembly to produce rigid preforms of near-net shape prior to resin injection.

Summary

At the present time the major driver in the civil aerospace market is cost reduction. The challenge for material and processing development is to provide the full benefits of composites at economically viable prices. The focus of technical development is changing to find cheaper methods of manufacture with less emphasis on improving properties. Resin transfer moulding or one of the other similar dry fabric processes have the potential for being the most economic route for many applications. A better understanding of composites is also required to improve the predictive design tools reducing the amount of development and testing required for structurally demanding components.

References

[1] McCarthy R F J, Haines G H and Newley R A 1994 *Composites Manufacturing* **5**(2) 83–93

[2] Hancox N L and Mayer R M 1994 *Design Data for Reinforced Plastics* (London: Chapman & Hall)

[3] Aluminium Federation 1993 *The Properties of Aluminium and its Alloys*

[4] McCarthy R F J 1990 *Plastics Metals Ceramics 11th International SAMPE Conference, Basel* 29-31.

[5] Reinfronider K L 1991 *Fatigue of Composite Materials*, vol. 4, *Composite Materials* (Amsterdam: Elsevier Science Publishers BV)

[6] Curtis P T 1998, Conference Designing Cost Effective Components

[7] Jackson K E, Kellas S and Morton J 1992 *J. Composite Materials* **26**(18) 2674–2705

[8] Williams C, Summerscales J and Grove S 1996 *Composites* Part A, **27A** 517–524

Chapter 7

Materials developments in aeroengine gas turbines

David Clarke and Steve Bold

Introduction

Aeroengine gas turbines can be optimized in different ways for different applications, and even in different ways for the same application. This optimization process is driven by a number of broad factors as diverse as the basic laws of physics and gas behaviour through to aircraft operational performance and the financial needs of both the engine manufacturer and the end user of the aircraft. Through all of this runs a common factor: materials—metals, polymers and ceramics—are key to balancing all these factors for achievement of the optimum engine design and, in turn, the design needs for future engines define the need for materials development.

Current engine design

Key factors driving engine design

In civil applications, fuel efficiency is a key factor but, with the three major engine manufacturers all offering engines with similar fuel burn performance, secondary factors such as noise, emissions, weight and reliability become the major product differentiators. Highest engine efficiency is achieved through a high pressure ratio in the compressor and a large temperature rise through the combustor. Both factors are limited by the temperature capability of the materials available and the cooling technology used. Propulsive efficiency also depends on matching the exit velocity of the gas stream to the speed of the aircraft. For this reason, civil engines use large bypass ratios so that work is put into moving a larger mass of air more slowly than the air moving

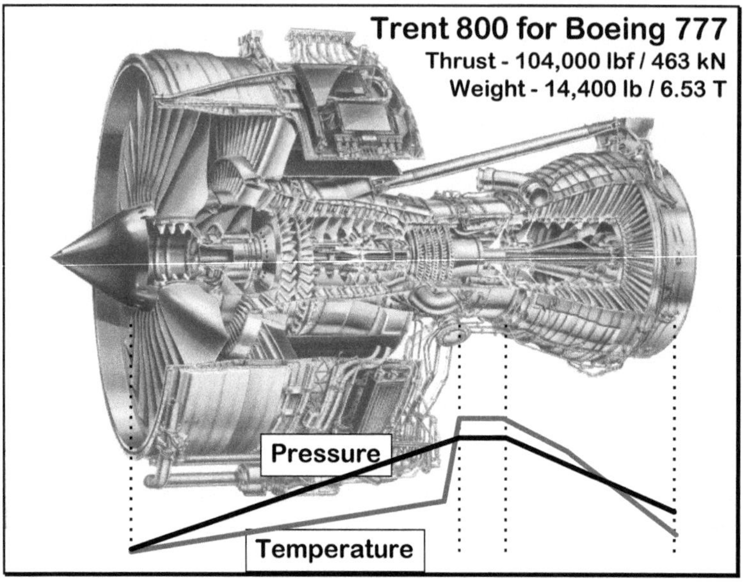

Figure 7.1. Pressure and temperature cycles through the high thrust, high bypass Trent 800 engine.

through the engine core. On take-off up to 80% of air goes through the bypass duct. Figure 7.1 shows pressure and temperature cycles through the high thrust, high bypass Trent 800 engine.

Whilst it is the obvious basis of all engine design, it is worth remembering that the first criterion in the design process is having the ability to deliver the necessary thrust to fly and manoeuvre the aircraft. This is particularly relevant in military applications where specific thrust (thrust/engine mass) is generally paramount, followed by fuel consumption, emissions (observability), reliability, maintainability and noise. Because of the higher aircraft speed and the need for additional responsiveness, military engines have low bypass ratios and very high gas exit velocities. Pressure ratios tend to be lower giving maximum specific work but generally not maximum fuel economy.

For both civil and military applications there is a common factor—cost. Product development costs and production unit costs are considered from the first stages of engine design and create as many challenges for materials development as the engine performance parameters.

Two engine designs

The vast majority of civil aircraft are available fitted with engines from more than one manufacturer. Figure 7.2 shows a real example of the results of two

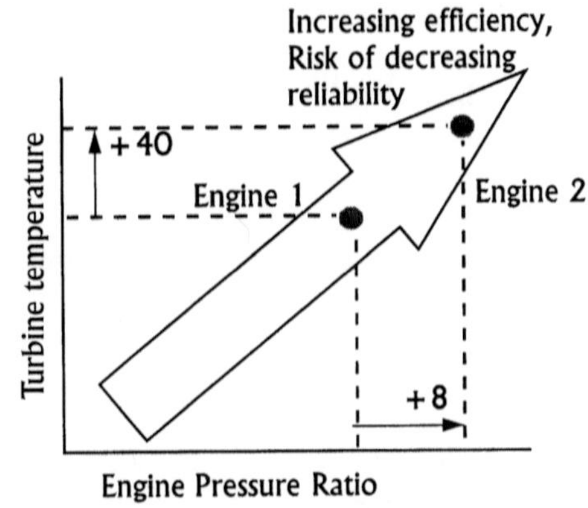

Figure 7.2. Comparison of turbine temperature and pressure ratio for two aeroengines designed for the same aircraft.

different manufacturers' design solutions for one particular aircraft. Both engines were designed at the same time for the same duty and whilst both use similar basic technologies, produce the same thrust and weigh similar amounts, their design philosophies are quite different.

Physics dictates that engine thermodynamic efficiency increases with increasing turbine gas temperature and increasing pressure ratio (combustor entry pressure to ambient air inlet pressure). Primarily as a result of running hotter and at higher pressures, engine 2 burns around 4% less fuel than engine 1 for the same take-off thrust. The impact of raising the gas temperature and pressure to achieve this higher efficiency, however, is that core engine components in engine 2 degrade much more quickly than those in engine 1, and engine 2 has to be removed from the aircraft twice as often for major maintenance. The financial impact of this to the operator is that engine 1, *the less technically efficient engine*, costs nearly 10% less to operate than engine 2. The financial impact on the engine manufacturer is that engine

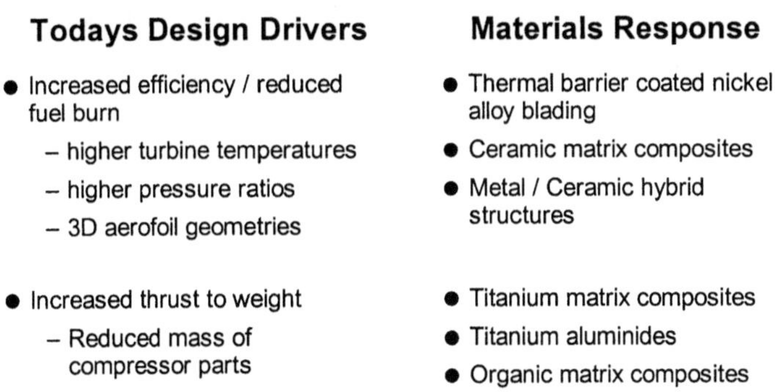

Figure 7.3. Current design drivers and materials responses.

1 now accounts for 80% of this market sector with over 1000 engines in service.

Engine design is thus clearly a complex optimization process and there are many solutions to one overall requirement. The impact of the different design styles and operational parameters of these two engines on the materials needed to manufacture them is equally dramatic. Figure 7.3 shows the current design drivers and material responses.

Materials

Turbine blade alloy development

Since the initial development of the gas turbine in the 1940s the temperature capability of the nickel alloys used for the highest temperature parts of the engine—high pressure turbine blades, discs and the combustion chamber—has increased by around 400°C to nearly 1200°C.

This 400°C increase has involved four major stages of development in material and manufacturing technologies, from forged alloys to cast systems, then directionally solidified (DS) castings to eliminate creep problems associated with transverse grain boundaries, and now to single crystal castings with control of both the longitudinal and the transverse crystal orientation. This development is summarized in figure 7.4. To achieve a further increase in temperature capability is now requiring a fifth stage of development, with the introduction of surface coatings to reduce oxidation and corrosion, and ultimately a complex system of thermal barrier coatings (TBCs) to reduce the

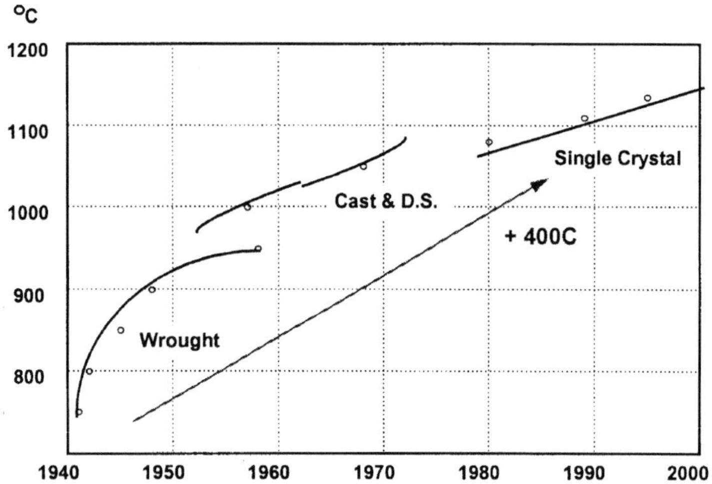

Figure 7.4. Temperature capability of different turbine blades since 1940.

rate of heat transfer from the blade surface to the internally cooled surfaces in its hollow core. This latter development is discussed in more detail later.

An increase of 40°C in turbine blade temperature as highlighted between the two engines in the example in figure 7.2 denotes a significant increase in materials technology level. Typically this represents the difference between two technology levels, e.g. a directionally solidified to a single crystal alloy, or a single crystal to a system of single crystal plus oxidation-resistant coatings. Whilst this level of operating temperature increase gives a significant fuel burn benefit (4% is around the maximum difference seen between different engines designed for a single airframe application) it is only of real value if this can be matched by comparable reliability between the two units. The need for a particular level of materials technology or a significant advance is thus created as the fundamental engine architecture and thermodynamic cycle are set. It is obviously possible to operate engines at low temperatures but only with a defined performance penalty and in an increasingly competitive market this is unacceptable. Continuous performance enhancements, only deliverable by ongoing materials developments, are essential.

Evolutionary versus revolutionary development

Historically the majority of materials developments for gas turbines have been evolutionary, as in the gradual evolution of each nickel alloy technology described above. Both nickel and titanium alloys, however, are now at a level of development where further small advances are of limited benefit and are increasingly expensive to achieve. Engine design now demands revolutionary developments in the fields of increased temperature capability and reduced component mass. The primary materials development programmes to achieve this are almost all based around either composite materials or, increasingly often, a composite structure, integrating the benefits and properties of various materials systems into a single component. The following sections discuss some specific examples of this approach.

Nickel based metal/ceramic structures

The majority of turbine components are cooled by air from the compressor. This increases component life but reduces the engine efficiency, since the engine pressure ratio is effectively being progressively decreased as air is bled off to these various sections. The effect of this is to increase the engine fuel burn compared with the theoretical minimum. In practice this results in increased cost for the operator. Whilst the reduction in efficiency as a result of these cooling bleeds is very small, the overall financial effect to a large fleet operator may be very significant. A 1% reduction in fuel burn to a typical long haul operator of ten Boeing 747-400s would reduce annual fuel bills by well over $1 million.

To reduce the need for cooling air, many turbine components now incorporate ceramic materials. The latest developments of high pressure (HP) turbine blades use nickel alloys coated with a graded series of layered materials to give a component which relies for its operation on the synergy of properties from the metal substrate and ceramic coating. The single crystal nickel base alloy is first coated with a plasma sprayed MCrAlY overlay coating (M represents Ni or Co). The chromium and aluminium provide oxidation resistance while the presence of yttrium improves scale adhesion. In addition the layer acts as a bond coat to prevent spalling of the outer coating layer—the thermal barrier coating. The thermal barrier coating is a low thermal conductivity ceramic which restricts the flow of heat from the gas stream to the metal blade. This maximizes the benefit obtained from blade cooling and offers a potential increase in operating temperature of over 100°C.

Thermal barrier coatings have been used in the combustion chambers of the RB211 since 1975 but it is only with advances in bond coat technology and ceramic deposition techniques that they can be reliably used on critical rotating parts where coating failure could lead to premature component removal. Coatings are applied by electron beam physical vapour deposition to develop the columnar grain microstructure necessary to resist thermal and mechanical strains, particularly around blade leading and trailing edges. Figure 7.5 shows a coated blade from the Trent 800.

Full exploitation of coating technology requires the coating to be seen as an integral part of the component from the start of the design process. The coating system must be compatible with the requirements of aerodynamics,

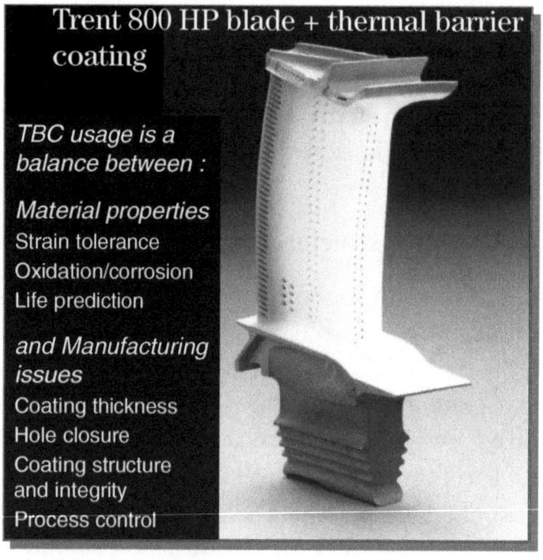

Figure 7.5. Trent 800 high pressure (HP) turbine blade, showing cooling holes and thermal barrier coating (TBC).

mechanical integrity and blade cooling and must be created in a cost-effective manner suitable for mass production. All of these considerations have led to the unique, cost-effective thermal barrier coating system designed for use in the Rolls-Royce Trent engines. Thermal barrier coatings are discussed in more detail in chapter 22.

Nickel/ceramic hybrid combinations are also being developed on a macro scale for mechanically integrated large static structures. Here a simple metallic unit forms the major load carrying element and a semi-structural ceramic shell, or more commonly a fibre reinforced ceramic composite, forms the high temperature surfaces of the component. This approach has been used in turbine blade tip seals on the Trent engine, and exhaust structures for military engines such as the EJ200.

Titanium aluminides and titanium metal matrix composites

Unlike turbine components, compressor components are not normally cooled. They operate at gas stream temperatures which reach around 650–750°C at the compressor exit. The primary design driver here is to increase pressure ratio and this largely drives further increases in this exit temperature.

Compressor discs and blades are largely formed from titanium alloys, with much of the historical alloy development emphasis having been on developing defect tolerance and increased temperature capability. The latter is exemplified in alloys such as IMI834 which are used in applications up to 630°C—almost twice the capability of the routine Ti-6Al-4V at 350°C. Moving to these higher temperature alloys has allowed weight savings in compressor modules of over 15%. Alternatives to titanium are primarily steel and nickel alloys with the consequent mass increases. Major performance improvements are now being made by either using intermetallic (titanium aluminide) materials or silicon carbide fibre reinforced titanium composites.

Gamma titanium aluminides are the most advanced of the intermetallics with half the density of current titanium alloys. Primary benefits are very high specific stiffness and inherent non-burning chemistry. Temperature capability is currently around 750°C, making the system suitable for back end compressor and turbine applications. Their non-burning nature makes them particularly suitable for stator vanes, where titanium alloy use is limited by the risk of titanium fires. Application of aluminides could lead to whole engine weight savings of up to 4%. These materials will also give reduced engine life cycle costs and improved engine/airframe functionality since reducing mass on one element of an engine structure has a consequent knock-on effect on surrounding components—lower-mass blades lead to lower-mass discs and shafts, reduced stiffness casings etc. Titanium aluminides are discussed further in chapter 17.

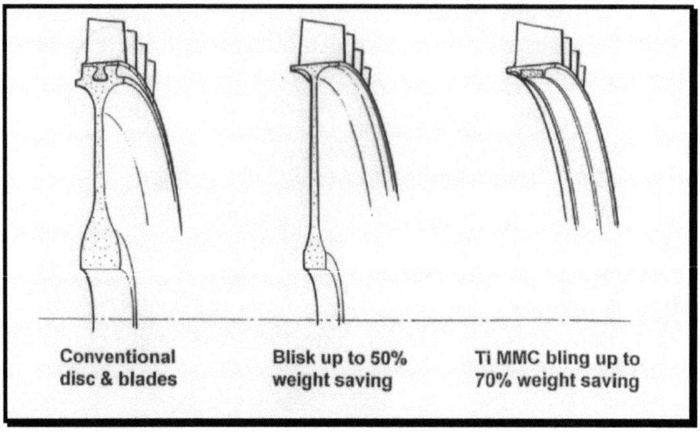

Figure 7.6. Weight savings with blisc and bling designs using titanium metal matrix composites (MMCs).

Knock-on effects in reducing mass of associated components are particularly significant in rotating bling (bladed ring) structures made in titanium metal matrix composite (Ti MMC). The bling is an integral structure of titanium blades on a titanium metal matrix composite. The reinforced ring increases hoop stiffness by 100% and strength by 50%. This removes the need for the heavy bore of the disc, giving weight savings of the order of 40% over a conventional titanium blisk (bladed disc) design, as shown in figure 7.6. Rolls-Royce Allison has successfully run the first demonstration of titanium metal matrix composite blings in an engine. Titanium metal matrix composites are discussed in more detail in chapter 18.

The weight savings in the bling itself are greatly increased by the options presented for major changes to the surrounding structures, e.g. the possibility of using a larger diameter (and hence stiffer) shaft, since the shaft diameter constraint imposed by the disc bores is effectively removed. This geometric change opens up the possibility of replacing the steel shaft with a titanium unit, with dramatic weight benefits. Steel shafts are currently used to ensure adequate torsional and bending stiffness, which must be achieved within the very low diameter allowed for a conventional shaft geometry.

Future design drivers

Future engine designs demand still further increases in the basic design parameters of temperature and pressure, but are increasingly requiring enhanced materials capabilities in more diverse fields. Engine design has evolved for the past 50 years around a largely unchanged group of mechanical technologies, and these have needed a standard set of materials with optimized mechanical

Tomorrows Design Drivers	**Materials Response**
• Increased accessory reliability / increased thrust to weight – mechanical drives to accessories (pumps etc.) replaced with electrical motors – mechanical bearings in engine core replaced with electromagnetic bearings	• Permanent magnets with retention of magnetic properties to >300 °C. (Limit today around 220 °C)
• Reduced cost / complexity control hardware / software – Centralised control electronics moved to distributed systems close to (hot) engine core	• Electronic materials operable at around 400 °C. Probably Silicon Carbide based rather than Silicon, (Si limit around 100°C).

Figure 7.7. Future design drivers and material responses.

and structural properties. Competitive new designs now require revolutionary changes in the engine mechanical design, and this is driving materials development in new fields (for the engine manufacturers) such as electrical rather than mechanical properties. Figure 7.7 shows the future expected design drivers and the material responses.

The more-electric engine (MEE)

Engines are heavily dependent on mechanical/hydraulic actuation and drive mechanisms to operate pumps, variable nozzles and vanes, etc. These all require mechanical drive systems from the engine core. To increase reliability and reduce cost and weight, these will be replaced with small electric motor drives. This will ultimately require an electrical generator mounted within the engine core in an environment of over 300°C. Permanent magnets necessary for use in this machine will currently only operate up to 220°C, so that the unit will have to be insulated and cooled with air, similar to current turbine structures, until new magnet capabilities are available. Again, this use of cooling air will detract from the self-same improvements in efficiency being made by the new technology.

Distributed control software

With the move to the MEE, control systems will also become more distributed around the engine into smaller units. All control software is currently mounted on electronics in a single heavily-insulated and fireproof case on the outside of the engine. Distribution of the systems around the engine will increase reliability and reduce cost and weight, but for the maximum benefit will require high temperature electronics based on materials such as silicon carbide (silicon based semiconductors cannot be reliably operated above around 100°C). Silicon carbide based electronics systems have been demonstrated at over 400°C for military applications but are currently not cost effective for commercial use.

Materials development for gas turbines is ongoing in these fields and the more conventional areas and is presenting a more diverse set of requirements than ever before. Many of these materials, however, will only ever fill limited niche applications in gas turbines and similar high speed, high temperature machines. As a result perhaps the biggest and most common materials challenge for the future will be the one currently faced by high temperature polymer composites and structural ceramic composites—how to develop and bring to production a highly specialized, low volume material quickly and cost effectively. The answer must lie in collaboration across the industry supply chain and the research and development institutes to simplify processes, to automate aspects of materials assessment, and above all to innovate and develop new, creative solutions in manufacturing, and to then apply these quickly with controlled levels of risk.

Chapter 8

Blading materials and systems in advanced aeroengines

Tasaduq Khan and Marie-Pierre Bacos

Introduction

Turbine blades are critical components in aerospace turbomachines since the safety and performance of the engine depend strongly on such parts. Turbine inlet temperatures have increased very significantly over the past 30 years and will approach 1650°C at maximum power for the large commercial turbofan engines over the next few years. The turbine inlet temperature attained in the M88 engine of the Rafale fighter plane is already of the order of 1580°C. This level of performance has been achieved by a combination of advances in air cooling design technology for stresses and airflow, and single crystal technology. During the past 15 years, ONERA has played a leading role in Europe in the development and study of new cost effective single crystal superalloys tailored for the specific needs of the engine makers. Among the various materials developed within France, alloys such as AM1 (M88 engine), AM3 (ARRIEL and ARRIUS helicopter engines) and MC-2 for advanced single crystal blades are worth mentioning. The MC-2 alloy developed by ONERA, which is one of the strongest low cost currently available single crystal superalloys without rhenium, is being engine tested by Turbomeca and by Pratt and Whitney for advanced engine applications.

First, second and now third generation single crystal superalloys have progressively been developed in order to exploit the full potential of these materials. The next leap forward in terms of turbine inlet temperature, performance and durability will be provided by the use of thermal barrier coatings (TBCs) on cooled aerofoils. The increase in turbine inlet temperature can be expected to be about 150°C, which is impossible to achieve with alloy chemistry modifications alone.

Evolution of single crystal superalloys

About 40 years ago, Versnyder *et al* [1] demonstrated the possibility of producing both columnar grained and single crystal superalloys by directional solidification. However, it was only at the end of the 1970s that superalloy chemistry was specifically designed for single crystal blade applications. These chemical compositions were not very innovative, since they were still based on polycrystalline nickel based alloys. One such example is the single crystal alloy NASAIR 100 derived from Mar-M247 essentially by suppressing grain boundary strengthening elements such as carbon, boron, zirconium and hafnium [2]. The resulting increase of the incipient melting temperature from 1240 to 1330°C allowed complete solutioning of the secondary γ' precipitates and the almost complete elimination of coarse primary interdendritic γ' particles by using a super-solvus high temperature solution heat treatment. The resulting increase of creep strength was shown to be equivalent to a 58°C temperature advantage at high temperatures and low stresses. This was the first significant leap towards introducing single crystal technology.

After the development of the so-called first generation single crystals (CMSX2, SRR99, AM1, AM3), several alloy designers showed that a significant improvement of the creep strength of single crystal superalloys may be obtained by the addition of rhenium at the expense of other refractory elements such as molybdenum or tungsten. Thus, a study carried out on modified Mar-M200 single crystal alloys showed that additions of rhenium substituting for tungsten substantially lower the γ' coarsening kinetics and result in large negative γ–γ' misfits. Atom-probe studies performed on CMSX-2 and PWA 1480 base alloys modified by additions of rhenium also showed the existence of rhenium atom clusters within the γ matrix of these alloys, that could be a more potent source of strengthening than the conventional solid solution effect. The introduction of 3 wt% rhenium to single crystal nickel based superalloys resulted in a temperature capability improvement of about 30°C [3]. On the other hand, these rhenium additions have induced some drawbacks compared with the first generation materials: (1) increased density; (2) increased propensity for precipitation of brittle intermetallic phases known as topologically close packed (TCP) phases; and (3) increased cost.

The first rhenium-containing alloys ($\sim$3 wt%) were labelled as second generation single crystal superalloys. Typical examples of this class of alloys are PWA1484, René N5 and CMSX-4. However, a rational criterion for the designation of these second generation alloys should be based on their high temperature creep strength instead of their rhenium content. In these conditions, the MC-2 superalloy developed by ONERA should be considered as an integral part of this group, in spite of the fact that it does not contain rhenium [4]. It is noteworthy that the absence of rhenium makes this alloy cheaper and lighter than the other second generation single crystal alloys.

The third generation single crystal superalloys contain a high level of rhenium, up to 6 wt%, in order to benefit from the significant strengthening effect previously demonstrated with the second generation single crystal alloys. The first developed third generation single crystal superalloys in the USA are René N6 from General Electric and CMSX-10 from Cannon-Muskegon. More recent development work conducted by General Electric was devoted to third generation alloys also containing some additions of ruthenium [5]. A new generation of alloys, a typical example of which is the MC-NG alloy, has also been developed in France by ONERA [6]. In these alloys rhenium is still associated with other refractory elements such as molybdenum, tungsten and tantalum which are also known to reduce the bulk diffusion rate that slows down the coarsening kinetics of the strengthening γ' precipitates. These elements also induced strong solid solution hardening effects due to their large atomic radii compared with those of nickel or aluminium.

The latest innovation in terms of chemical composition for third generation superalloys is the partial replacement of rhenium by ruthenium. The idea to incorporate some amount of ruthenium in the nickel base superalloys is quite old, but it is only recently that designers developed alloys where they clearly exploit the advantages of this alloying element. Such developments were conducted by General Electric and ONERA. The more evident advantages of ruthenium compared with rhenium are its lower atomic weight and its lower propensity to promote the precipitation of topologically close packed phases [5, 7]. These ruthenium containing third generation alloys exhibit excellent creep strength, especially at high temperatures, but complementary investigations are needed to evaluate and understand the influence of ruthenium on other properties. The more recent idea to introduce iridium in superalloys was exploited by Kobayashi *et al* [8] in TMS-80. The most evident effect of iridium is to stabilize the alloy with respect to the topologically close packed phases, but no significant effect was demonstrated on other properties. Microstructure and creep of superalloys are discussed in detail in chapters 19–21.

Thermal barrier coatings

In spite of the improvement in terms of creep strength of the most recently developed single crystal alloys, one cannot expect to achieve an increase in their temperature capability beyond 1140–1150°C. The only real possibility to obtain a very significant increase in blade operating temperatures is therefore to develop single crystals coated with thermal barrier coatings. The use of such coatings in conjunction with alloy development can provide the designer with more freedom to:

1. further increase the turbine inlet temperature;
2. increase life by reducing component temperatures; and
3. reduce cooling air required at a given turbine inlet temperature.

Plasma spraying is currently the most common means for depositing thermal barrier coatings. It results in the deposition of platelet-shaped particles incorporating porosity and microcracks parallel to the interface that is beneficial in term of thermal conductivity. Unfortunately, the plasma spray process is inadequate for depositing ceramic coatings on to high pressure turbine blades due to the problem of obstructing cooling holes. For critical components such as rotating blades the electron-beam physical vapour deposition (EBPVD) technique is the most appropriate technique because it provides extremely fine columnar structures, which are highly tolerant of thermal strains induced by differential expansion of the underlying substrate. The advantages of electron beam deposition over plasma sprayed coatings include improved thermal cycle life, higher erosion resistance and improved surface finish.

The electron beam deposition technique is a line of sight process and is limited to the coating of simple-shaped aerofoils. The deposition of thermal barrier coatings by electron beam deposition shadowed regions of multiple aerofoils is much more difficult [9]. A non-directive alternative deposition process such as chemical vapor deposition (CVD) presents definite advantages in this respect. Based on this, ONERA has developed a plasma enhanced chemical vapour deposition (PECVD) technique, as shown in figure 8.1, for the deposition of an yttria partially stabilized zirconia [10], which produces coatings with a columnar morphology but with high deposition rates (up to 150 μm/h) compared with conventional chemical

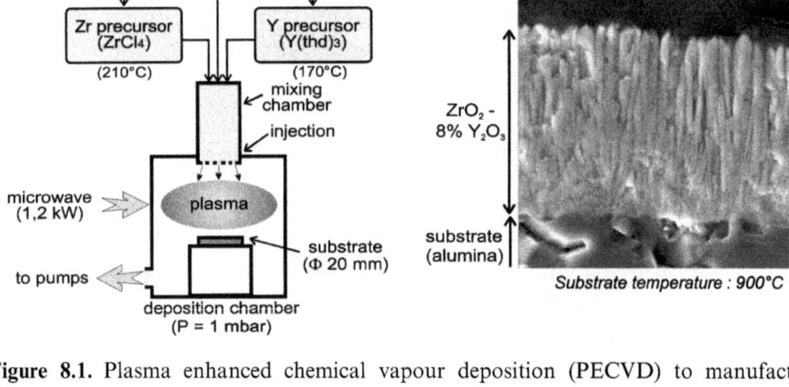

Figure 8.1. Plasma enhanced chemical vapour deposition (PECVD) to manufacture thermal barrier coatings (TBCs).

vapour deposition processes. The performance of the plasma enhanced chemical vapour deposition coatings is comparable with that of the electron beam deposited coatings.

The main function of thermal barrier coatings is to isolate the coated superalloy components (blades and vanes) thermally from the hot gases circulating in the turbine. Recently, significant attention has been increasingly directed towards characterizing, understanding and lowering the thermal conductivity of the ceramic coating. Yttria partially stabilized zirconia coatings have a low thermal conductivity (typically between 1 and 2 W/mK) that can be further decreased either by substituting zirconium with heavier metallic ions [11, 12] or by increasing the yttrium content [13, 14]. Thermal conductivity can also be lowered by modifying the morphology of the coating. Indeed, pores, voids or microcracks constitute obstacles against the heat transfer propagation. It is desirable to optimize the defect distribution in order to lower the thermal conductivity. A variety of models [15–18] have been developed to estimate the thermal conductivity of multiphase solids, in particular porous materials. However, most of these models assume morphological parameters that are somewhat idealized and removed from the real microstructure. ONERA has developed and evaluated a finite-difference model for computing the thermal conductivity of thermal barrier coatings using, as input data, digitized images of the real materials [19], as shown in figure 8.2. This new approach takes into account the actual complex morphology of the ceramic coatings.

Yttria partially stabilized zirconia coatings are not impermeable to oxygen transport, which leads to a significant oxidation of the underlying alloy. In order to provide oxidation protection, an oxidation-resistant metal (called bond coat) is applied beneath the ceramic coating, which also serves to anchor it to the underlying superalloy. Typically this bond coat is an MCrAlY (M = Ni, Co) or nickel aluminide. Pre-oxidation of the bond coat promotes the formation of a thin alumina film (called thermally grown oxide TGO) that acts as a glue for the thermal barrier ceramic coating to be deposited. It is universally acknowledged that bond coats for thermal barrier coatings have a strong effect on the thermal fatigue life of the ceramic. Bond coat oxidation has been clearly linked to spallation of the ceramic coat [20, 21], but the mechanism by which oxidation causes failure has not been clearly identified due to the differences in alloy chemistry. Indeed the degree of spallation depends upon numerous factors including the composition of both the alloy substrate and the bond coat. In fact the different components of the structure—superalloys, MCrAlY bond coat, NiAl bond coat, ceramic—have specific and complex chemical interactions with each other. This obviously has an impact on the performance of the overall coating. Impurities such as sulphur, which segregate from the base alloy to the interface, can weaken the interface bonds [22–24]. Addition of reactive elements (such as yttrium), either in the superalloy or in the bond coat,

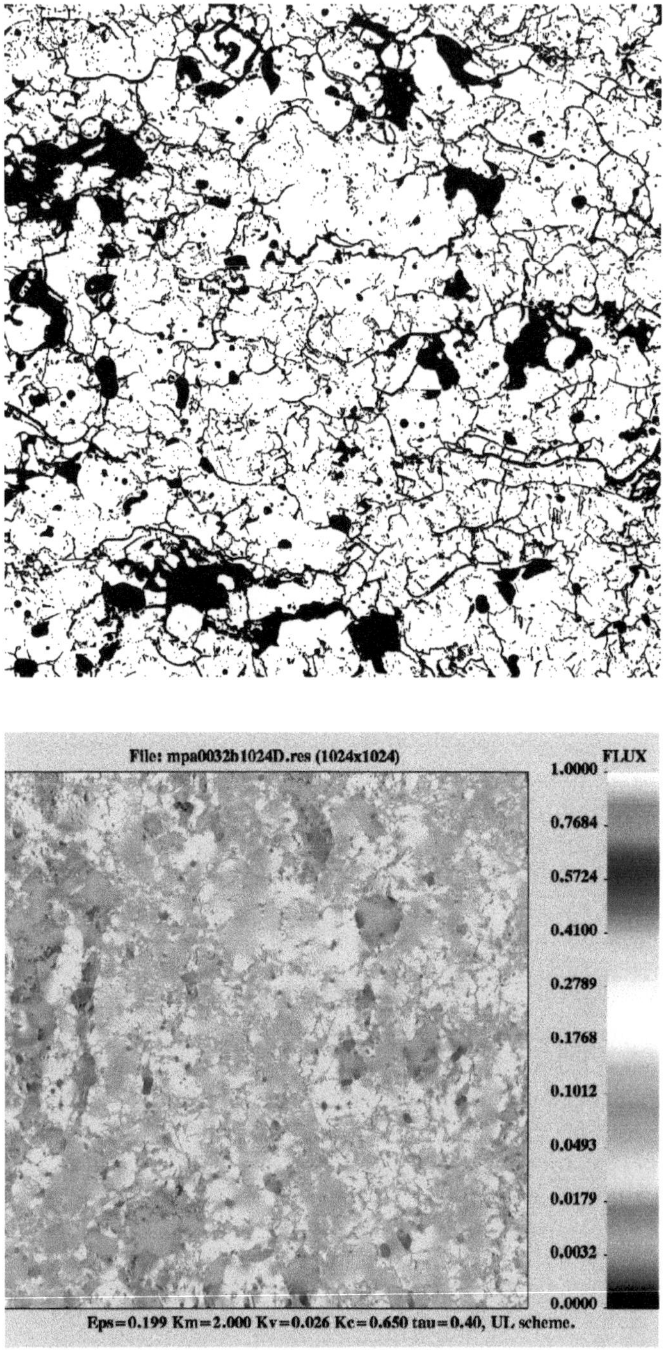

Figure 8.2. Columnar structure of a thermal barrier coating.

can increase the adhesion of the thermally grown oxide by gettering the indigenous sulphur, or by modifying the oxidation mechanisms [25, 26]. Therefore a fine microstructural characterization of the layers (superalloy/bond coat/thermally grown oxide/ceramic coating) is necessary to design thermal barrier coatings with optimal coating performances. Thermal barrier coatings are discussed further in chapter 22.

Summary

Turbine inlet temperatures have increased very significantly over the past 30 years. This has been achieved by a combination of advances in materials (improved alloy compositions and improved processing) and air cooling technology. Significant improvements have been brought about by second and third generation single crystals. The latter alloys all contain rhenium additions. The creep and fatigue temperature advantage of the third generation alloys over the second generation single crystals is estimated to be about 30°C.

It is clear now that superalloys are reaching their temperature asymptote. The only way to increase significantly the temperature capability of such materials is, therefore, to coat them with thermal barrier coatings. However, the major difficulty in using thermal barrier coatings in a cyclic high temperature environment is spalling of the ceramic coating. It is now well known that this spallation is linked to the chemistry of both the superalloy and the bond coat. Considerable research effort will therefore be focused on the superalloy/thermal barrier coating system rather than on the superalloy alone. No single bond coat–thermal barrier coat combination will provide optimal performance on all substrate alloys.

References

[1] Versnyder F L and Guard R W 1960 *Trans. ASM* **52** 485

[2] Strangman T E, Hoppin G S III, Phipps C M, Harris K and Schwer R E 1980 *Superalloys 1980* ed J K Tien *et al* (Metals Park, OH: American Society for Metals) p 215

[3] Erickson G L 1995 *J. Metals* **47**(4) 36

[4] Caron P and Khan T 1990 *Advanced Materials and Processes* vol 1 ed H E Exner and V Schumacher (Oberursel, Germany: DGM Informationsgesellschaft mbH) p 333

[5] O'Hara K S, Walston W S, Ross E W and Darolia R, General Electric Company, US Patent 5,482,789

[6] Caron P and Khan T 1998 *Materials for Advanced Power Engineering* (Liège, Belgium) in press

[7] Caron P unpublished work (ONERA)

[8] Kobayashi T, Koizumi Y, Nakazawa S, Yamagata T and Harada H 1997 *Advances in Turbine Materials, Design and Manufacturing* ed A Strang *et al* (London: The Institute of Materials) p 766

[9] DeMasi-Marcin J T and Guppka D K 1994 *Surf. Coat. Technol.* **68/69** 1–9

[10] Chevillard S, Drawin S and Vidal-Setif M-H 1998 *AGARD Report 823* 11-1/11-9

[11] Kingery W D, Bowen H K and Uhlmann D R 1996 *Introduction to Ceramics* (New York: Wiley)

[12] Madarasz F L and Klemens P G 1997 *Internat. J. Thermophysics* **8**(2) 257

[13] Youngblood G E, Rice R W and Ingel R P J 1988 *Am. Ceram. Soc.* **71**(4) 255

[14] Hasselman D P H, Johnson L F, Bentsen L D and Syed R 1987 *Am. Ceram. Soc. Bull.* **66**(5) 799

[15] Murabayashi M 1969 *J. Nucl. Sci. Tech.* **6** 47

[16] Koh J C and Fortini A 1971 NASA Report NAS3-12012, CR 1200854

[17] Cunningham M E and Peddicort K L 1991 *Internat. J. Heat and Mass Transfer* **24** 1081

[18] McLachan D S, Blaszkiewicz M and Newnham R E 1990 *J. Am. Ceram. Soc.* **73**(8), 2187

[19] Dorvaux J-M, Lavigne O, Mévrel R, Poulain M, Renollet Y and Rio C 1998 AGARD Report 823, 13-1/13-10

[20] Lee E Y and Sisson R D Jr 1994 *Proceedings of the 7th National Thermal Spray Conference* 20–24 June, Boston, Massachussetts

[21] Freborg A M, Ferguson B L, Brindley W J and Petrus G J 1997 *Proceedings of the 85th Meeting of the RTO Structures and Materials Panel* 13–17 October, Ålborg, Denmark

[22] Smialek J L 1996 Microscopy of Oxidation—3, in *Proceedings of the Third International Conference, Trinity Hall, University of Cambridge* p 127 eds S B Newcomb and J A Little, 16–18 September

[23] Smeggil J G, Shuskus A J, Bornstein N S and DeCrescente M A 1989 *The Role of Active Elements in the Oxidation Behavior of High Temperature Metals and Alloys* ed E Lang pp 271–285

[24] Grabke H J, Kurbatov G and Schmutzler H J 1995 *Oxidation of Metals* **43**(1/2) 97

[25] Pint B A, Nagaraj B A and Rosenzweig M A 1996 in *Elevated Temperature Coatings: Science and Technology II* ed N B Dahore and J M Hampikian (The Minerals, Metals & Materials Society) pp 163–174

[26] Andersson C A, Bratton R J, Lau S K and Lee S Y 1980 *Thin Solid Films* **73** 481

SECTION 2

LIGHTWEIGHT MATERIALS

The developments in lightweight materials for aerospace technologies, principally for use as structural components in airframes, are centred upon two major classes of material: aluminium alloys and polymer matrix composites. In the case of aluminium alloys, considerable attention has been focused upon aluminium–lithium alloys, where recent improvements in damage tolerance, and the advantages of weldability and formability can overcome disadvantages such as the inherent expense of the lithium. Polymer matrix composites are largely based around carbon fibre/epoxy resin structures. Their inherent high strength/low density properties make them attractive, but there are important hurdles to overcome in terms of manufacturing costs, reliability (confidence) and temperature resistance, before they can be more widely exploited. This section discusses many of the issues at the forefront of the development of these two classes of materials, which to some extent are competing for exploitation in the aerospace industry. In addition, the section will discuss the other important lightweight materials systems that are or may be exploited, such as novel metastable aluminium alloys, aluminium based metal matrix composites and novel polymeric systems.

Chapter 9 gives an overview of the subject of lightweight materials, discussing the various materials classes and their application, and also presenting other materials requirements that interface with the low component mass requirement, such as high temperature stability and manufacturing route. Chapter 10 focuses on the key area of fatigue in aluminium alloys. It presents an extensive survey of the current knowledge base in this subject and shows how fatigue analysis is being generalized. Chapters 11 and 12 focus on two other lightweight alloy developments. The former considers aluminium alloys with amorphous or very fine crystallite structures where the novel strengthening mechanisms in these non-equilibrium phases allow very high strength materials to be formed. An alternative method for strengthening aluminium alloys is to include a ceramic reinforcement. Chapter 12 describes the work that has been done towards overcoming the problems of low toughness in these materials.

The latter four chapters of the section describe developments in polymer matrix composite materials for aerospace application. Chapter 13 gives an overview of some of the materials issues in fibre and matrix development, and the two chapters following focus on two major issues in the application of such materials: chapter 14 describes developments in thermoplastic toughening of the composite structures, and chapter 15 the synthesis of hydrophobic materials to allow the use of these materials in higher temperature environments in which they are exposed to humidity. The final chapter of the section analyses the cost implications for the application of these lightweight materials, making a comparative study of various composite systems and their manufacturing routes and the costs associated with application of aluminium alloys.

Chapter 9

Advances in aerospace materials and structures

Chris Peel

Introduction

Examples are given in this chapter of some of the factors driving research programmes aimed at materials developments, that subsequently control their embodiment into commercial projects, in particular affordability. This will be achieved by specific reference to the status of developing aerospace materials, including new light alloys, metal matrix composites, polymeric composites and trends towards more affordable structures. Finally, a speculative look is taken at future aerospace structures and the consequential requirements for aerospace materials.

It is first necessary to consider the driving forces for the application of new materials in aerospace. These can be simplified into four interdependent issues namely:

1. improvements in aircraft performance, e.g. range, payload, speed;
2. reductions in cost, e.g. cost of acquisition and operation;
3. improvements in survivability, crashworthiness and ballistic performance; and
4. enhancement and maintenance of the environment.

The translation of new concepts, materials and ideas into commercial reality is currently dominated by cost-effectiveness, in terms of the cost of acquisition, including qualification and manufacture, and in terms of the subsequent operating cost associated with application of the end product. It seems clear that, although there is a continuous drive in the aerospace world for new materials developments, there are many negative aspects and significant costs associated with them and they are undertaken at risk.

Improvements in aircraft performance

Previously, it has been the case that the performance of military systems has been the dominant factor, there being no place for second best. There are real arguments for the maintenance of technology at the leading edge for systems that have a critical impact on military performance. This transposes into lighter, higher performance materials and those with higher temperature capabilities. In addition to these real incentives for the selective enhancement in performance of military combat and transport aircraft, there are different but equally demanding ones for the civil transport field.

Reduction in structural mass

For combat aircraft, the speed of climb, flight and turn, the fuel and payload masses, rates of fuel burn, visibility and survivability are obvious issues. Many, but not all, of these are determined by total allowable aircraft mass and, hence, reduction in parasitic structural mass is a major driver. For engines, the thrust to mass ratio is a key issue and, remembering that thrust is limited by the maximum thermal capability of the materials employed, the current drive combines mass reduction by the use of advanced materials with increases in operating temperatures and hence increased thrust.

Transport aircraft performance can be measured in terms of the cost of transport per passenger-mile or load-mile. In reviewing the impact of new technologies on the likely gains in reduced fuel consumption for a large transport fleet, Swihart [1] revealed that advanced materials and structural designs had a significant part to play. However, the impact of the new materials is relatively small in comparison with the impact of improved engine performance or enhanced aerodynamics, as shown in figure 9.1.

Higher temperature airframe developments

Enhancement of performance envelopes can be illustrated by the application of new materials at elevated temperatures for airframe, aeroengine and advanced missiles. The instantaneous allowable temperatures for conventional aluminium alloys or polymeric composites may be limited to approximately 150°C but there is a significant prospect for the development of higher temperature capabilities for both material types that may be critical for certain projects. For example there is now a case for the development of a second generation supersonic transport aircraft, a Concorde replacement. It has already been indicated that increases in payload and range are critical to the success of a new supersonic transport, just as could be an increase in an aircraft's operating envelope. There are simultaneous requirements for reduced airframe mass and a longer structural life at elevated temperatures,

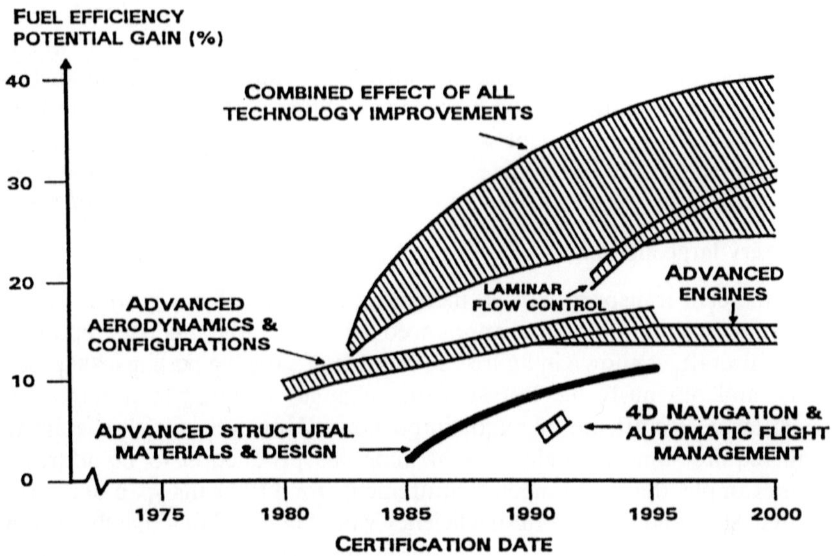

Figure 9.1. The effects of advancing technologies on specific fuel burn showing the relative contributions of advanced materials and structures (after Swihar [1]).

higher than those experienced by Concorde. Considering these aspects, advanced materials will be of direct benefit in extending the range of an aircraft with a prescribed payload, as shown in figure 9.2. For example [2], for the next generation of supersonic transport flying at Mach 2.5, conventional

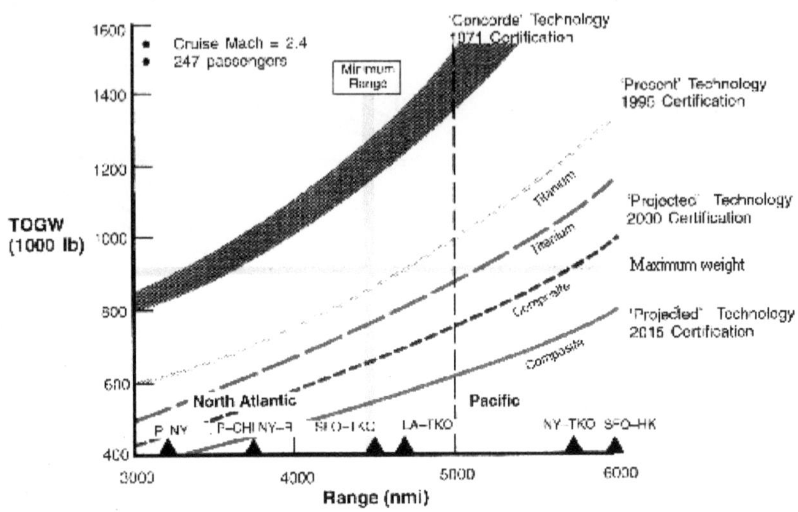

Figure 9.2. The impact of materials performance on potential range and mass of a second-generation supersonic transport aircraft (after Quist [2]).

aluminium or titanium alloys may not produce a structure light enough to allow sufficient range to cover the Pacific routes. New materials, whether polymeric or metal composites or advanced alloys, must simultaneously produce weight reduction and higher temperature performance to produce a viable aircraft. If insufficient weight saving is achieved or high temperature performance proves inadequate, the aircraft may not be commercially viable.

The very large airframe

For the civil transport market, increasing aircraft size has become increasingly important. Plans are announced to design and build exceptionally large aircraft, as shown in figure 9.3, capable of carrying perhaps 800 passengers, and obviously the infrastructure changes required to accommodate such large aircraft are of major importance. However, such large aircraft will require some critical issues in materials production to be addressed. For example, the simultaneous requirements for a full wing span of approximately 80 m and for structural efficiency would demand that ideally, whether metallic or composite, pieces of material approximately 40 m long would be required, or more than one wing joint will have to be employed. Currently half wing covers are typically constructed from single lengths of rolled or extruded aluminium alloy approximately 30 m long. This increase in length

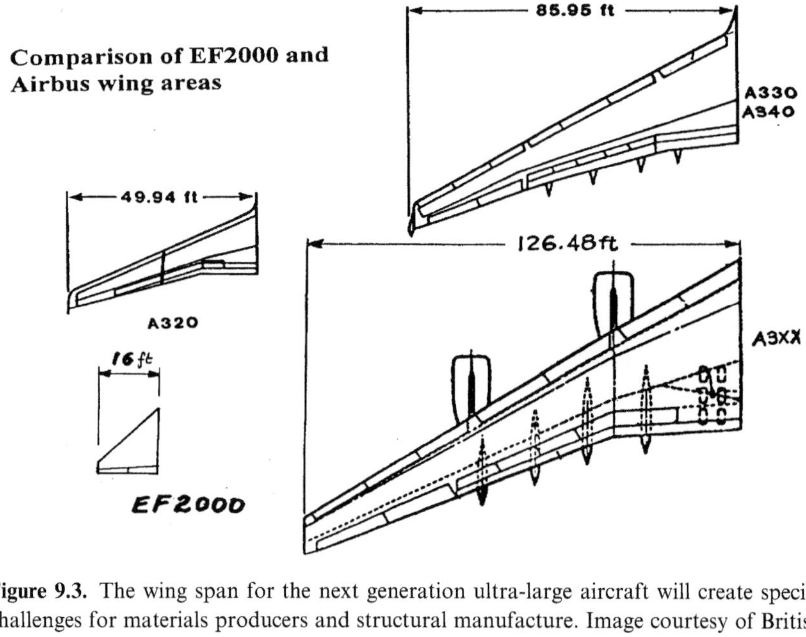

Figure 9.3. The wing span for the next generation ultra-large aircraft will create special challenges for materials producers and structural manufacture. Image courtesy of British Aerospace Airbus.

will require significant capital investment whether the choice is an enlarged aluminium alloy plant or extremely large composite facilities. Composites can produce lighter structures when specific stiffness is the issue, and metallic materials are more efficient for strength requirements, so it may well be that future large wings will be hybrid.

This major demand for new materials in large structures should provide a strong incentive for the materials industries. Similar trends can be found in the military transport aircraft where the wing structures of large transports are similarly constructed from exceptionally large planks.

Developing requirements for aerospace gas turbines

The requirements for materials employed in gas turbine engines parallel those applicable to the airframe construction, with a continuing drive to reduce parasitic mass but now with a distinct emphasis on the need to derive new materials that will operate at ever increasing temperatures. This is because the thermodynamic efficiency of a gas turbine engine is increased as the working temperatures and pressures are increased, reducing specific fuel consumption and increasing thrust to weight ratios. Current military gas turbine engines achieve a thrust to weight ratio approaching 10:1 but this ratio needs to be doubled as a longer term goal. Targets for improvement include combinations of weight reduction, reduced engine diameter (figure 9.4), reduced number of stages to achieve lower mass and cost, increased pressure ratio and higher turbine entry temperature to effect thrust improvements. Since the increasing operating temperature required to increase thrust will be limited by the high temperature performance of existing and new materials, weight reduction by materials development must remain a major issue.

The property requirements for components of the gas turbine are varied. Cooler compressor areas will tend to be dominated by corrosion, erosion, impact and fatigue considerations. Turbine blades operating at high

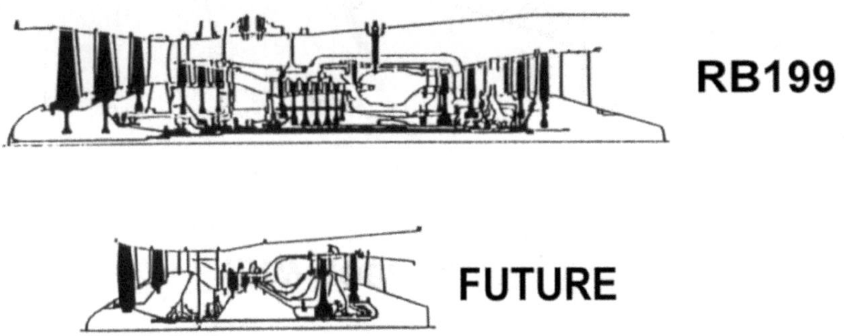

Figure 9.4. Targets for future military engines include increased thrust to weight ratios and reduced engine cross sections (after Ruffles [3]).

temperatures will be dominated by creep and corrosion resistance, primarily because the stresses, limited by creep resistance, are too low to cause fatigue problems. However, the turbine discs on which they are mounted are cooler in operation, more highly stressed and therefore are again fatigue critical.

To meet these challenges, two activities run in parallel with attempts to increase operating temperatures or to reduce parasitic mass. These include the derivation of new nickel superalloys and tough engineering ceramics, to operate efficiently at increasingly elevated temperatures, and the derivation of new alloys, metal composites and ceramics designed to reduce parasitic mass in components operating at higher temperatures than may be achievable currently. The development of high temperature nickel alloys is also discussed in chapter 8.

Reduction in costs

Costs of acquisition and ownership

A current trend with all new materials being embodied in military and civil projects is the minimization of both initial cost and cost of ownership. A problem of increasing magnitude is the escalating cost of new military equipment, as shown in figure 9.5. Higher initial purchase costs can be offset to a certain extent by lower operating costs. Weight savings achieved by the use of modern materials can reduce fuel burn, or can increase the proportion of the total structural mass that is payload or can increase periods between maintenance.

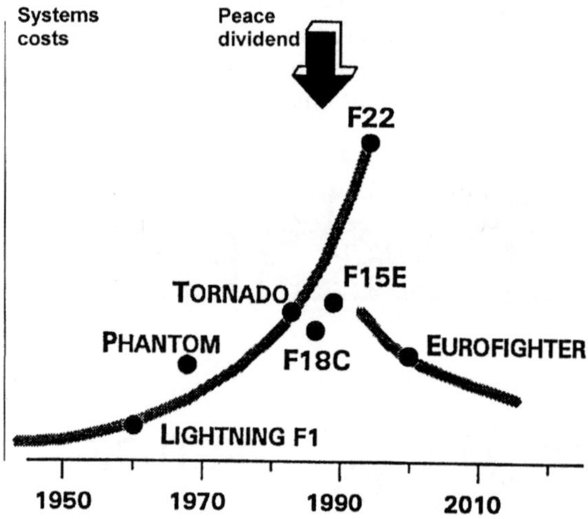

Figure 9.5. Schematic illustration of the escalation in the cost of military aircraft (after Jagger).

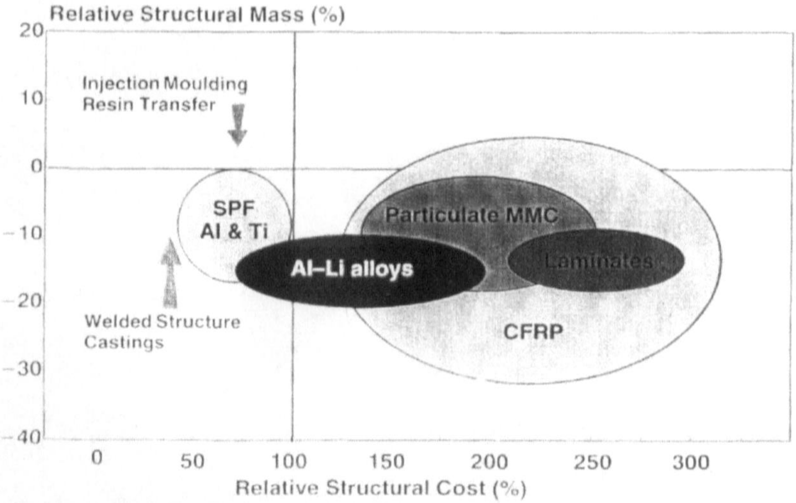

Figure 9.6. Relative structural mass savings and costs for structure embodying new materials.

Whilst these improvements may lead to a reduction in the cost of ownership, any premium associated with an initial increased cost of new materials must be taken into account. Frequently, new materials offer weight savings at increased cost, as shown in figure 9.6. For example, fibre reinforced polymeric composites generally offer higher weight savings than new metallic materials but at a potentially higher cost. Superplastic forming (SPF) coupled with diffusion bonding (SPF-DB) of titanium or aluminium alloys appears to combine cost and weight saving but is limited in the size of component that can be manufactured. Considerable average weight savings can be achieved with improved aluminium–lithium alloys and carbon fibre reinforced polymers.

The impact of maintenance costs

For both civil and military aircraft, the costs associated with maintenance and operation will be at least two to three times greater than initial purchase values, and this trend will be magnified by the need to extend future operational lives for military equipment, as shown in figure 9.7. A distinct difference may be found between the distribution of operating costs for military and civil aircraft because, with relatively low utilization of military aircraft, fuel burn becomes of less importance than systems up-grade, maintenance and repair. Indeed, in the context of the impact of materials and structures on the cost of new combat aircraft, it can be seen in figure 9.7 that the cost of materials has a very small influence on the overall cost, but the cost of maintenance of the airframe and engine structures is most significant.

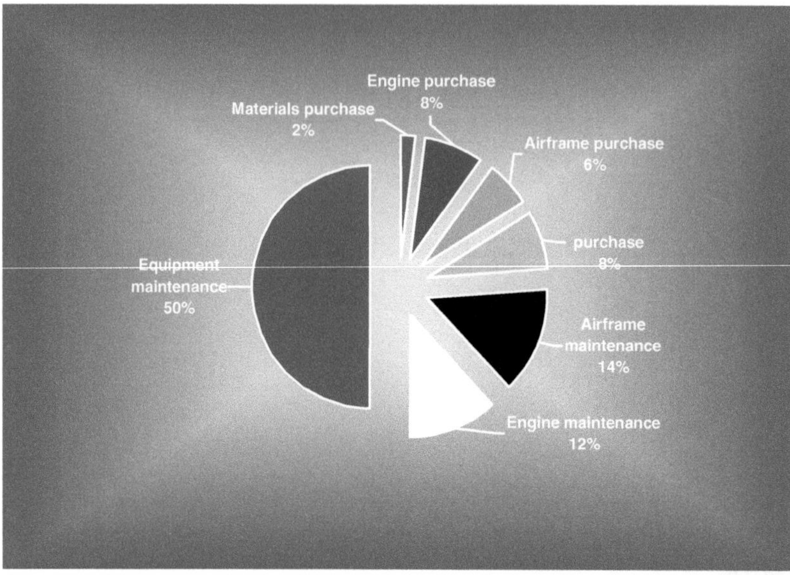

Figure 9.7. Relative distribution of costs for a military aircraft showing the small dependence upon material costs.

Additional to simple arguments related to fuel burn or payload, mass reduction in an airframe or the rotating parts of a gas turbine can reduce the acceleration-induced loads, which in turn can lead to an increase in fatigue life. Moreover, new materials such as Al-Li alloys or metal matrix composites can be employed purely because of their improved resistance to fatigue cracking and enhanced levels of damage tolerance, whilst polymeric composites can be used to offset the damage caused in corrosive environments. For example, in general terms, lightweight alloys have been employed very extensively in aerospace to reduce parasitic mass, and the developments of the past two decades have been oriented towards the enhancement of performance in service by the reduction in fatigue damage rates or sensitivity to corrosion problems. With Al-Li alloys there appears to be the possibility to reduce the parasitic mass of the structures to which they are applied, and to improve the performance in service. For example, Al-Li alloys are being used in the fuselage frames of military aircraft, with up to a fivefold increase in fatigue life being achieved for this critical application, and with a concomitant mass reduction as a further bonus.

Repair technology

Further examples of the influence of materials choice on operating costs can be found. For example, during the long development of carbon fibre reinforced polymeric composites, insufficient attention was given in the early stages to the effect on the operating costs, with special reference to the cost of repair. Following the paper presented by Thorbeck [4], it can be found that repair costs for composite structures can exceed those for conventional metals by a factor of at least two. In the event of the need to replace entire items containing significant damage levels, such as a complete fin, the re-investment required to replace the damaged item cannot be sufficiently offset by the relatively modest fuel burn reductions. This has resulted in a regression towards the cheaper metallic structures, at least whilst fuel prices remain low or while emphasis is placed on more affordable maintenance costs. Analogous comments can be made on the potential difficulties in repairing Al-Li alloys and particularly on the impact of the appearance of lithium in the aluminium scrap cycle.

The status of current materials

New materials aimed at military aerospace application, such as Al-Li alloys or fibre reinforced composites, are continuously being adopted. These also include particle and fibre reinforced metal matrix composites based on aluminium, magnesium and titanium alloys, and more exotic mechanically alloyed and powder materials. The application of titanium alloys in combat aircraft is showing a resurgence, although a drive towards cost effective manufacture tends to limit alloy choice to traditional variants. Nickel superalloys and ceramics continue to make incremental improvements. New polymer matrices are appearing with enhanced toughness and higher temperature performance, whilst new fibres and fibre concepts continue to be derived.

Developments in aluminium alloys

Although aluminium alloys have dominated the construction of military and civil airframes since the inception of the stressed metallic airframe in World War Two, new developments continue. These include improved high strength Al-Zn based alloys (7000 series), better damage tolerant Al-Cu and Al-Mg-Si alloys (2000 and 6000 series) and the more novel Al-Li and Al-Be variants.

High strength aluminium–zinc alloys

There has been a steady development of the Al-Zn-Mg-Cu alloys (7000 series), mainly used in structures loaded compressively, such as upper wing skins, but also for internal ribs, frames and landing gear. The drive for increased compressive strength in upper wing skin material for civil aircraft leads to ever increasing levels of applied stresses to maximize weight saving, such that good fatigue strength and fracture toughness become increasingly important.

Additionally, problems that beset the early use of the 7000 series alloys with poor exfoliation and stress corrosion resistance have to be kept in check. Techniques such as double ageing practices, controlled combinations of heating rates and mechanical deformation between quenching and ageing, and the use of the so called reversion ageing practices has enabled strength, fracture toughness and corrosion resistance to be increased simultaneously. In Europe, initiatives in France and the UK have developed variants of ISO 7449 alloy with improved strength and fracture toughness that maintain an appropriate level of corrosion resistance.

The damage tolerant aluminium–copper alloys

Traditionally, the 2000 series Al-Cu and Al-Cu-Mg alloys have been used for applications requiring damage tolerance. Damage tolerance can be considered as the beneficial combination of very high fracture toughness and very low rates of fatigue crack propagation, particularly in natural environments under loading spectra containing the occasional high tensile load. From early development and steady evolution, naturally aged 2000 series alloys have come to dominate this type of application in both fuselage skin and lower wing covers. In both, the strength of the material has generally been less important than its damage tolerance, and stress levels are, in the main, set to achieve satisfactory fatigue lives. In the case of the fuselage sheet skin material, sufficient fracture toughness has been required to enable the fuselage to withstand a crack two frame bays in length at maximum service pressure. The improvement achieved in increasing the plane stress fracture toughness of 2524-T3 is illustrated in figure 9.8, using data from Rioja [5].

Damage tolerance has improved progressively by minimizing levels of impurity elements, by improving fabrication practice to optimize grain structures, and by cold working of fastener holes to maximize fatigue strength. Working stress levels have as a consequence been increased to save weight. This has resulted in a need to increase the yield strength of materials used in lower wing skins to offset yielding, illustrated by the application of 2324-T39.

Aluminium–lithium alloys

Recent years have seen the emergence of Al-Li alloys offering typically up to 10% density reduction and 15% increase in stiffness, coupled with significant improvements in damage tolerance, particularly achieved by improvements in resistance to the growth of fatigue cracks. The first generation of new Al-Li alloys, such as the Al-Mg-Li alloy 1420 developed in Russia, and the

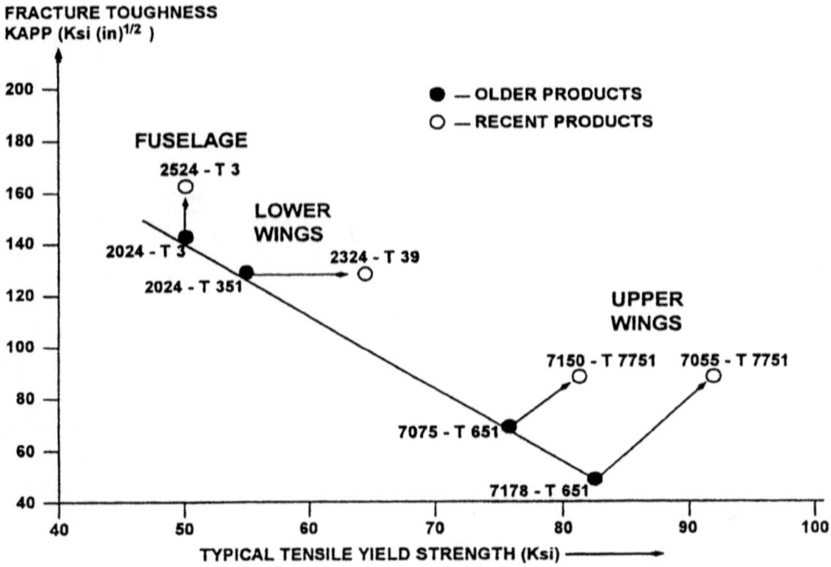

Figure 9.8. Improvements in strength and fracture toughness (Alcoa data [5]).

Al-Li-Cu-Mg alloy 8090 developed in the UK, are now in service in major military and space structural applications. Studies in the USA are centred on the use of alloys more dilute in lithium to impart improved fracture toughness and weldability such as the Al-Cu-Li alloys 2095/2195 and 2097. Applications include space shuttle fuel tank frames and skins, where weight reduction is at a premium, and aircraft fuselage frames. The use of Al-Li alloys in helicopters is discussed in chapter 4.

Improvements have been achieved over the past decade in the performance of Al-Li alloys by the eradication of some of the persistent problems hampering their application. These have been achieved partly with the development of a second generation of more dilute alloys and partly by improved control of microstructure. Research has concentrated on factors causing scatter in short transverse fracture toughness in thick sections, the maximization of in-plane toughness for sheet products and improvements to thermal stability.

Since growing use of these new alloys is being achieved, work now proceeds on a second generation of alloys for further application. Two themes are emerging, one being property enhancements, the other extensive reduction in the cost of structures embodying Al-Li alloys. It may be noted that the damage tolerant variants of the Al-Li alloys now being studied in the UK have higher fracture toughness and fatigue crack growth resistance in sheet than even improved 2000 series alloys, but at a slightly lower level of strength.

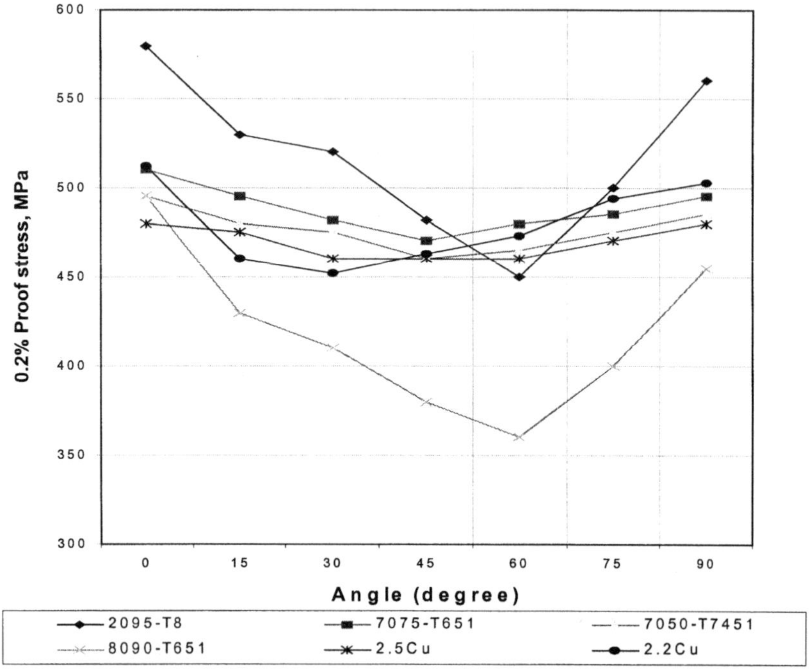

Figure 9.9. Improvements in isotropy have been achieved with the DERA high strength Alloy X Al-Li-Cu-Mg compositions.

Attention in the UK is concentrated on the derivation and optimization of a high strength variant with maximum weight saving potential in military applications. In particular, material with high compressive performance is sought for undercarriage, spar and wing plank applications. Progress to date has been good. In particular material with isotropic properties is being developed to alleviate problems with anisotropic values in strength that reduce the design allowables for these alloys, as shown in figure 9.9.

Lower cost alloys and production methods

To reduce the cost of aluminium alloy structures, two approaches may be considered. Of course cheaper alloys may be selected but, as a general trend, technical requirements are increasingly exacting and consequently attract price increases. A second approach is to develop lower cost semi-fabrication and final assembly techniques. For example, stringers or stiffeners may be integrally extruded in sheet, as shown in figure 9.10, to obviate built-up structures, or alternatively welding may be employed. These changes in fabrication may require alloys to be derived and optimized to enable the technology. For example, the high performance aerospace variants of the

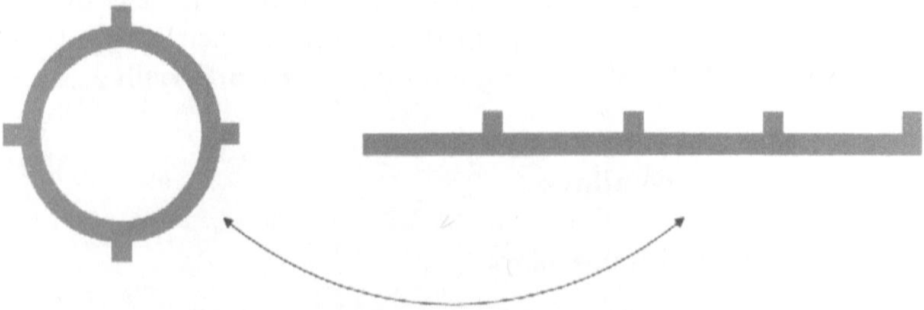

Figure 9.10. The use of integrally stiffened extruded panels may significantly reduce the cost of fabrication for thin section stiffened shells.

Al-Cu and Al-Zn series have traditionally been regarded as poor in terms of weldability. This issue is being addressed especially with the emergence of improved welding technology.

As an example of a potentially cheaper material, aluminium alloys precipitation-hardened by Mg_2Si have been in general use (including aerospace) for decades. Recently, interest in this system has increased for several reasons. First, the alloys appear to offer potential for reduction in initial cost, being potentially cheaper than Al-Cu alloys without the expensive copper addition. Second, being weldable, they appear to offer the potential to reduce manufacturing costs. Third, variants with improved property levels have appeared typified by the ISO 6013 alloy. Somewhat surprisingly, the damage tolerance of the 6000 series alloys has been shown at least to match that of the incumbent 2000 series alloys in terms of sheet fracture toughness and resistance to fatigue crack growth.

Cost reduction using Al-Li alloys may seem a difficult task because of the inherently high cost of lithium, until certain critical features are revealed. For example, the alloys can be successfully welded and have been applied extensively to the structure of combat aircraft and to the space structures in the welded state. Whilst 50% cost reductions for the initial structure may be achieved by welding, the impact of this type of structure on sustainability and cost of ownership must be carefully considered.

A second major feature for cost reduction in Al-Li involves the use of creep forming technology typically used on metal combat aircraft wings with conventional Al-Cu alloys. Creep/age forming involves pre-setting the wing plate material to the wing curvature and then using the artificial ageing cycle of the aluminium heat treatment to creep the material permanently to shape. Alternative methods with metallic wing planks involve either complex machining operations or the use of shot blasting. Creep forming is efficient in that it involves two operations simultaneously and in particular it can produce the deep curvatures in very large structures required in modern high lift wing forms. However, Al-Li alloys of the 8090 type have a particular advantage in that they have been shown to possess the correct balance of damage tolerant properties when artificially aged, whilst damage tolerant Al-Cu alloys are not normally aged artificially.

Higher temperature alloys

Higher temperature airframe alloys

There persists a barrier to the application of materials at elevated temperature. When operating temperatures of airframe or missile skins exceed approximately 150° to 200°C, aluminium alloys or polymeric composites tend to be replaced by titanium alloys, increasing cost significantly. If light alloys or composites can be improved in this respect, the use of titanium or other heavier materials can be obviated, saving cost and weight, as shown in figure 9.11.

For this reason, powder metallurgy aluminium alloys are being evaluated at temperatures up to 450°C. Close control of the composition of the atomizing gas coupled with sieving and gas classification techniques in powder atomization allows the production of fine spherical powders to pre-selected sizes that can be readily mixed, whilst the relatively rapid quenching rates ($\sim 10^4\,\mathrm{K\,s^{-1}}$) of the powder process allows the retention of supersaturation of normally insoluble elements. Perhaps more importantly, the

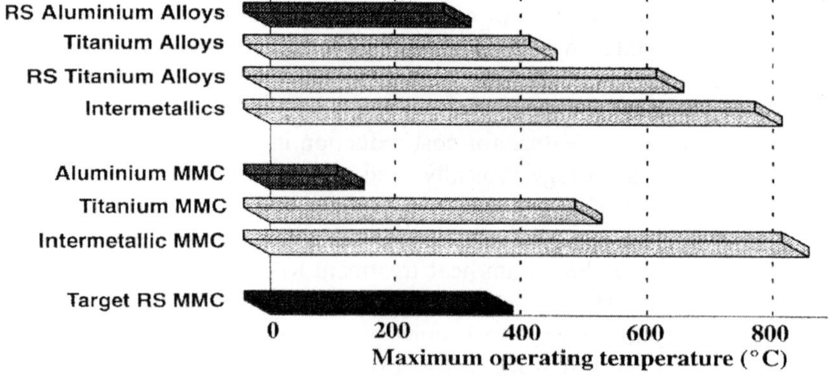

Figure 9.11. Relative performance for contending high temperature materials with targets for maximum operating temperatures.

powder route lends itself to the use of mechanical alloying techniques for material fabrication.

Alloy compositions currently being pursued are demonstrating thermal stability at temperatures as high as 450°C with satisfactory levels of strength at ambient temperatures but, still with inadequate strength at temperature (figure 9.11). Research is aimed at the derivation of compositions and microstructures that optimize the need for high temperature ductility during fabrication with maximum creep strength and strength at service temperatures. A strong tradition persists in the use of compositions that produce fine dispersions of thermally stable precipitates typified by the combination of iron, vanadium and silicon in the ISO 8009 alloy. However, many of these systems can show a temperature range in which embrittlement occurs by the precipitation of very fine phases in matrix and grain boundaries.

If temperatures higher than can be sustained with light alloys or polymer composites are envisaged, the traditional choice has been an aerospace titanium alloy. An issue of some importance is that titanium is electrochemically more compatible with carbon fibre reinforced polymers than aluminium alloys and is therefore being chosen for hybrid metal–composite structures to circumvent corrosion problems associated with the use of aluminium.

Higher temperature engine alloys

Titanium alloy usage has centred on the $\alpha + \beta$ alloys such as Ti-6Al-4V or Ti-6Al-2Sn-4Zr-2Mo in wrought forms, and on commercially pure titanium material or Ti-6Al-4V in cast forms. Although there is a large range of titanium alloys potentially available [6], considerations of cost, castability, fabricability and weldability have tended to limit the choice for most applications. Features such as the need to heat-treat the more complex alloys after welding or diffusion bonding have had a critical influence on materials selection.

The substantial incremental increases in performance of the titanium alloys and titanium based intermetallic alloys in the UK should not be underestimated. Maximum service temperatures are illustrated in figure 9.12. It may be noted that the step change from the standard Ti-6Al-4V alloy to Ti-4Al-4Mo-2Sn-0.5Si (IMI 550) not only adds strength at room temperature but also gives a 100°C increase in maximum operating temperature. The further step to the near α alloys like Ti-5.8Al-4Sn-3.5Zr-0.7Nb-0.5Mo-0.3Si-0.05C (IMI 834) illustrates the complex alloy chemistry required to increase maximum operating temperature to approaching 600°C.

Intermetallic alloys appear to offer a further extension in operating temperature, although the maturity of these materials is not as advanced as the complex titanium alloys. Intermetallic systems are based, in the main, upon titanium aluminides centred on hexagonal close packed super $\alpha 2$, tetragonal Ti-Al and orthorhombic Ti_2Al-Nb compounds. These have

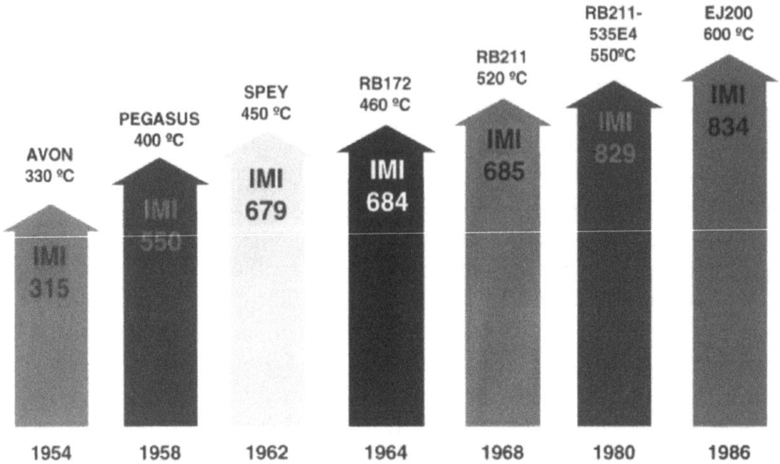

Figure 9.12. The steady progression in maximum operating temperatures and performance achieved with titanium alloys [6].

complex chemistries, adjusted to attempt to impart satisfactory room temperature ductility and toughness, coupled with high temperature strength and creep resistance, by the generation of fine duplex or complex microstructures. Titanium aluminides are discussed further in chapter 17.

The derivation of new nickel superalloys for high temperature turbine application and the associated manufacturing technologies can be traced from the inception of the gas turbine using Nimonic 80 in the 1940s. Step changes can be seen with the introduction of cast alloys during the 1960s, to be followed by the use of directional solidification and then single crystal developments. These developments are described in more detail in chapter 8. The principal aim has been to improve resistance to creep damage at temperatures approaching 1100°C. Hence the recent emphasis on control of grain structure to maximize creep strength through the casting processing. Target operating temperatures beyond the currently achievable maxima of approximately 1100°C would appear to require the use of ceramic-based materials probably employing fibre reinforcement, discussed below. The engineering technology has also kept pace with the materials developments, for example the introduction of internally cooled blades allows their operation in hot gas streams, at temperatures above the melting point of the blade alloy.

Affordable titanium structures

The application of titanium based materials is closely linked to cost effective manufacturing technologies. The inherently high price of titanium, based on current extraction, alloying and fabrication technologies, makes it unattractive for general structures. However, when the ability to form the material superplastically, with or without diffusion bonding, and the excellent weldability of the simpler alloys are taken into consideration, titanium structures become highly competitive. Recent trends have seen a resurgence in the use of titanium alloys, especially when weight reduction is required by replacement of steel components, or strength is required at temperatures higher than can be sustained by an aluminium alloy or polymer composite. The wide chord fan blade, flying control surfaces and internal structure of combat aircraft provide good examples of this technology reaching application. One exciting trend that may develop is the qualification of sources and extraction methods for reduced cost titanium. Several approaches are being adopted in the UK but these also include the potential use of materials from Russia and the East.

Composites

Metal matrix composites

To extend further the performance of metals, work has continued on the derivation and extension of a family of metal matrix composites. In a metal matrix composite, use is made of ceramic reinforcements in the form of fine particles, fine fibres or continuous monofilaments to reinforce matrices typically of aluminium, magnesium, titanium, nickel or iron. The inclusion of the ceramic reinforcement imparts much increased stiffness to the composite, with a significant increase in strength and in certain cases very much enhanced strength at temperature. The use of particulate or randomly aligned multi-filament tows produces a composite with isotropic properties. Alignment can be achieved with multi-filament and monofilament variants, to tailor properties and to enable greater optimization of the structure to save weight. To fabricate the composites, use is made of liquid metal processes, such as squeeze casting or liquid metal injection, powder technologies and mechanical alloying with hot isostatic pressing for consolidation. In the case of alloys reinforced with SiC monofilaments, a key element is the ability to form superplastically and diffusion bond the matrix material.

There are major benefits for aircraft structure in the use of material with increased specific stiffness because approximately half of the structural mass may be limited by stiffness requirements. Composites based on aluminium alloys reinforced by ceramic particles or chopped fine fibres, typically of silicon carbide or alumina, show increasing stiffness and strength with increasing levels of reinforcement, but this is at the cost of reducing toughness and ductility. Maximum additions for aerospace engineering products tend to be approximately 25% to 30% by volume although additions of up to 60% can be achieved for other types of application.

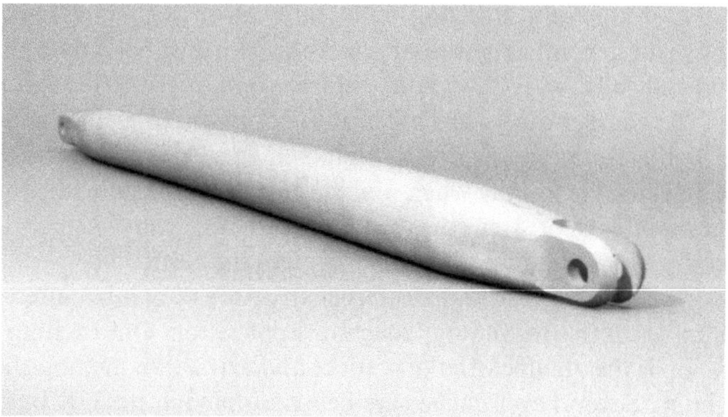

Figure 9.13. One piece particulate reinforced aluminium alloy metal matrix composite strut. Courtesy of Aerospace Metal Composites and Robert Cruzet.

Recent detailed studies and worked examples have revealed that particulate reinforced composites based on aluminium alloys can provide matching specific strengths, stiffness and damage tolerance to polymeric composite components at competitive cost. As a result there has been a growing interest in their application to stiffness-critical components, typified by transverse wing and floor support struts for a transport aircraft (figure 9.13), rotor head components, crashworthy crew seat struts, drive shafts and struts etc. These are basically small components where the combination of strength, stiffness and wear resistance can be put to good purpose. Early production applications have been achieved in the USA with, for example, particulate reinforced aluminium alloy for limited skin replacement on updated combat aircraft.

Multi-filament metal matrix composites

There remains further potential to enhance metal composites performance with the use of randomized or aligned fibre reinforcement. Multi-filament composites, typically consisting of tows of fine fibres of alumina or carbon in aluminium, have been developed. Currently, manufacturing technology favours the use of a squeeze casting technique ensuring the consolidation of the solidifying metal under significant pressure but also embodying the need to control carefully the delivery of molten metal through fibre preforms. Accurate fibre placement and minimization of fibre swim is achieved by proprietary techniques. The use of randomly oriented alumina fibres incorporated in a series of aluminium casting alloys is now being adopted progressively for components such as pistons for military diesel engines, where the demand for stability at elevated temperatures, control of thermal expansion, wear resistance and high temperature fatigue strength matter. For this type of application, rapid production methodologies are essential to maintain high levels of cost effective manufacture. The squeeze casting technique lends itself to this form of automated production but is naturally limited to relatively small components.

For a potentially higher level of structural optimization, emphasis has been placed on the use of aligned fine fibre tows containing alumina, silicon carbide and graphite fibres, with typical individual diameters below 10 μm and with reinforcement again incorporated by liquid metal and squeeze casting techniques. Many potential applications are being addressed for small components such as missile parts including control fins.

Monofilament reinforced metal matrix composites in aeroengines

There are potential applications for aluminium and titanium alloy components selectively reinforced with continuous monofilaments, especially when there are requirements for use at high service temperatures in aeroengine components. In these composites, a single fibre or monofilament, typically of silicon carbide, 100 μm or more in diameter, is distinguished from a tow of fine fibres described above. Monofilaments up to 25 km in length are currently produced by chemical vapour deposition, typically of boron or silicon carbide ceramic around either a tungsten wire or a single carbon fibre, as opposed to spinning tows of fine ceramic multi-filaments. The monofilament composites are capable of very high specific strengths and stiffness in the direction of fibre alignment especially at increased operating temperatures. However, the current price of the monofilament alone is in the region of £5000/kg.

Several fabrication routes can achieve selective reinforcement by monofilaments of aluminium alloys, titanium alloys and intermetallic compounds. These include the diffusion bonding of packs of aligned monofilaments between sheets of alloy foil or similar packs in-filled with alloy wires. Alternatively, filament wound structures can be produced using monofilament coated with the requisite matrix alloy and consolidated again by diffusion bonding. The two most common approaches for coating monofilament are based upon plasma spraying during filament winding of the fibre or the use of physical vapour deposition of matrix material around individual fibre before filament winding and consolidation. A sustained drive for high temperature metal matrix composites based on silicon carbide monofilaments in titanium alloys is aimed at achieving higher engine thrust to weight ratios, on military engines typically from circa 10 : 1 to approaching 15 : 1. Inclusion of these metal composites into structures such as an integrally bladed compressor discs (bliscs or blings) is predicted by Doorbar [7] to reduce rotating mass by up to 70% (figure 9.14) and depends critically upon the high temperature performance of the new composite material. Here the term bling is used to describe an integrally bladed ring, differentiating it from blades mounted conventionally on discs using fir tree fittings.

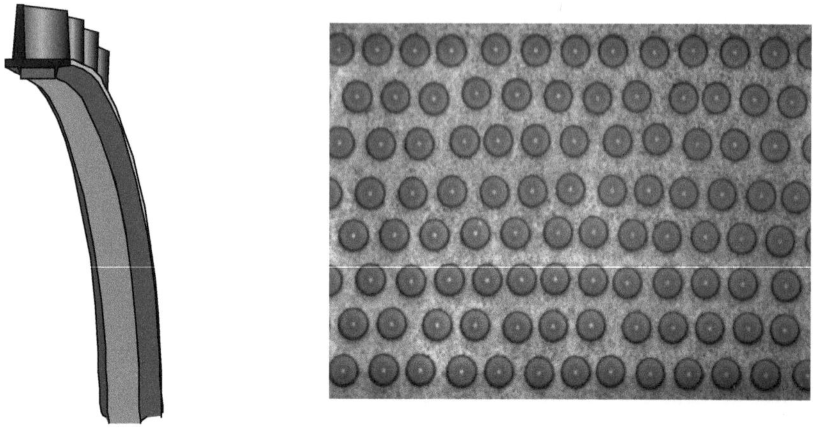

Figure 9.14. Selectively reinforced titanium alloy metal matrix composite used to reinforce aeroengine blings.

Composites of this form are intended to be used for the selective reinforcement of structure since, presently, the high cost of the base material would preclude large scale usage or general application within a component except perhaps for the most expensive weight critical space structures. However, significant potential for application can be found in the aeroengines of the future, where weight reductions and performance increases have been identified for a range of components such as low pressure compressor blades and vanes, compressor and turbine blings, turbine shafts, compressor casings etc. At present the performance of monofilament reinforced metal composites is being assessed in prototype engine components with cost of manufacture very much in mind.

Ceramics and ceramic composites

Monolithic ceramics and composites based on ceramic matrices offer a powerful balance of properties stemming from their covalent bonded structures. These have exceptional capability at high temperature in terms of strength, stiffness, hardness and wear resistance coupled with relatively low density, making their choice seemingly inevitable for high temperature applications, typically in gas turbine components. However, lack of ductility, fracture toughness and resistance to shock held back structural application until the advent of toughened materials during the 1980s, although use of ceramic coatings has long been made for critical wear and temperature resistant surfaces.

Typical monolithic ceramics include those based on alumina and alumina–silica mixtures, with toughening achieved by the inclusion of up

Table 9.1. Relative properties for selected ceramics in sintered form.

Ceramic	Density ($g\,cm^{-3}$)	Elastic modulus (GPa)	Flexural strength at 25°C (MPa)	Estimated max temp (°C)
Alumina	4.0	400	550	900
Silicon carbide	3.1	300	380	1400
Silicon nitride	3.1	240	420	1000
Toughened zirconia	5.7	210	600	900
Alumina/SiC whiskers	3.7	380	640	900

to 20% by volume silicon carbide as fine whiskers to produce a composite. Fracture toughness levels approaching $10\,MPa\,m^{1/2}$ have been achieved. However, inconsistency in properties and fears of the health risks associated with fine whiskers are limiting structural application.

A key step towards inherently higher toughness has been the development of fine grained polycrystalline material whether in alumina or zirconia. A second key step has been to toughen zirconia with stabilizing oxides such as magnesia or yttria that prevent high temperature phase changes. This has formed the basis for a wide range of engineering ceramics with mixtures of tetragonal zirconia in cubic zirconia (PSZ) or alumina (ZTA) or fine grained polycrystalline zirconia itself (TZP). Mechanical properties are given in table 9.1 but it should be noted that the strength and toughness of these toughened materials deteriorates as temperatures of 1000°C are approached, limiting high temperature applicability.

A further class of engineering ceramic materials is loosely described as 'non-oxide ceramics', comprising systems based on silicon carbide, silicon nitride and the Sialons. Toughening is achieved by whiskers or short fibres of silicon carbide/nitride. Having lower coefficients of thermal expansion than the oxides, these materials appear more attractive for high temperature engine applications, where thermal shock is critical.

It should be noted that fibre and monofilament reinforced ceramic matrix composites may extend the toughened ceramic concept by allowing anisotropic strengthening, with the use of aligned fibres typical of a metal or polymer matrix composite. Glass or silicate based composites, carbon/carbon and silicon carbide/silicon carbide combinations all appear to offer exceptional properties at temperature, with increased toughness achieved by crack deflection and fibre pullout mechanisms imparted by the fibre reinforcements. Although the large scale application of these developing ceramic materials is still awaited in airframes and engines, the significant contribution of early systems such as carbon/carbon in aircraft disc brakes and carbon/carbon or silicon carbide/silicon carbide rocket engine nozzles points to the potential for extensive use in the future.

Polymers

Polymers and polymeric composites

Fibre reinforced polymer composites will undoubtedly find extensive application in future aerospace structures. Now regarded as conventional materials, in the sense that it is more than 30 years since the early forms of carbon fibre reinforced epoxy were evaluated for aerospace applications and were first adopted for use in sports goods, fibre reinforced polymeric materials have been continually improved over this period.

Major phases of significance can be recognized in the use of glass fibre composites for radar transparent covers, and the early use of boron fibres in epoxy resins typified by application in tail structures, followed by the adoption of high strength and stiffness carbon fibre reinforcement in combat aircraft wings. Separately, the use of aramid intermediate and high modulus fibres has become commonplace in many secondary structure applications, where toughness and durability have been the key issues. Fibre properties are compared in table 9.2. Carbon fibre based on pyrolysed polyacrylonitrile precursor has been adopted for structures of high mechanical integrity, typified by wing structures for current combat aircraft. Offering similar stiffness to aluminium alloys in a typical near isotropic composite, the primary advantage of the polymer-based material has been reduced density, affording weight savings in structures dominated by stiffness requirements.

The past decade has seen increased utilization of hybrid materials in which fibre types are selected and mixed to balance the requirements for strength, stiffness, toughness and impact resistance and, most recently, control of signature. Moreover, the past decade has also seen the emergence of new fibres with further improved mechanical properties such as increased modulus and failure strains, reflected by increased usable tensile strengths. Indeed it should not be assumed that fibre and polymer developments have ceased.

Table 9.2. Relative fibre properties.

Fibre	Tensile modulus (GPa)	Tensile strength (GPa)	Density ($g\,cm^{-3}$)	Specific modulus ($10^6\,m^2\,s^{-2}$)	Specific strength ($10^6\,m^2\,s^{-2}$)
Ultrahigh modulus carbon	725	2.2	2.15	336	1.02
Intermediate modulus carbon	310	5.2	1.8	170	2.9
Medium modulus carbon	235	3.8	1.8	130	2.1
R/S Glass	85	4.5	2.52	35	1.8
E Glass	69	2.4	2.54	27	1.0
Boron	400	3.5	2.6	155	1.4
Silicon carbide	400	4.0	3.4	120	1.2
Aramid	125	3.6	1.45	85	2.5

Figure 9.15. The first example of a continuous hollow carbon fibre.

For example, new fibres appear regularly, with extended mechanical properties. Typical new arrivals could be illustrated by the ultrahigh performance Nexel alumina fibres, ultrahigh molecular weight polyethylene, hollow glass and carbon fibres, ultrahigh performance silicon carbide fibres and monofilaments and ultimately, perhaps, diamond fibre. Some of these developments have been driven purely by mechanical property improvement, but others such as the hollow carbon and glass fibres have interesting possibilities for smart material applications in which the hollow core of the fibre may be ultimately used for the inclusion of strain sensing and reactive materials and potentially self-repairing matrix materials [8], as shown in figure 9.15.

The greater application of fibre reinforced composites has continued to be hampered by a few key problems. These can be simplified to fundamental difficulties in the use of what is currently, essentially, a two-dimensional material, limitations imposed by minimum gauge requirements, and the all-pervading issue of cost effectiveness.

Affordable composite structures

Since improvements in the technical performance of the polymer composite may be constrained by third axis properties, emphasis is being currently placed on the cost-effective manufacture of structures with three-dimensional capability. Progression from mechanically fastened composite skins attached to metallic and pre-cured composite substructures, through adhesively bonded structures, to the current co-cured combinations of skin and substructure has produced steady improvements, although mechanically fastened removable covers will remain a requirement for many applications. Further developments already taken to demonstrator stages include the use of stitched three-dimensional structures, three-dimensional woven fibre architectures and the use of woven fibre laminates bonded or mechanically fastened.

Cost reduction developments have recognized that the use of autoclave bonding of pre-preg laminates is unlikely to be affordable for transport applications or large structures in general. Developments are therefore being directed at techniques such as resin transfer moulding, resin injection, the use of adhesive films and non-crimp fibre fabrics. Automation of fibre placement with tape and tow laying techniques may prove most significant. Thermoplastic matrix composites may offer possibilities for techniques such as diaphragm forming and the use of superplastic metal membrane forming at least for smaller components. An issue in all of these techniques will be the cost of tool design and manufacture and, in particular, the need to predict material performance accurately and to minimize production development costs.

It was illustrated earlier that adoption of polymer based composites may produce significant increase in the cost of new structures but a reduction in subsequent service costs. Achieving an accurate appreciation of both initial and lifetime costs against a background of ever changing application requirements is a very major challenge. However, the adoption of increasing levels of sophistication in electronic methods of product definition and life modelling may allow designers to simulate the effects of selections and changes very rapidly, aiding the optimization process.

Polymers and higher temperature polymers

The two basic forms of polymers used monolithically and as matrices for polymer composites are based on either thermosetting or thermoplastic principles. Thermosetting materials require polymerization to be generated in the base material by addition of a curing agent and, usually, application of an elevated temperature. The material so formed is set in its final shape. Typical thermosetting matrices include epoxy, polyimide (bismaleimide), polyester and phenolic resins. Thermoplastic materials are fully polymerized *ab initio* and are set to shape at elevated temperatures where they are softened or even melted, being capable of repeated thermal cycles. Typical thermoplastic materials include polyether ether ketone (PEEK), polyether imide (PEI), polyphenylene sulphide (PPS) and polyarylene sulphone.

There are materials that are thermoplastic in nature but that require a chemical reaction initially to polymerize such as the semi-thermoplastic polyimides. Many of the thermoplastics are semi-crystalline in nature and possess better toughness and chemical resistance than thermosetting materials. A recent trend has been the derivation of hybrid mixtures of thermosetting and thermoplastic materials to achieve ease of application and increased toughness simultaneously.

A major issue of significance to future application, especially in the gas turbine, is the limiting temperature of operation achieved by the

Table 9.3. Typical maximum hot–wet operating temperatures.

Matrix polymer	Type	Max hot/wet limit (°C)
Epoxy	Thermoset	120
Polyimide (bismaleimide)	Thermoset	220
Polyimide	Semi-thermoplastic	260
Polyether ether ketone (PEEK)	Thermoplastic	120
Polyetherimide	Thermoplastic	120

matrix of polymer composites, typified by the maximum hot–wet operating temperature allowable. Table 9.3 shows the typical maximum operating temperatures of selected matrix polymer systems. Notable are the thermosetting polyimides and the semi-thermoplastic polyimides, with maximum operating temperatures in excess of 200°C. The key problem is to balance a high glass transition temperature, necessary for high service temperature, against the minimization of thermally induced micro-cracking during initial setting. New materials are emerging from the USA that have significant promise in this respect.

As a general trend, the matrix materials for high temperature application are significantly more expensive than those used in general low temperature applications. However, they may meet a requirement that otherwise would require the use of an expensive titanium based metallic solution.

Discussion

It has been the intention in this chapter to review the state of development of emerging materials for military aerospace applications, with specific comment being addressed to cost of manufacture and ownership. Costs of acquisition and ownership are relative and the following two distinct trends emerge.

If a new material is more expensive than the incumbent that it is intended to replace, to be cost effective, improvements must be made in utilization or in reduction of costs associated with fabrication of the material into the structure. Research has identified means to increase utilization and to embody more effective manufacturing, for example in the current re-investigation of the welded airframe.

However, there is the further possibility that if a new material can match the technical performance of a more exotic material then it can be cost effective. Thus, if expensive polymer or metal matrix composites with the appropriate stiffness and high temperature strength can match the performance of the intrinsically more expensive titanium alloys, both performance increase and cost reduction can be achieved.

Future trends

It is extremely difficult to extrapolate current trends in aerospace materials development much beyond the next decade and any such extrapolation must be regarded as speculation. In this light, the more obvious trends dictating the immediate future can be separated from longer-term issues. Immediate trends may be summarized as:

1. improving performance, including survivability;
2. reducing costs;
3. improving the environment; and
4. sharing commercial risks.

Airframes and engines will continue to benefit from reduction in parasitic mass. Systems such as landing gear will require to be reduced in weight and if possible volume simultaneously, posing major challenges for engineering with improved materials including, perhaps, a move away from steel towards a titanium alloy base. Undoubtedly a major challenge for improved performance is raised by the requirements of future airframes and engines to accommodate increased operating temperatures. Large structures such as a new supersonic transport aircraft are contemplated that, from their sheer size, costs and risks, must be produced in near conventional materials and must be optimized for the high temperature requirement. Combat aircraft and missiles will fly faster or employ vectored thrust techniques demanding the derivation and application of weight efficient materials with enhanced high temperature performance. Engines continue to reduce in size but increase in thrust. To achieve the 15 : 1 thrust to weight ratios in the compact engines required for military aircraft of the future requires the combination of *all* the new materials technologies outlined in this chapter.

Cost effective manufacture of both transport and combat airframes, aerospace structures and aeroengines is a major current theme moving towards more affordable structures. This issue will have a major influence on the potential for utilization of materials such as fibre-reinforced composites and Al-Li alloys. Unless the costs of built-up structures utilizing these expensive materials can approach those using the incumbent materials there is little hope for their embodiment on a large scale. To reach this objective, demonstrator programmes investigating low cost manufacturing routes for large composite and metallic structures are now in planning, if not under way. Cost effective manufacturing processes have been considered in passing throughout this chapter, but these are matters in which the UK capability has been traditionally underdeveloped. Present trends are towards modular or unitized structures maximizing commonality in terms of materials, process and units of built-up structure. Techniques that minimize parts counts are being researched for both metallic and polymer composite systems exemplified by the current interest in welded metallic airframes.

Reduction of service lifetime costs is a major issue. Since the lifetime service and repair costs of a military airframe may be two to three times those of the initial purchase, more will need to be done to model and control lifetime performance and to select materials with improved fatigue and corrosion resistance, and this aspect can reverse the trend towards materials with lower initial and manufacturing costs. Protecting and conserving the environment is an issue of ever increasing importance. The effects of emissions from aircraft engines may well limit speeds and altitudes in the short term and will continue to provide pressures for higher temperature materials to enable turbine entry temperatures to be further increased, reducing fuel burn and emission levels simultaneously. More attention is being applied currently to scrap and waste matters such as the replacement of the use of cadmium and chromate for corrosion protection, the reduction of the levels of volatile substances in surface finishes and paints and the complete eradication of halon fire suppressants following the Montreal Protocol. An ability to recycle the materials used in aerospace structures may even become a requirement that will affect the use of current polymers and alloys. In the same vein, the method of ultimate disposal of a material will become a requirement for consideration during development.

For the far future, a trend continues to emerge in which there is a steady separation of the needs of military equipment from those for general transportation. To be efficient in reducing service costs, transport aircraft will continue to grow in size and range. In contrast, for example, the development of unmanned military systems may be an extension of present capability. Without the encumbrance of the human physiology, sensitive as it is to acceleration and environment, and the continuous need to train flight crews, speeds and turn rates may be increased significantly, structural loads increased and system flight lifetimes reduced dramatically to enable much more efficient use of the airframe materials. Selectively these structures themselves will become smaller and smaller, with increased thrust to weight engines compacted in reducing volumes, and with miniaturized electronic and control systems. Whole body castings, welded metallic structures and injection moulded thermoplastics or derivatives thereof would seem to offer cost effective solutions. Manufacturing trends in the cost-conscious automobile industry may be adopted. Even the re-utilization of welded pressed sheet structures may occur as these are consistently cost and weight efficient.

Application of aerospace materials will become a holistic process, combining materials and process routes with prediction of engineering and market performance, with developments predicted by modelling. Whatever the trends that emerge, whether it is the greater application of more robust marketable materials to aerospace structures, the greater use of predictive modelling, or the continued derivation of specialized materials for highly optimized structures, the critical role of the materials scientist will persist in the aerospace industry.

References

[1] Swihart J M 1980 'The next generation of commercial aircraft: the technological imperative' *12th Congress of the International Council of Aeronautical Sciences* ICAS, October

[2] Quist W E 1990 'Advanced materials for a high speed civil transport' *Capital Metals and Materials Forum*, November

[3] Ruffles P C 1996 Innovation in aero engines *Aeronautical J.* **100**(1000)

[4] Thorbeck J 1991 'Economical view on composite structures maintenance' *12th International European Chapter of SAMPE* (Elsevier Materials Science Monographs)

[5] Rioja R J 1996 Private communication on 'ALCOA data for aluminium alloy development'

[6] Farthing T 1988 'Designing with titanium' *6th World Conference on Titanium, Cannes, France* (Les Editions de physique) pp 37–48

[7] Doorbar P J 1993 'Metal matrix composites in aero-engines' *Aerotech 1993* (Institute of Mechanical Engineers)

[8] Curtis P T 1996 *Multifunctional polymer composites: advanced performance materials* vol 3 (Boston: Kluwer Academic Publishers) pp 251–262

Chapter 10

Fatigue optimization in aerospace aluminium alloys

Ian Sinclair and Peter Gregson

Introduction

Fatigue is a critical design consideration in many airframe components. As improvements are sought in o'verall performance, maintenance requirements and total service life, the demand for fatigue resistant materials and structures is increased. In this respect it is important to recognize that fatigue performance is highly sensitive to applied load levels, with the fatigue life of a built-up airframe structure containing stress concentrations (particularly fasteners) being inversely proportional to the load range raised to a power of about seven, i.e. small increases in working load levels place disproportionate demands on fatigue resistance. It is telling that whilst tremendous effort is expended in predicting fatigue performance, and significant progress has been made in the scientific understanding of fatigue processes, aircraft structures are still verified by expensive structural tests at every design level (i.e. ranging from simple single panels and joints, through to entire airframes). This may, at least in part, be related to the complex, multi-mechanistic character of fatigue failure, where numerous competitive and synergistic interactions arise from changes in load parameters such as applied stress state (e.g. proportional and non-proportional multi-axial loads), load history (simple transients, block loads, etc.) and chemical environment. Unlike primary structural properties such as elastic modulus and yield strength, there is no guarantee that any one fatigue performance parameter established in the laboratory will be meaningful or accurate under service conditions. In terms of detailed microstructure-based fatigue research and understanding, most reports within the literature have been biased towards relatively simple load environments, rendering detailed quantification of microstructural effects on fatigue in realistic performance environments difficult [1]. Whilst a full discussion of all the associated issues is beyond the scope of this chapter, it is clear that quantitative *ab initio* modelling and optimization of realistic commercial microstructures for all aspects of in-service fatigue performance are as yet largely untenable. This chapter considers the underlying physical principles associated with this problem, identifying important aspects for future scientific assessment, along with general technical issues relevant to the future aluminium airframe design.

Current status

In understanding fatigue failure it is well established that the associated physical processes fall into regimes of distinct and often contradictory microstructural dependencies. The regimes are primarily: initiation, short crack growth and long crack growth, with long crack growth further breaking down into near-threshold, Paris and static mode regimes. These regimes do not define individual controlling mechanisms, but rather a broad set of mechanical and microstructural interactions.

Crack initiation

Fatigue crack initiation is essentially promoted by strain localization. This is manifest through a range of initiation sites in conventional alloys, including transgranular slip bands, intermetallic particles (cracked, decohered or intersected by local slip bands), and porosity, varying with material type, condition and applied load conditions [2–20]. In terms of basic materials character, it is recognized that shearable strengthening precipitates and the associated propensity for heterogeneous planar deformation may accelerate initiation via increased strain localization within slip bands, with corresponding detrimental effects on fatigue life being augmented by the environmental sensitivity of such slip bands [20–24], as shown in figure 10.1. Conversely, microstructural characteristics promoting homogeneous deformation may be expected to moderate the effects of shearable precipitates, namely decreasing grain size, increasing grain boundary misorientations, the presence of a sub-structure and/or non-shearable dispersoid particles [20, 22, 25–29]. Attempts to quantify the relationship between slip heterogeneity and microstructure have been put forward, with more recent work by Starke and co-workers addressing both shearable precipitate and available slip length effects [30, 31]. By equating the degree of local work softening due to precipitate shear, and the work hardening that occurs as a result of dislocation pile ups at major slip barriers (grain boundaries and/or non-shearable particles), an explicit expression is derived to describe slip heterogeneity.

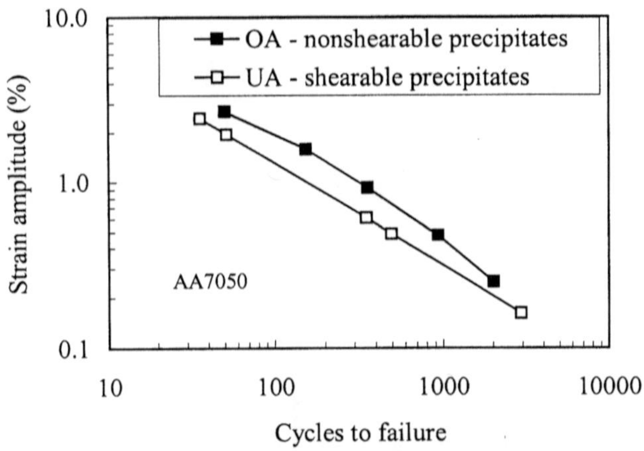

Figure 10.1. Precipitate shearability influence on low cycle fatigue performance (after Starke and Lutjering [20]).

Correlations between such results and monotonic mechanical behaviour have been identified [30, 32]; however, quantitative links to fatigue behaviour are not well established as yet (e.g. see reference [33]).

For moderate to high cycle fatigue lives, intermetallic particles represent the most commonly reported initiation site for commercial alloys. Crack initiation at intermetallic particles is known to be promoted by increasing particle size, aspect ratio, alignment with the loading direction, increasing dimensions of the immediately surrounding grain, prior particle damage from processing and lamellar/clustered particle morphologies [2, 9, 11–15, 34], e.g. figure 10.2. An intrinsic size dependence for particle initiation may be related to the energy balance of crack formation and/or the strength variability of a given volume of brittle material [12, 35, 36]. Kung and Fine [9] also suggest that increasing particle size may increase the probability of particle/slip band intersection. Particle elongation and alignment in the direction of loading may be expected to amplify any local stress and strain partitioning effect between the matrix and particle phases due to their compliance mismatch [16, 35]. In terms of the effects of surrounding large grains on crack initiation from particles, James and Morris [12] identify in some detail the preferential cyclic straining, and hence the increase in particle stress levels, that occurs in individual large grains due to the locally increased slip distances.

The incidence of grain boundary crack initiation in peak and overaged materials has been identified with strain localization within precipitate free zones (PFZs), although boundary precipitates and slip band/grain boundary intersections are also known to promote failure [20–22, 25, 26, 37–40]. In keeping with monotonic failure mechanisms, fatigue failure from precipitate free zones may be inhibited by decreasing grain dimensions (reducing average precipitate free zone strain levels as the number of boundaries per

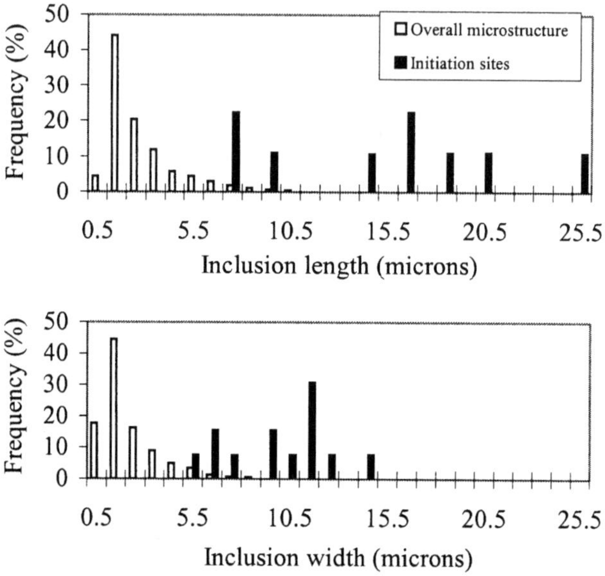

Figure 10.2. Crack initiating and overall inclusion size distributions in AA2024-T3 (after Laz and Hilbery [14]).

unit volume at which strain is accommodated increases, as shown in figure 10.3) and preferential boundary alignment parallel or perpendicular to the loading direction (precipitate free zone strain concentration being favoured by shear parallel to the boundary plane due to the constraint of the surrounding hardened material) [20, 26]. Detailed quantification of the influences of

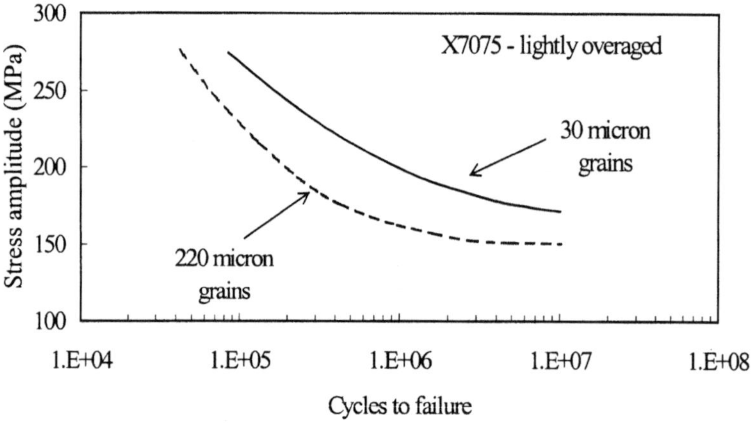

Figure 10.3. Grain size effect on high cycle fatigue in PPZ containing material (after Lutjering *et al* [26]).

individual boundary microstructure parameters on precipitate free zone fatigue failure does not appear to have been widely reported [41], although Welpmann *et al* [40] identify a beneficial effect of grain boundary ledges on boundary initiation due to the constraint of precipitate free zone shear. Park and Nam [41] report a systematic set of experiments on a laboratory Al-Zn-Mg alloy close to AA7017, where precipitate free zone widths were varied from 0.07 to 0.13 μm without significantly affecting the yield stress or matrix and grain boundary precipitate distributions. An increase in low cycle fatigue life was identified with increasing precipitate free zone width. The authors suggest that this was associated with an increasing degree of stress relaxation at slip band/grain boundary intersections as precipitate free zone width increases, but may also be attributed to a reduced average precipitate free zone strain as the volume of precipitate free zone material in which preferential deformation occurs is increased.

Whilst information on various microstructural effects exists regarding crack initiation and early growth, there is significant scope for detailed simultaneous quantification and separation of individual effects within specific systems. Various advances in materials analysis methods may be valuable in this respect. Electron backscattered diffraction (EBSD) [42, 43] in particular provides a convenient method for obtaining crystallographic information on individual grains and boundaries associated with failure, and may be extended to measuring local plastic strain levels. Given the generally recognized role of favourable slip orientations during crack initiation, it may be seen that local orientation distributions represent a critical parameter for any crack initiation study, with recent work on AA8090 [43], for example, quantifying the link between the Schmid factor for individual grains and crack initiation, as shown in figure 10.4. Local boundary orientation effects are also noted by

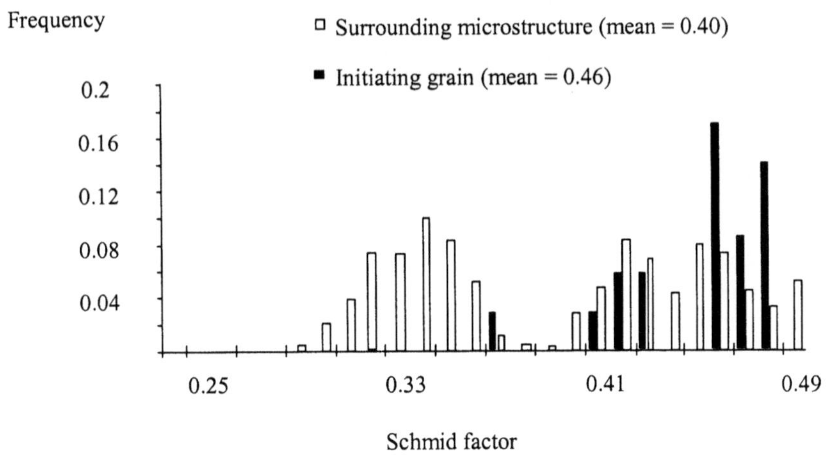

Figure 10.4. Schmid factor distributions in initiating and non-initiating grains in AA8090 (after Taylor *et al* [43]).

Feng *et al* [27], showing that increasing texture sharpness may favour initiation as effective slip distances are increased beyond individual grain dimensions. Kuo and Starke [25] identify various correlations between crack initiation and grain boundary types, with initiation within grains having low coincident site density boundaries being identified with the increasing availability of boundary sliding. Few works have deconvoluted microstructural effects to this level. Given the underlying correlations that exist between key microstructural elements (local grain size and texture links to local constituent and dispersoid phase distributions) this is clearly required for a proper quantitative description of material behaviour.

Short crack growth

The early propagation (or indeed, non-propagation) of freely initiating cracks within the short crack regime may be identified with conditions where: (1) crack length and associated plastic zone sizes are of the order of the local microstructural scale (i.e. grain boundaries and/or secondary phase particle separations); (2) the relative scales of crack-tip plasticity and crack length limit the applicability of linear elastic fracture mechanics; (3) crack-tip shielding processes associated with conventional quasi-steady state macroscopic crack growth tests are limited due to the limited size of the crack wake; and (4) environmental interactions in the crack-tip region are not significantly affected by transport along the crack path. Fatigue within the short crack regime is often characterized by a crystallographic slip band (i.e. Stage I) propagation mechanism (e.g. see reference [44]), with conventional $\mathrm{d}a/\mathrm{d}N$–ΔK plots exhibiting considerable scatter as acceleration and deceleration in growth rate occur at individual slip blocking features, as shown in figure 10.5.

The characteristically high growth rates associated with short crack propagation are evident in figure 10.6 (cf. the long crack curves), attributable to reduced closure levels associated with a short crack wake, the break down of small scale yielding/continuum mechanics descriptions, and the limited constraint associated with crack-tip deformation within individual grains at a free surface (e.g. see references [45, 46]). Various detailed models of short crack growth behaviour have been developed, a number explicitly accounting for the local deformation and crack-tip interactions with individual slip blocking features [47–52]. Extensive work by de los Rios and co-workers [50–52], for example, relates local growth rates to crack-tip displacements derived from a Bilby, Cotterell and Swinden description of crack-tip plasticity and the blocking of dislocation motion at successive boundaries. Using this approach successfully reproduces the initial discontinuous nature of short crack growth, and may incorporate materials parameters such as grain size, the effective resistance of boundaries to slip transfer, and work hardening.

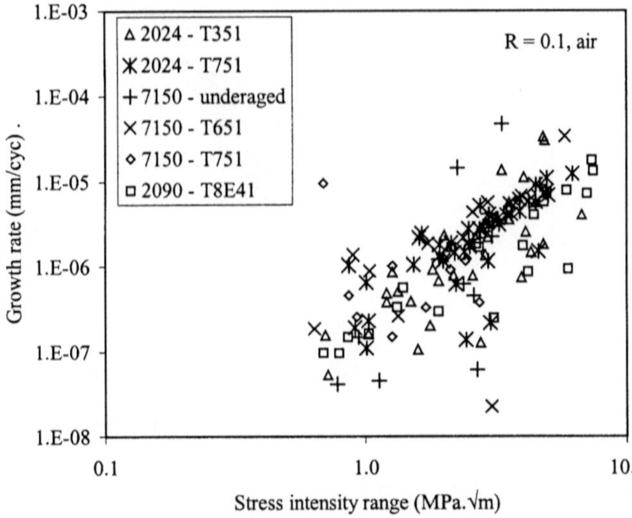

Figure 10.5. Short crack growth rate comparison for various heat treatable alloys (after Venkateswara Rao *et al* [53]).

Considerable overlap between short crack growth results for a range of alloys and heat treatments is generally reported in the literature, e.g. see references [53–56] and figure 10.5, although reports of generally lower growth rates in underaged materials compared with overaged materials have been made and attributed to increased reversibility of heterogeneous

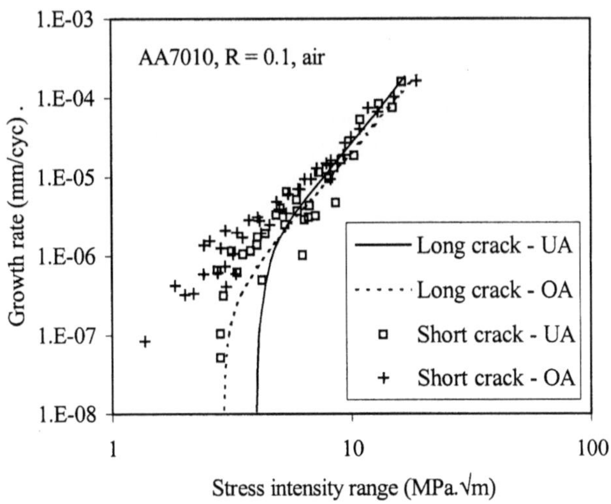

Figure 10.6. Heat treatment effect on short crack growth rates (after Bolingbroke and King [54]).

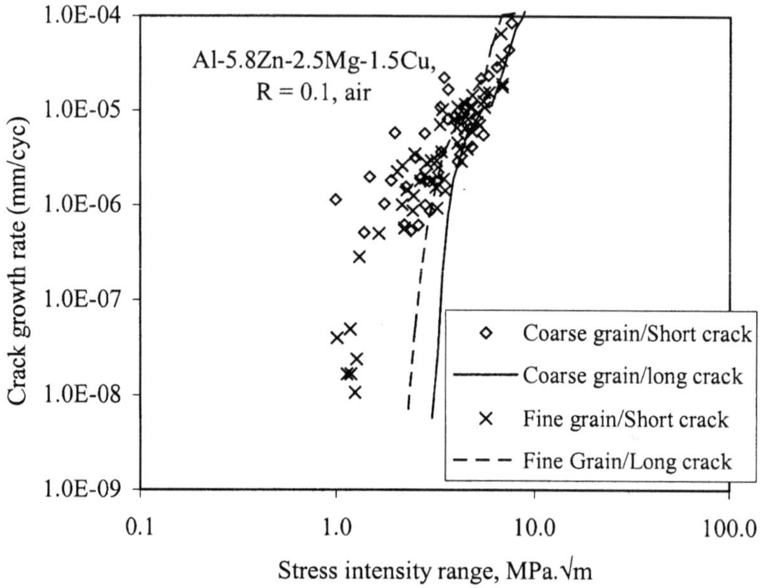

Figure 10.7. Grain size effect on short crack growth rates (after Edwards and Zhang [55]).

planar deformation, as shown in figure 10.6 [54]. Broadly speaking, however, growth rates are dominated by local grain boundary decelerations/arrest events. Fatigue resistance may therefore be expected to increase with decreasing grain size, as seen by the shift in the small crack data in figure 10.7 [55]. At extremely small grain dimensions, the characteristic acceleration of short crack growth compared with long cracks may in fact be suppressed, with propagation rates following closely the corresponding long crack growth curve [57]. Whilst the results shown in figure 10.7 are consistent with boundary control of growth rates, contradictory results [6] have been interpreted in terms of competitive influences of grain size on plasticity-induced closure and fatigue-induced surface residual stress levels. In terms of the mechanics of crack arrest at boundaries, growth rate decelerations are directly attributable to the constraint of crack-tip deformation, with growth rates being seen to fall as crack-tip plastic zones first reach a neighbouring boundary, and the size of the plastic zone then becomes limited to the distance between the boundary and the crack-tip [49, 55, 58, 59].

In addition to the slip blocking effect of grain boundaries it should also be noted that grains within which growth first occurs may be expected to experience particularly high resolved shear stress levels on the crack growth plane, and hence exhibit relatively high Stage I growth rates compared with neighbouring grains [43]. Given the role of boundary misorientation in the transferral of slip and the mechanical compatability

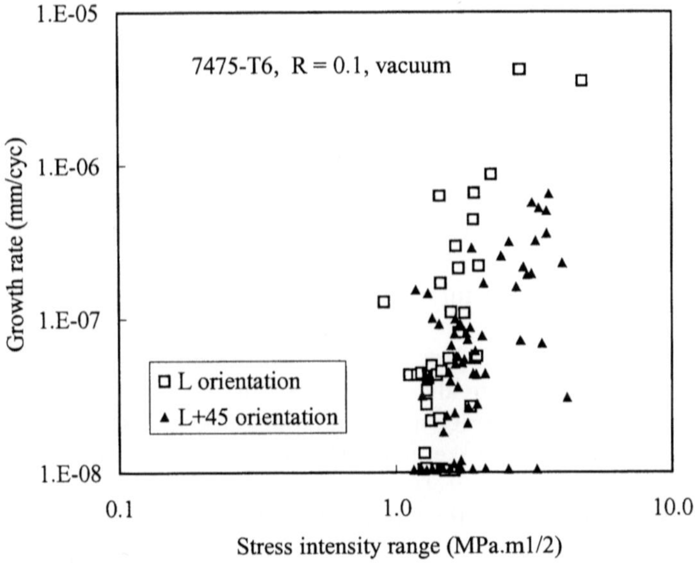

Figure 10.8. Influence of test orientation on short crack growth in strongly texture plate (after Gregory *et al* [60]).

between neighbouring grains, local and global texture may then directly influence short crack growth, with e.g. Gregory *et al* [60] seeing a marked enhancement in grain boundary crack arrest in a highly textured 7475 material when tested in an L + 45° orientation compared with the L orientation, as shown in figure 10.8. This was linked to the presence of two orientation variants in the underlying brass texture, and the different Schmid factors experienced by each variant for the two orientations tested. Li *et al* [61] have confirmed that the severity of individual grain boundary crack decelerations may be linked to local boundary misorientation, whilst Zhang and Edwards [44] have quantified this in terms of differences in resolved crack-tip shear stress between the available slip systems within successive grains (based on simple elastic crack stress fields).

Varying constituent particle effects on short crack growth have been identified, potentially increasing crack growth resistance by blocking slip like grain boundaries, and/or decreasing crack growth resistance by offering easy crack paths if particles themselves fail. The frequency of crack–particle interactions may of course be limited in higher purity materials, although the location of particles along grain boundaries may also be critical, with Mulvihill and Beevers [39] and Edwards and Gungor [62] identifying the arrest of *intergranular* short cracks at boundary particles (as well as triple points). Zabett and Plumtree [11] identify a detrimental influence of boundary particles during transgranular short crack growth, where particle cracking effectively bridges growth between grains, suppressing the normal grain boundary/crack arrest process.

Long crack growth

Overall, resistance to long fatigue crack growth may be differentiated between: (1) intrinsic crack-tip behaviour, which may vary with basic alloy condition, environment and the extent of crack-tip plasticity in relation to characteristic microstructural dimensions such as grain size, secondary phase particle separation and/or dislocation cell size [63, 64]; and (2) sensitivity to crack shielding processes, and particularly crack closure at low mean stress levels due to the relatively small crack-tip opening distances (cf. fracture surface asperity sizes and oxide thickness) and the presence of a macroscopic crack wake [65]. The separation of intrinsic crack-tip behaviour and shielding effects (or extrinsic behaviour) has of course constituted a major theme of fatigue research in the past 20 years, with shielding being used to rationalize a wide range of microstructural and mechanical influences on growth rate behaviour (e.g. see reference [66]).

Whilst the accurate and meaningful measurement of crack closure levels is open to discussion (e.g. the identification of the crack opening point and observed variations between global and near-tip measurements) [64, 67–69], and the significance placed on crack closure has been strongly criticized by Vasudévan and co-workers in recent years [70, 71], an extensive body of direct and indirect physical evidence and theoretical analysis does exist to support the role of closure in crack growth and the corresponding need to separate extrinsic and intrinsic influences on crack propagation.

Mechanisms

In terms of intrinsic crack growth behaviour, long crack growth in polycrystalline materials generally occurs along macroscopically tensile paths by either localized shear (extended Stage I) or tensile (Stage II) mechanisms. Shear growth is of course closely related to initiation/short crack propagation via single slip bands, and is characterized by tortuous kinked and twisted paths and multiple crack-tip bifurcation as different slip plane orientations are followed [72]. Non-crystallographic and crystallographic Stage II growth modes have been identified. The former may be considered to occur via classical continuum crack-tip blunting and resharpening processes [73, 74], whilst the latter is commonly associated with environmental interactions (particularly hydrogen-producing environments such as laboratory air). Crystallographic tensile growth in air is commonly identified with $\{001\}$ planes, but may also occur in $\{110\}$ orientations, both of which may be linked to symmetrical {111} slip activity with or without environmental influence [74–80]. Slavik and co-workers [33, 72] note, however, that such growth has never been identified under true high vacuum conditions (they indicate the need for test vacuums significantly better than 1 mPa), and, as such, may be entirely an environmental effect.

Conventional wisdom indicates that slip band crack growth is associated with near-threshold growth conditions, where critical crack-tip processes occur in a quasi-single crystal manner within individual grains at the crack-tip, whilst propagation in the Paris regime occurs via a Stage II mechanism as deformation extends from the crack-tip over many grains at a time to produce more continuum behaviour. The intrinsic incidence of such behaviour is indeed apparent in some cases, including single crystal materials tested in vacuum, where transitions in growth mode are clearly independent of the slip constraint of grain boundaries [81, 82]. Intrinsic deformation behaviour and environment may significantly modify this classical progression in growth mechanisms, however, with extended Stage I failure being particularly promoted by heterogeneous planar slip and inert/vacuum conditions [22, 25, 27, 81, 83–90]. The influence of intrinsic deformation character may be linked to the generally acknowledged propensity for crystallographic crack growth in underaged materials containing shearable precipitates. In Al-Li materials tested in moist air, highly heterogeneous planar slip may promote intensive slip band propagation up to very high stress intensity levels, with environmentally-induced {001} and {110} failure occurring at low/near-threshold stress intensities, e.g. see references [79, 91]. It is significant to note that whilst crystallographic failure modes are widely discussed in relation to long crack growth resistance, quantitative identifications of surface crystallography and correlations with underlying microstructure are not widely carried out.

Closure-free behaviour

For instances of the clear predominance of single growth mechanisms, Petit and co-workers [82, 92] have posited the existence of just two unifying long crack growth resistance curves for a range of medium and high strength aluminium alloys when closure is accounted for, as shown in figure 10.9. The authors specifically consider crack growth in vacuum, with the incidence of either extended Stage I or Stage II crack growth then exerting a controlling influence on growth rates. Whilst the increased crack growth resistance associated with extended Stage I growth compared with Stage II growth in figure 10.9 is commonly supported by the literature and attributed to enhanced slip reversibility when crack extension is dominated by single slip, the absence of microstructural influences is contradicted by other investigations [25, 33, 83], as shown in figure 10.10. As such, intrinsic microstructural influences on

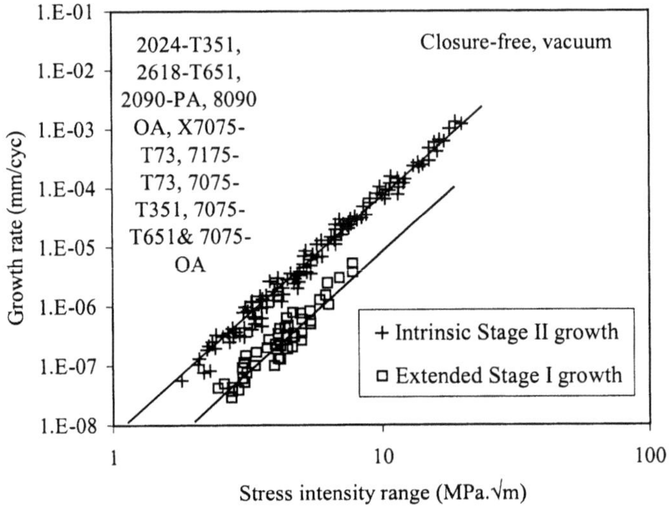

Figure 10.9. Stage II and extended Stage I growth rates for various heat treatable alloys (after Petit and Mendez [92]).

growth rates are unclear, even when the complexities of environmental interactions and closure are removed. Detailed consideration of the data in figure 10.10 in terms of flow stress and elastic modulus effects on crack-tip opening levels, or potential improvements in slip reversibility with increasing slip planarity, shows no simple correlation with observed growth rate behaviour [33].

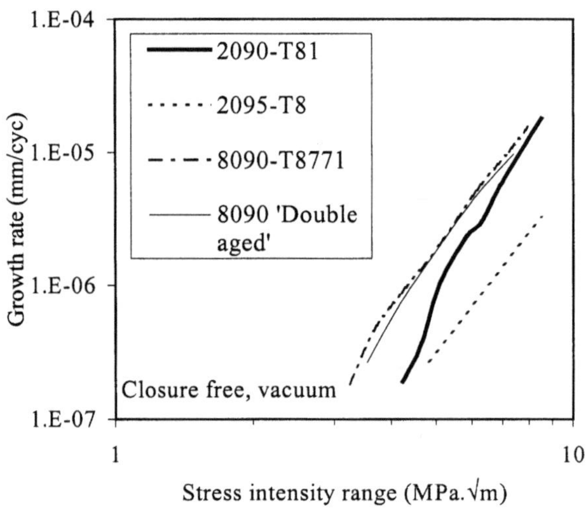

Figure 10.10. Closure-free extended Stage I growth rates for various Al-Li alloys (after Slavik *et al* [33]).

Observed improvements in closure-free extended Stage I crack growth resistance with increasing grain size have also been discussed in other works in terms of the reduced average strain experienced throughout an individual grain ahead of a crack as the grain size increases [25, 93]. It is important to note, however, that extended Stage I growth rate results, such as those shown in figures 10.9 and 10.10, are not intrinsic, since crack deflection, twisting and bifurcation shielding [65, 82] are necessarily associated with extended Stage I growth, even without addressing the validity of any closure correction procedure that has been used. Furthermore, whilst microstructural influences on long crack growth may be considered to be averaged through the bulk of the material (cf. short crack growth), some degree of local microstructural interaction may also give rise to ligamented crack fronts when microstructural propagation modes operate, e.g. slip band or grain boundary failure [83, 94]. Regions of locally high or low crack growth resistance give rise to pinning and dragging of the crack front. This may in fact be linked to the apparent incidence of macroscopic crack growth rates below one lattice spacing per cycle in the near-threshold region, although intrinsically discontinuous growth modes (i.e. damage accumulation processes) may also contribute to this phenomenon [95]. The formation of crack front ligaments by local dragging may be expected to reduce growth rates [96], with Grau *et al* [83, 97] highlighting the beneficial effects of ligamented crack fronts on fatigue resistance in coarse grained underaged materials in particular. Edwards *et al* [94] also consider complex crack front effects on growth, identifying a semi-cohesive zone at the crack-tip as independent ligaments are left behind the main crack front. In this case increasing dispersoid content was identified to increase both the number and load life of these ligaments. Ligament/semi-cohesive zone formation may not be accounted for in closure measurements and as such may still influence data such as that shown in figures 10.9 and 10.10.

The potential magnitude of three-dimensional crack deflection effects on closure-free crack growth rates is not generally recognized, but is illustrated by recent work by Liu *et al* [98, 99]. Here the strong crystallographic texture associated with AA8090 plate in a commercial unrecrystallized form was exploited in controlling crack paths in both conventional mode I and controlled mixed mode I and II tests. Specimens were specifically taken from the centre section of plate materials where a strong single brass-type texture occurred, and were tested under moderate to high mode II conditions in a nominal transverse/longitudinal (LT) orientation such that the crack starter notch lay parallel to the very strong {111} plane orientation that is common to both orientation variants of the brass texture, see figure 10.11. As such it was possible to produce extremely flat, slip band dominated crack paths under known mixed mode I/II opening conditions. Tortuous extended Stage I crack growth was observed in the same materials under

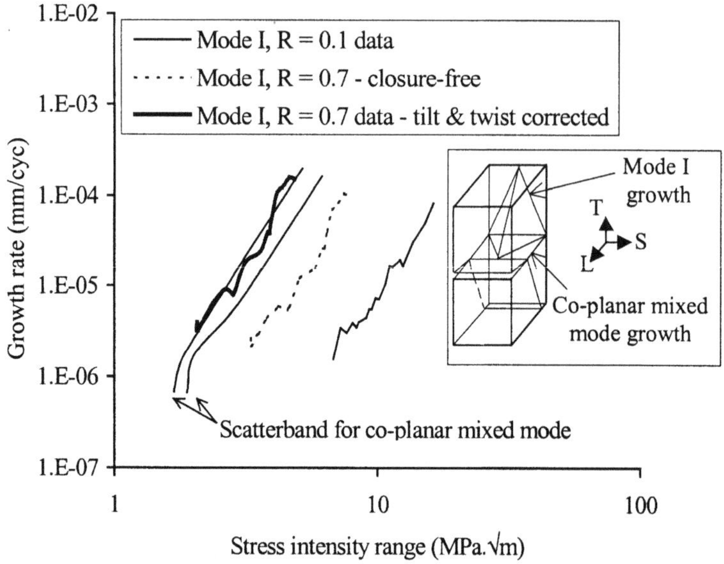

Figure 10.11. Nominal mode I and co-planar mixed mode growth rates in 8090-T8151 plate, in air. Schematic shows nominal specimen orientation and preferred {111} plates for crack growth.

conventional mode I testing. Given the strong texture of the material, kink and twist angles associated with extended Stage I crack growth in the mode I tests were relatively constant (i.e. essentially fixed by the preferred slip plane orientations [79, 100]), with local mixed mode I, II and III crack-tip conditions from the associated deflection then being readily estimated from the resolved tensile and shear stresses ahead of a nominal mode I crack [65, 101]. Figure 10.11 compares growth rates from the mixed mode and closure-corrected conventional mode I tests using an equivalent strain energy release stress intensity factor, where it may be seen that mixed mode crack growth is considerably faster than nominal mode I growth. Calculations of deflected crack-tip driving forces for the closure-free ($R = 0.7$) mode I tests exhibit a close correlation with the mixed mode test results, as shown in figure 10.11.

The high crack growth rates associated with the mixed mode tests in figure 10.11 are consistent with the results of Petit and Mendez [92] in figure 10.12, comparing the mixed mode behaviour of single crystal high purity Al-4.5 wt% Zn-1.25 wt% Mg material, with Stage II and extended Stage I growth in polycrystalline samples in mode I. The results of Petit and Mendez, furthermore, contradict the assertion that improved slip reversibility particularly contributes to improved intrinsic crack growth resistance during extended Stage I crack growth compared with Stage II, as the true Stage I results are to the left of the Stage II data independent of ageing condition.

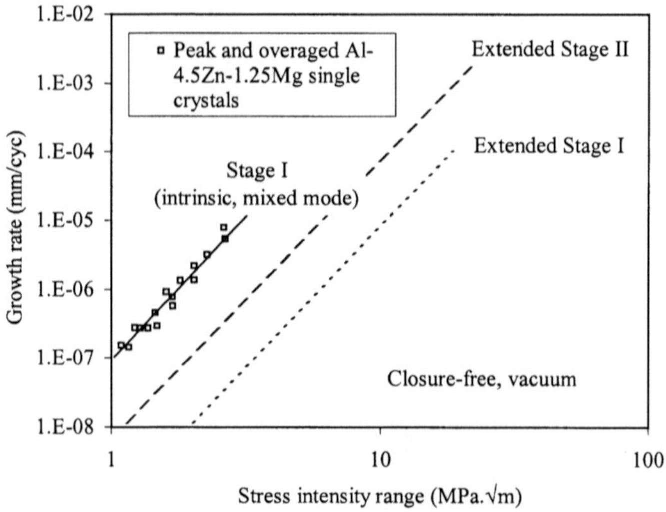

Figure 10.12. Intrinsic Stage II, extended Stage I and mixed mode Stage I growth rates (after Petit and Koshe [82], and Petit and Mendez [92]).

Closure effects

At lower stress intensities and *R*-ratios, crack growth may of course be strongly influenced by crack closure effects. Whilst a range of closure mechanisms may be identified (e.g. oxide- and plasticity-induced [102, 103]), roughness-induced closure represents the most significant microstructurally controlled mechanism, being accentuated by crystallographic crack paths, and is therefore promoted by equivalent microstructural characteristics as noted previously [22, 66, 67, 83, 84, 87–89]. Increasing crack closure effects have been identified with increasing grain dimensions in terms of both changing underlying crack growth mechanism (i.e. favouring crystallographic growth) and increased fracture surface asperity sizes produced during crack growth (e.g. see references [67, 83, 89, 104]). Minakawa *et al* [67], for example, identify the complete disappearance of crack closure effects at low positive *R*-ratios in powder metallurgy materials when the grain size falls from 5 to 1 μm. Increased asperity sizes and hence closure levels are also identified with strong crystallographic textures, permitting the effective extension of individual facets across many grains at a time [22, 27, 79, 83, 84, 100].

Qualitative correlations are widely drawn between increasing fracture surface roughnesses and increasing crack closure levels, with Al-Li-based alloys often exhibiting particularly clear increases in closure levels over conventional alloys and a marked test orientation dependence in strongly textured Al-Li materials due to variations in preferred crystallographic

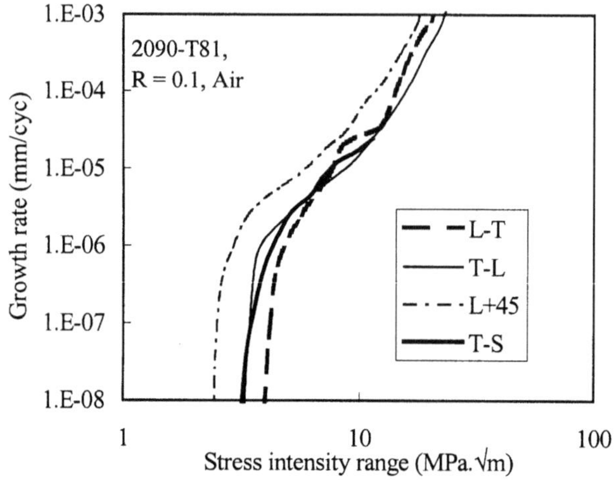

Figure 10.13. Orientation effect on long crack growth in texture Al-Li plate (after Venkateswara *et al* [84]).

growth plane orientations with loading direction, as shown in figure 10.13. Grau *et al* [83] suggest that increasing texture intensity may enhance net mode II displacement levels (and presumably, by implication, irreversible mode II displacement levels) at the crack-tip, increasing crack closure levels, although this effect has not been explicitly proven.

Correlations between closure levels and simple asperity height and path roughness parameters have been suggested (e.g. see references [89, 105, 106]), however, detailed descriptions and models are not well established. The early two-dimensional modelling of crack paths used by Suresh [107] achieves a reasonable first-order description of measured closure effects, and implies that the frequency and angle of crack deflection control closure levels (rather than absolute asperity sizes), with more recent finite difference modelling by Llorca [108] broadly justifying this approach. Llorca also notes a significant influence of deflection regularity (potentially controllable via texture intensity), with changes in successive deflection angles particularly increasing closure levels.

With increasingly sophisticated roughness-induced closure models becoming available [105, 106, 109–111], further clarification and separation of realistic microstructural influences on three-dimensional surface contact characteristics may be possible. As a general point, it should be noted that if a true, physically-based understanding and optimization of fatigue is to be taken, it would appear that the measurement and understanding of crack closure processes is essential. Given the very large body of physical effects attributed within the literature to closure (particularly including variable amplitude effects), it is perhaps surprising and disturbing that no truly accepted and entirely objective measurement procedure exists. This may be attributable to the complexity of closure processes, with important factors such as the through-thickness variation in closure level that may be expected going from surface plane stress to bulk plane strain conditions in a sample, and the fact that the closure and the build-up of load transfer across crack surfaces may be a gradual process rather than an instantaneous event at a specific load point.

Additional considerations

Scientific understanding of physical processes may be developed for individual aspects of fatigue in aluminium alloys; however, important concatenation of test conditions requires further investigation, e.g. environmental and/or variable amplitude load effects on short crack growth. In terms of the current potential for fatigue optimization of materials, it would appear that, in the absence of comprehensive scientific understanding, an increased appreciation of more complex, representative performance conditions for materials of interest and their incorporation into earlier stages of development programmes is required, with the information gained then also contributing to an improved generic basis for rational fatigue performance optimization.

Environment

It is well established that fatigue mechanisms in aluminium alloys are highly sensitive to environmental effects, particularly degradation in crack growth resistance on going from vacuum to wet gaseous to aqueous halide conditions. In terms of the more aggressive halide solutions, controversy remains as to the significance of the processes that may occur (hydrogen embrittlement, anodic dissolution and surface film formation and rupture, etc.), although in relatively mild moist air environments, hydrogen embrittlement is reasonably established as the major environmental interaction (atomic hydrogen being released from water on reacting with aluminium) [24, 34, 72, 74, 77–80, 91, 95, 113–128]. Embrittlement processes may then include changes in lattice cohesion and/or enhanced dislocation mobility, reversibility, localization or injection from free surfaces.

As noted earlier, crystallographic {001} and {110} propagation is commonly associated with hydrogen interactions, although detailed measurements are generally lacking, with recent observations by Slavik and Gangloff [72], for example, identifying a range of facet orientations in AA2090 in aqueous 1% NaCl lying around {521} (depending on the texture of the material). Such faceting was tentatively attributed to crack-tip hydride formation or preferential hydrogen embrittlement of dislocation cell walls. The kinetics of airborne moisture effects on crack growth have been extensively examined by Wei and co-workers [78, 124–126]. They particularly identify rate-limiting mass transport effects along the crack path (subsequent reactions at the crack-tip being relatively rapid given the high reactivity of aluminium) and the controlling influence of environmental exposure, given as the ratio of the water vapour pressure and the test frequency. Based on impeded molecular flow characteristics along the crack path, environmental interaction kinetics may then be enhanced with increasing mean crack opening (i.e. increasing *R*-ratio) and decreasing path tortuousity (microstructure/micromechanism dependent).

Overall, environment effects are known to be a complex function of microstructure and loading conditions, the full details of which are not established. 7000 series alloys are generally recognized as more susceptible to environmental fatigue interaction compared with copper-containing 2000 series materials, in keeping with the susceptibility of 7000 materials to stress corrosion. Planar slip may enhance environmental effects [86, 129], with Starke and co-workers [23, 129] particularly identifying an increase in environmental sensitivity in a range of Al-Zn-Mg-Cu alloys where the copper content was increased to limit precipitate shearability in the peak aged condition, as shown in figure 10.14. This result is also broadly consistent with the reduced environmental sensitivity of the copper-containing 2000 series alloys materials, although the implied significance of slip planarity does not appear to be borne out in undue environmental sensitivity of Al-Li-Cu alloys [22, 113].

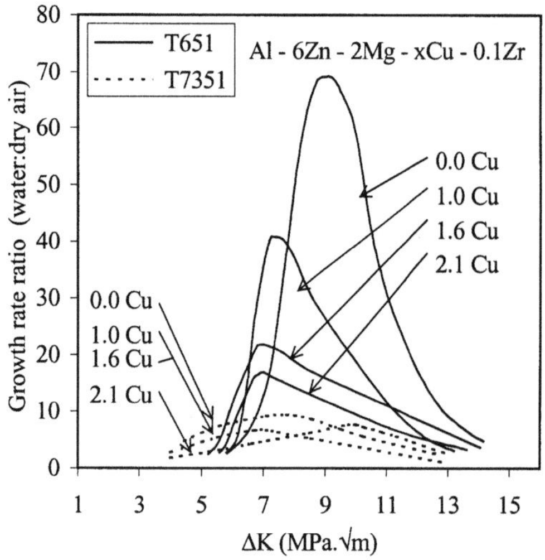

Figure 10.14. Heat treatment and Cu content effects on environmental sensitivity in 7xxx (after Lin and Starke [129]).

Grain boundary and subgrain boundary failure may also occur, depending on the microstructure, with Grau *et al* [83] and Gingell and King 130, for example, confirming an increasing susceptibility to boundary failure with decreasing grain size in 7000 materials in NaCl solutions. Boundary microstructure may then be significant, with the presence of anodic T_1 (Al_2CuLi) at boundaries being identified as detrimental in Al-Li materials [77, 131], and η ($MgZn_2$) in 7000 alloys [121, 127]. Gao *et al* alternatively highlight the detrimental effect of magnesium being segregated to grain boundaries in 7000 materials [78]. The scale of crack-tip plasticity may also have direct mechanistic influences on environmental interactions. Piasick and Gangloff [77] identify the onset of intersubgranular failure with increasing ΔK in AA2090 with the cyclic plastic zone size encompassing the subgrain size, potentially enhancing hydrogen transport to these boundaries, whilst Kemp *et al* [91] identify the transition to extensive slip band crack growth in AA8090 with increasing stress intensities with the plastic zone size beginning to exceed the extent of hydrogen diffusion ahead of the crack-tip. In terms of intrinsic changes in crack growth behaviour with increasing plastic zone sizes in relation to key microstructural dimensions, as identified by Yoder *et al* [63] and Wanhill [64], it is significant to note that the corresponding test data had all been obtained in air, with only the work of Wanhill addressing crack closure influences in detail: it has been pointed out that the da/dN–ΔK curve transitions identified by these authors do not occur in high vacuum tests, and as such due consideration of environmental effects on their results is necessary [33, 77]. Extrinsic influences on crack growth should also be noted where changes in crack path occur due to environment. As such the efficacy of deflection shielding and roughness-induced closure may be influenced by the inhibition of slip band propagation in planar slip materials, or by variations in grain size when intergranular failure occurs [130]. The incidence of oxide induced closure has been discussed in relation to high strength Al alloys, being identified with overaged conditions in 7000 materials in air in some instances [102, 114]. This does not appear to be a particularly prevalent effect (cf. low strength steels), but may be promoted in more aggressive environments, e.g. see reference [132].

Variable amplitude

It is clear that extrapolating conventional constant amplitude fatigue characteristics to variable amplitude situations representative of real structural applications represents a considerable problem, combining complex microstructural and micromechanical interactions with bulk material and structural mechanics. In terms of the ranking and development of new materials in practice, problems encountered over the past decade may be highlighted by table 10.1 [1], where the merits and failures of a number of

Table 10.1. In-service inspectable crack growth ranking of new and established alloys [1].

Material comparison		Constant amplitude	Flight simulation: Gust	Flight simulation: Fighter	Changes in behaviour under simulation loading
PM 7091-T7E69 versus IM 7xxx-T7xxx		+/○	– –	– –	Oxides in 7091-T7E69 initiate voiding during severe flights.
2324-T39 versus 2024-T351		+/○	–		Under MINITWIST the higher yield stress 2324-T39 limits retardation during peak overloads, particularly in thin gauges.
2nd generation Al-Li alloys versus equivalent 2xxx and 7xxx alloys	High strength	++/+	–	+/○	Certain spectra inhibit the particular faceting/tortuousity of Al-Li fracture surfaces and hence shielding levels.
	Medium strength	+	○/–	+	
	Damage tol. sheet	+	+/○	○/–	

+ = better, – = worse, ○ = equal.

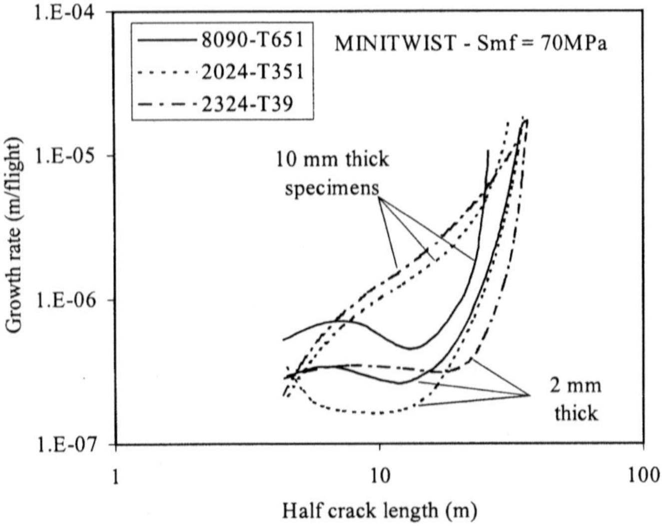

Figure 10.15. Comparison of spectrum crack growth rates in AA8090, 2024 and 2324 (after Wanhill [152]).

advanced materials systems are summarized with respect to nominal and service load conditions. As with constant amplitude conditions, variable amplitude loading behaviour may be separated into different regimes, extending the number of basic constant amplitude load variables with additional critical parameters such as transient amplitude(s), sign (i.e. tension or compression), sequence and periodicity.

In many practical situations, thickness and yield stress may be critical, particularly as achieving plane stress conditions at the crack-tip may markedly enhance post-overload crack retardations and the extent to which successive peak transients may overlap in influence (broadly scaling with the crack-tip plastic zone size, which increases going from plane strain to plane stress conditions). This effect is illustrated in figure 10.15, where MINITWIST spectrum load growth rates are dominated by the shift from 2 mm to 10 mm thick samples, particularly for the 2000 materials shown. It is interesting to note the change in ranking of the materials in figure 10.15, as the AA8090 material shows the same increase in growth rates with increasing thickness, but to a much lesser extent and while retaining a dip in initial growth rates due to successive overload interactions. This is attributable to the in-plane delamination that may occur in this material during load peaks, and the associated development of local plane stress conditions [1, 133].

As a first step in addressing the influence of variable loading effects, simple single and block overload/underload studies are commonly carried

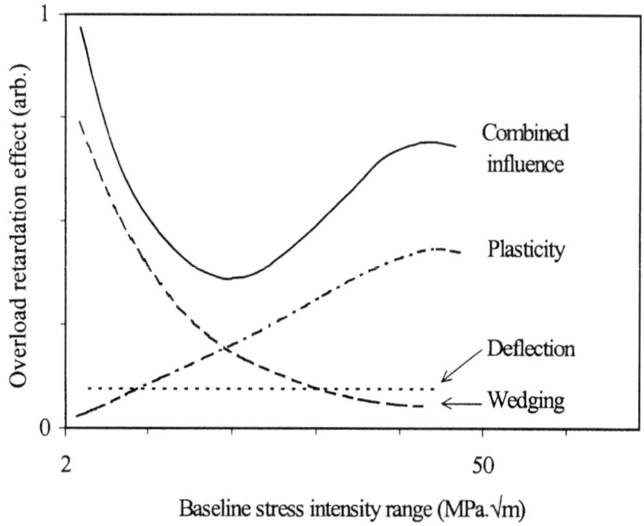

Figure 10.16. Schematic illustration of overload effects with changing baseline loads (after Venkateswara Rao and Ritchie [134]).

out. Such tests do not give a direct assessment of more complex spectrum performance, although a valuable mechanistic insight may be obtained. The physical processes that have been shown to influence variable amplitude crack growth rate behaviour are complex and remain somewhat controversial, and include plasticity-induced closure, residual stress generation, crack deflection and bifurcation, roughness-induced closure, crack-tip blunting, activation of near-threshold growth mechanisms, strain hardening of crack-tip material, static failure mode activation and crack wake ligament formation [34, 128, 132, 135–154].

Crack closure effects are emphasized by a number of authors, with, for example, Venkateswara Rao and Ritchie [134] separating the influence of different mechanisms during simple single overloads as a function of the baseline stress intensity range, as shown in figure 10.16. Here the primary effects are identified as roughness-induced closure at low stress intensities (cf. constant amplitude near-threshold behaviour), giving way to plasticity-induced closure at higher K levels. The effective delay in growth rates characteristic of single peak overloads then exhibits a minimum at intermediate stress intensity levels as the two mechanisms cross-over in influence. In this work and others [134, 136, 143] marked post-overload growth rate retardation in Al-Li materials compared with conventional 2000 and 7000 alloys was identified with the characteristically tortuous crystallographic crack growth mode of the Al-Li materials and associated roughness-induced closure effects. However, decreasing the number of baseline load cycles between single peaks, or changing to certain spectrum load conditions, may inhibit crystallographic crack growth (in Al-Li alloys and other materials), compromising any gains in crack growth resistance [34, 138, 139, 148, 152]. In keeping with this effect, the very planar fracture surfaces associated with fine grained powder metallurgy materials have been identified as limiting post-overload crack growth retardation [137, 140, 147, 148]. Bray and Starke [140] specifically use this effect to isolate the influences of plasticity-induced closure on near-threshold overload behaviour, noting that the increasing near-threshold influence of roughness-induced closure shown in figure 10.16 is not necessary to explain the overload response in this region, as the increasing sensitivity of near-threshold growth rates to crack-tip opening conditions will augment the influence of whatever closure mechanism operates.

At higher stress intensity levels the activation of static failure modes may also become significant, particularly during peak load excursions. At moderate baseline stress intensities it is found that the increased intermetallic content of lower purity alloys may in fact enhance variable amplitude fatigue performance, as static failure at intermetallics during load peaks promotes crack deflection and bifurcation [135, 150]. Whilst a range of detailed micromechanical influences on variable amplitude fatigue behaviour is evident from the above discussion, corresponding quantitative understanding and modelling of growth rate behaviour has been generally limited, with the exception of plasticity-induced closure, where a degree of predictive success has been achieved (e.g. see references [154, 169]).

Future developments

There are clearly several elements in the optimization of fatigue in airframes, such as the physical behaviour of the materials used (as discussed above), changes in design characteristics and developments in lifing methodologies. With the advent of novel fabrication methods, such as age-forming, laser welding, friction stir welding etc., all these aspects may be touched upon to different extents. In using any material (novel or otherwise) it is important to recognize the requirements and limitations of the design process. This process has been built up by the experience of manufacturers over many years, and aspects may constitute part of the airworthiness certification process, with any significant deviations from proscribed methods requiring careful validation.

Whilst fatigue design methods may be well established (the reader is referred to any number of extensive works in the area: see reference [168]), in most cases they involve significant simplifications of the fatigue failure process. For example, even a relatively sophisticated variable lifing fatigue approach such as Newman's 'Fastran' approach [169] only considers the action of one crack closure mechanism (plasticity), with associated implications for the relationships between baseline crack growth rates, plastic flow behaviour and predicted performance (a degree of fitting, in the form of the plastic constraint factor, is available in this approach). As such, a combination of material performance criteria, and how they may be designed with, is required for the successful substitution of advanced alloy systems in a given structure. Overall it should be recognized that structural performance optimization is a complex integration of cyclic properties over a range of failure regimes, with any rational materials optimization process necessarily being linked to an understanding of design, lifing and manufacturing issues.

In terms of manufacturing, age forming represents an attractive forming process for shaped airframe components (e.g. aerofoil surfaces and associated stringers [155]), avoiding involved forming operations such as peen forming and hand finishing. The process is known to be applicable to higher strength artificially aged materials, but may be expected to compromise damage tolerance in materials such as 2024-T351 and 2324-T39. As such, the optimization of artificially aged damage tolerant materials, which are then amenable to age forming, is of interest. A range of materials may be identified, such as the Al-Mg-Si alloys 6013 and 6056, lower strength 2000 series alloys, and Al-Li materials such as 8090 and 2x95 [156–159]. It is of course widely reported that Al-Li alloys can exhibit attractive fatigue crack growth characteristics, but this depends on the exact load regime of interest. From the point of view of damage tolerant usage of the Al-Li materials, fatigue crack deviation and long term thermal stability of toughness remain generic causes for concern, although approaches to these problems are available [160–163].

Significant developments in the welding of high strength aluminium alloys have occurred in the past five to ten years, with the identification of alloys that are amenable to welding (e.g. 6000 series alloys, and Russian 1420-type alloys [156, 157, 163]) and the development of techniques that may be applied to alloys that are traditionally not considered weldable (e.g. laser welding and friction stir welding [164–167]). In the use of welds, it is clear that welded joints in airframes represent mechanically and microstructurally novel crack initiation and propagation features, e.g. the dynamically recrystallized, fine-grained microstructures associated with friction stir welds may in fact be similar to those associated with mechanical alloying—as such, initiation/short crack behaviour in these regions may be intrinsically quite good. Overall, however, there appears to be little information on the fatigue characteristics of advanced airframe weld locations and the controls that may be exerted on damage tolerance. Understanding and optimizing the overall damage tolerance of welded airframe structures is in itself a key issue, as mechanical joints offer a significant crack stopping effect within a structure (in a welded joint cracks may of course propagate directly from one part to the next). Implications for the balance between different fatigue and fracture characteristics for welded airframes, and indeed other mechanical characteristics, requires clarification as welded designs are developed.

Summary

Aluminium alloys remain a key airframe material, particularly for civil aircraft. In terms of fatigue optimization, it is clear that a combination of materials and design/lifing improvements are required, with improved understanding and controlling of physical processes guiding the development of improved analysis and design tools (e.g. in the predictive treatment of variable amplitude behaviour). Whilst the physical processes contributing to fatigue characteristics in any given situation may be basically understood (e.g. see table 10.2), the competition and interaction that may occur between different mechanisms in commercial microstructures under service

Table 10.2 Basic microstructural influences on fatigue resistance in high strength monolithic Al alloys.

Regime	Relevant mechanisms	Desirable microstructural features
Initiation	Slip band formation	Non-shearable precipitates, dispersoids, reduced grain size, reduced texture
	Particle failure	Uniform spatial distributions, small equiaxed particles
	Porosity	Improved processing procedures
	GB failure	'Remove' boundary PFZs and particles, limit slip band activity
Short crack growth	Slip band/Stage I	Small grain size, high angle boundaries
	GB growth	Triple points, large boundary particles
Long crack growth	Slip band/extended Stage I	Shearable precipitates, fewer dispersoids, increased grain size, increased texture
	Roughness-induced closure	Shearable precipitates, fewer dispersoids, increased grain size, increased texture—texture effects vary with orientation
	Deflection shielding	Shearable precipitates, fewer dispersoids, increased grain size, increased texture—texture effects vary with orientation
	Plasticity-induced closure	Lower strength, increasing cyclic work hardening

conditions requires significant clarification and quantification if explicit fatigue optimization of airframe materials and structures is to be realized. With the ongoing demand for cost-effective performance improvements and the development of innovative materials/fabrication processes such as laser and friction stir welding, age forming and integrated component extrusions, high strength aluminium materials may be expected to maintain a competitive position in the next five to ten years. Such developments produce their own fatigue issues, as both materials and structural factors influence the processes of failure.

References

[1] Wanhill R J H 1994 *Fatigue* **16** 99
[2] Wanhill R J H and Schra L 1992 *Short Fatigue Cracks* eds K J Miller and E R de los Rios (London: MEP) p 3
[3] Blom A F, Hedlund A, Zhao W, Fathulla A, Weiss B and Stickler R 1986 *The Behaviour of Short Fatigue Cracks EGF1* eds K J Miller and E R de los Rios (London: MEP) p 37
[4] Morris W L 1979 *Mater. Trans. A* **10** 5
[5] Venkateswara K T, Yu W and Ritchie R O 1988 *Mater. Trans. A* **19** 563
[6] Zurek A K, James M R and Morris W L 1983 *Mater. Trans. A* **14** 1697
[7] Gerold V and Meier B 1987 *Fatigue 87* eds R O Ritchie and E A Starke (Warley: EMAS) vol 3 p 1517
[8] Pearson S 1975 *Eng. Frac. Mech.* **7** 235
[9] Kung C Y and Fine M E 1979 *Metall. Trans. A* **10** 603
[10] Langford J 1982 *Fat. Eng. Mater. Struct.* **5** 233
[11] Zabett A and Plumtree A 1995 *Fat. Frac. Eng. Mater. Struct.* **18** 801
[12] James M R and Morris W L 1982 *Mater. Sci. Eng.* **56** 63
[13] Bowles C Q and Schijve J 1973 *Int. J. Frac.* **9** 171
[14] Laz P J and Hillbery B M 1996 *Fatigue 96* eds G Lutjering and H Nowack (Oxford: Pergamon) vol 2 p 1293
[15] DeBartolo E A, Laz P J and Hillbery B M 1996 *Fatigue 96* eds G Lutjering and H Nowack (Oxford: Pergamon) vol 2 p 1311
[16] Petton G, Rinaldi C and Fougeres R 1994 *Proc. ICAA4* eds T H Sanders and E A Starke (Atlanta: GIT) vol 1 p 701
[17] Heinz A and Schelb W 1994 *Proc. ICAA4* eds T H Sanders and E A Starke (Atlanta: GIT) vol 1 p 733
[18] Magnusen P E, Bucci R J, Hinkle A J, Artley M E and Rolf R L 1989 *Proc. ICF7* eds K Salama *et al* (Oxford: Pergamon) vol 2 p 999
[19] Grosskreutz J C and Shaw G G 1969 *Fracture 1969* ed P L Pratt (Oxford: Pergamon) p 620
[20] Starke E A and Lutjering G 1979 *Fatigue and Microstructure* (Metals Park: ASM) p 205
[21] Sanders T H and Starke E A 1982 *Acta Metall.* **30** 927
[22] Venkateswara K T and Ritchie R O 1992 *Int. Metall. Rev.* **37** 153

[23] Starke E A and Williams J C *Microstructure and Fracture Mechanics of Fatigue Crack Propagation ASTM STP 1020* eds R P Wei and R P Gangloff (Philadelphia: ASTM) p 184
[24] Lin F-S and Starke E A 1979 *Mater. Sci. Eng.* **39** 27
[25] Kuo V W C and Starke E A 1985 *Metall. Trans. A* **16** 1089
[26] Lutjering G, Hamajima T and Gysler A 1977 *Advances in Research on the Strength and Fracture of Materials Proc. ICF4* ed D M R Taplin vol 2A p 7
[27] Feng W X, Lin F S and Starke E A 1984 *Aluminium–Lithium Alloys II* eds E A Starke and T H Sanders (Warrendale: TMS-AIME) p 235
[28] Sanders T H and Starke E A 1978 *Metall. Trans. A* **9** 1087
[29] Edwards L and Martin J W 1982 *Proc. Strength of Metals and Alloys* ed R C Gifkins (Oxford: Pergamon) vol 3 p 873
[30] Blankenship C P, Hornbogen E and Starke E A 1993 *Mater. Sci. Eng. A* **169** 33
[31] Duva J M, Daeubler M A, Starke E A and Luetjering G 1988 *Acta Metall.* **36** 585
[32] Blankenship C P and Starke E A 1993 *Metall. Trans. A* **24** 833
[33] Slavik D C, Blankenship C P, Starke E A and Gangloff R P 1993 *Metall. Mater. Trans. A* **24** 1807
[34] Swain M H, Newmann J C, Phillips E P and Everett R A 1990 *Fatigue '90* eds H Kitigawa and T Tanaka (Birmingham: MCEP) vol 3 p 1079
[35] Tanaka K and Mura T 1982 *Metall. Trans. A* **13** 117
[36] Knott J F and King J E 1991 *Materials and Design* **12** 67
[37] Dhers J, Driver J and Foureux A 1986 *Aluminium–Lithium Alloys III* eds C Baker *et al* (London: Institute of Metals) p 233
[38] Welpmann K, Lutjering G and Bunk W 1974 *Aluminium* **50** 263
[39] Mulvihill O and Beevers C J 1986 *The Behaviour of Short Fatigue Cracks EGF1* eds K J Miller and E R de los Rios (London: MEP) p 203
[40] Welpmann K, Lutjering G and Bunk W 1977 *Advances in Research on the Strength and Fracture of Materials Proc. ICF4* ed D M R Taplin vol 2A p 105
[41] Park D S and Nam S W 1995 *Mater. Sci. Tech.* **11** 921
[42] Quested P N, Henderson P J and McLean M 1988 *Acta Metall.* **36** 2743
[43] Taylor C J, Zhai T, Wilkinson A J and Martin J W 1998 to be published in *J. Micros.*
[44] Zhang Y H and Edwards L 1994 *Mater. Sci. Eng. A* **188** 121
[45] *Small Fatigue Cracks 1986* eds R O Ritchie and J Langford (Warrendale: TMS-AIME)
[46] *The Behaviour of Short Fatigue Cracks 1986 EGF1* eds K J Miller and E R de los Rios (London: MEP)
[47] Tanaka K, Akiniwa Y, Nakai Y and Wei R P 1986 *Eng. Frac. Mech.* **24** 803
[48] Tanaka K, Kinefuchi M and Yokomaku T 1992 *Short Fatigue Cracks ESIS13* eds K J Miller and E R de los Rios (London: MEP) p 351
[49] Zhang Y H and Edwards L 1993 *Fatigue '93* eds J-P Bailon and J I Dickson (Warley: EMAS) vol 1 p 383
[50] Navarro A and de los Rios 1988 *Fat. Frac. Eng. Mater. Struct.* **11** 383
[51] de los Rios E R, Xin X J and Navarro A 1994 *Proc. Roy. Soc. Lond. A* **447** 111
[52] Xin X J, de los Rios E R and Navarro A 1992 *Short Fatigue Cracks ESIS13* eds K J Miller and E R de los Rios (London: MEP) p 369
[53] Venkateswara Rao K T, Yu W and Ritchie R O 1988 *Eng. Frac. Mech.* **31** 623
[54] Bolingbroke R K and King J E 1986 in reference [39] p 101
[55] Edwards L and Zhang Y H 1994 *Acta Metall. Mater.* **42** 1413

[56] Brown C W and King J E 1986 in reference [38] p 73
[57] Langford J and Davidson D L 1986 in reference [38] p 51
[58] Zhang Y H and Edward L 1992 *Scripta. Metall. Mater.* **26** 1901
[59] Morris W L 1980 *Metall. Trans. A* **11** 1117
[60] Gregory J K, Gysler A and Lutjering G 1984 *Fatigue '84* ed C J Beevers (Warley: EMAS) vol 2 p 847
[61] Li C, Zhang P and Zhang T 1994 *Mater. Sci. Eng. A* **183** 23
[62] Edwards L and Gungor S 1992 *Small Fatigue Cracks ESIS13* eds K J Miller and E R de los Rios (London: MEP) p 469
[63] Yoder G R, Cooley L A and Crooker T W 1982 *Scripta. Metall.* **16** 1021
[64] Wanhill R J H 1988 *Eng. Frac. Mech.* **30** 233
[65] Ritchie R O 1988 *Mater. Sci. Eng. A* **103** 15
[66] Ritchie R O, Yu W, Blom A F and Holm D K 1987 *Fat. Frac. Eng. Mater. Struct.* **10** 343
[67] Minakawa K, Levan G and McEvily A J 1986 *Metall. Trans. A* **17** 1787
[68] Venkateswara K T and Ritchie R O 1988 *Acta Metall.* **36** 2849
[69] Chen D L, Weiss B and Stickler R 1996 *Mater. Sci. Eng.* **208** 181
[70] Louat N, Sadananda K, Duesbery M and Vasudévan A K 1993 *Metall. Trans. A* **24** 2225
[71] Vasudévan A K, Sadananda K and Louat N 1992 *Scripta. Metall. Mater.* **27** 1673
[72] Slavik D C and Gangloff R P 1996 *Acta Mater.* **44** 3515
[73] Laird C 1979 *Fatigue and Microstructure* (Metals Park: ASM) p 149
[74] Nix K J and Flower H M 1982 *Acta Metall.* **30** 1549
[75] Pelloux R M N 1969 *Trans. ASM* **62** 281
[76] Garrett G G and Knott J F 1975 *Acta Metall.* **23** 841
[77] Piascik R P and Gangloff R P 1993 *Metall. Trans. A* **24** 2751
[78] Gao M, Pao P S and Wei R P 1988 *Metall. Trans. A* **19** 1739
[79] Sinclair I and Gregson P J 1994 *Proc. ICAA4* eds T H Sanders and E A Starke (Atlanta: GIT) vol 2 p 436
[80] Lynch S P 1988 *Acta Metall.* **36** 2639
[81] Lafarie M C and Casc C 1983 *Fat. Eng. Mater. Struct.* **6** 329
[82] Petit J and Koshe K 1992 *Small Fatigue Cracks ESIS13* eds K J Miller and E R de los Rios (London: MEP) p 135
[83] Grau F-J, Gysler A and Lutjering G 1994 *Proc. ICAA4* eds T H Sanders and E A Starke (Atlanta: GIT) vol 1 p 709
[84] Venkateswara Rao K T, Bucci R J, Jata K V and Ritchie R O 1991 *Mater. Sci. Eng. A* **141** 39
[85] Ruch W and Starke E A 1986 *Aluminium–Lithium Alloys III* eds C Baker *et al* (London: IOM) p 121
[86] Suresh S, Vasudevan A K and Bretz P E 1984 *Metall. Trans. A* **15** 369
[87] Lindigkeit J, Gysler A and Lutjering G 1981 *Metall. Trans. A* **12** 1613
[88] Zaiken E and Ritchie R O 1985 *Mater. Sci. Eng.* **70** 151
[89] Carter R D, Lee E W, Starke E A and Beevers C J 1984 *Metall. Trans. A* **15** 555
[90] Brett S J, Cantor B and Doherty R D 1977 *Advances in Research on the Strength and Fracture of Materials Proc. ICF4* ed D M R Taplin vol 2B p 77
[91] Kemp R M J, Wilson R N and Gregson P J 1992 *Fat. Frac. Eng. Mater. Struct.* **15** 291
[92] Petit J and Mendez J 1996 *Fatigue '96* eds G Lutjering and H Nowack (Oxford: Pergamon) vol 1 p 15

[93] Starke E A, Lin F S, Chen R T and Heikkenen H C 1984 *Fatigue Crack Growth Threshold Concepts* eds Davidson D and Suresh S (Warrendale: TMS-AIME) p 43

[94] Edwards L, Busby A K and Martin J W 1986 *Mater. Sci. Tech.* **2** 823

[95] Langford J and Davidson D L 1983 *Acta Metall.* **31** 1273

[96] Bower A F and Ortiz M 1990 *J. Mech. Phys. Solids* **38** 443

[97] Grau F-J, Gysler A and Lutjering G 1993 *Fatigue '93* eds J-P Bailon and J I Dickson (Warley: EMAS) vol 2 p 623

[98] Liu Y, Sinclair I and Gregson P J 1998 *Engineering Against Fatigue* eds Baynon J *et al* (Rotterdam: Balkema) in press

[99] Liu Y, Sinclair I and Gregson P J 1998 unpublished data

[100] Yoder G R, Pao P S, Imam M A and Cooley L A 1989 *Scripta Metall.* **23** 1455

[101] Lawn B R 1993 *Fracture of Brittle Solids* (Cambridge: Cambridge University Press)

[102] Vasudevan A K and Suresh S 1982 *Metall. Trans. A* **13** 2271

[103] Budansky B and Hutchinson J W 1978 *J. App. Mech.* **45** 267

[104] Venkateswara K T and Ritchie R O 1989 *Mater. Sci. Tech.* **5** 896

[105] Wasen J, Hamberg K and Karlsson B 1988 *Mater. Sci. Eng. A* **102** 117

[106] Jung H Y and Antolovich S D 1996 *Eng. Frac. Mech.* **54** 307

[107] Suresh S 1985 *Metall. Trans. A* **16** 249

[108] Llorca J 1992 *Fat. Frac. Eng. Mater. Struct.* **15** 655

[109] Tong J, Yates J R and Brown M W 1995 *Eng. Frac. Mech.* **52** 599 and 613

[110] Sehitoglu H and Garcia A M 1997 *Mater. Trans. A* **28** 2263 and 2267

[111] Mendelsohn T S, Gross T S and Zhang Y 1995 *Acta Metall. Mater.* **43** 899

[112] Ngyun D, Thompson A W and Bernstein I M 1987 *Acta Metall.* **35** 2417

[113] Gangloff R P 1990 *Corrosion Fatigue Crack Propagation in Metals* NASA contractor report CR-4301

[114] Vasudevan A K, Suresh S and Bretz P E 1984 *Metall. Trans. A* **15** 369

[115] Ricker R E and Duquette D J 1988 *Metall. Trans. A* **19** 1775

[116] de los Rios E R, Sun Z Y and Miller K J 1994 *Fat. Frac. Eng. Mater. Struct.* **17** 1459

[117] Suresh S, Palmer I G and Lewis R E 1982 *Fat. Frac. Eng. Mater. Struct.* **5** 133

[118] Holroyd N J H and Hardie D 1983 *Corr. Sci.* **23** 527

[119] Piascik R S and Willard S A 1994 *Fat. Frac. Eng. Mater. Struct.* **17** 1247

[120] Magnin Th 1996 *Mater. Sci. Forum* **217–222** 83

[121] Niegel A, Gudladt H-J and Gerold V 1987 *Fatigue '87* eds R O Ritchie and E A Starke (Warley: EMAS) vol 3 p 1229

[122] Davidson D L and Lankford J 1983 *Fat. Eng. Mater. Struct.* **6** 241

[123] Rangahathan N, Bouchet B and Petit J 1987 *ASTM STP 948* eds J E Masters and J J Au (Philiadelphia: ASTM) p 424

[124] Wei R P, Pao P S, Hart R G, Weir T W and Simmons G W 1980 *Metall. Trans. A* **11** 151

[125] Wei R P and Gangloff R P 1989 *ASTM STP 1020* eds R P Wei and R P Gangloff (Philadelphia: ASTM) p 233

[126] Shih T-H and Wei R P 1983 *Eng. Frac. Mech.* **18** 827

[127] Christodoulou L and Flower H M 1980 *Acta Metall.* **28** 481

[128] Trockels I, Lutjering G and Gysler A 1996 *Mater. Sci. Forum* **217–222** 1599

[129] Lin F-S and Starke E A 1980 *Mater. Sci. Eng.* **43** 65

[130] Gingell A D B and King J E 1996 *Mater. Sci. Forum* **217–222** 1605

[131] Blankenship C P and Starke E A 1994 *Acta Metall. Mater.* **42** 845

[132] Zuidema J, Mense P J M and Edwards R A H 1987 *Eng. Frac. Mech.* **26** 927

[133] Kumai S and Higo Y 1996 *Mater. Sci. Eng. A* **221** 154

[134] Venkateswara Rao K T and Ritchie R O 1988 *Acta Metall.* **36** 2849

[135] Sanders T H and Staley J T 1979 *Fatigue and Microstructure* (Metals Park: ASM) p 467

[136] Petit J, Suresh S, Vasudevan A K and Malcolm R C 1986 *Aluminium–Lithium Alloys III* eds C Baker *et al* (London: IOM) p 257

[137] Reynolds A P 1992 *Fat. Frac. Eng. Mater. Struct.* **15** 551

[138] Kiese J, Gysler A and Lutjering G 1990 *Fatigue '90* eds H Kitigawa and T Tanaka (Birmingham: MCEP) vol 3 p 1523

[139] Kiese J, Gysler A and Lutjering G 1993 *Fatigue '93* eds J-P Bailon and J I Dickson (Warley: EMAS) vol 3 p 1569

[140] Bray G H and Starke E A 1993 *Fatigue '93* eds J-P Bailon and J I Dickson (Warley: EMAS) vol 3 p 1587

[141] Trockels I, Lutjering G and Gysler A 1996 *Fatigue '96* eds G Lutjering and H Nowack (Oxford: Pergamon) vol 1 p 571

[142] Trockels I, Lutjering G and Gysler A 1996 *Mater. Sci. Forum* **217-222** 1599

[143] Alexander D J and Knott J F 1987 *Fatigue '87* eds R O Ritchie and E A Starke (Warley: EMAS) vol 1 p 395

[144] Knott J F and Pickard A C 1977 *Metall. Sci.* **11** 399

[145] Vasudevan A K and Suresh S 1989 *Treatise on Materials Science and Technology* **31** 445

[146] Blankenship C P, Bray G H, Kaisand L R and Starke E A 1995 *Fat. Frac. Eng. Mater. Struct.* **18** 551

[147] Telesman J and Antolovich S D 1986 *Eng. Frac. Mech.* **24** 463

[148] Trockels I, Gysler A and Lutjering G 1994 *Proc. ICAA4* eds T H Sanders and E A Starke (Atlanta: GIT) vol 1 p 717

[149] Yu W and Ritchie R O 1987 *J. Eng. Mater. Tech.* **109** 81

[150] Bucci R J, Thakker T H, Sanders T H, Sawtell R R and Staley J T 1980 *ASTM STP* **714** 41

[151] Fleck N A 1985 *Acta Metall.* **33** 1339

[152] Wanhill R J H 1994 *Fatigue* **16** 3

[153] Lankford J and Davidson D L 1981 *Advances in Fracture Research* ed D Francois (Oxford: Pergamon) vol 2 p 899

[154] Newman J C 1992 *Fatigue of Aircraft Materials* (Delft: Delft University Press) p 83

[155] Newman J, Goodyear M D, Witters J J, Veciana J and Platts G K 1991 *Proc. Al-Li 6* eds M Peters and P-J Winkler (Oberursel: DGM) vol 2 p 1371

[156] Sato H, Lutjering G and Gysler A 1998 *Proc. ICAA5* eds T Sato *et al* (Tokyo: JILM) vol 3 p 1421

[157] Ronan D, Brechet D, Warner T and Ribes H 1998 *Proc. ICAA5* eds T Sato *et al* (Tokyo: JILM) vol 3 p 1991

[158] Vine W J, Sutton G R and Price H J 1998 *Proc. ICAA5* eds T Sato *et al* (Tokyo: JILM) vol 3 p 1973

[159] Blankenship C P and Starke E A 1994 *Acta Metall. Mater.* **42** 845

[160] Blankenship C P and Starke E A 1993 *Metall. Trans. A* **24** 833

[161] Bowen A W 1990 *Mater. Sci. Eng.* **6** 1058

[162] Sinclair I and Gregson P J 1996 *J. Aero. Eng.* **210** 117

[163] Kolobnev N I, Grushko O E, Cherkasov V V, Dolzhansky Y M, Miller W S and Couch P D 1994 *Proc. ICAA4* eds T H Sanders and E A Starke (Georgia: GIT) vol 1 p 305

[164] Norman A F and Prangnell P B 1998 *Proc. ICAA6* eds T Sato *et al* (Tokyo: JILM) vol 3 pp 1435 and 1501

[165] Mahoney M W, Rhodes C G, Flintoff J G, Spurling R A and Bingel W H 1998 *Metall. Mater. Trans. A* **29** 1955

[166] Murr L E, Liu G and McClure J C 1998 *J. Mater. Sci.* **33** 1243

[167] Nicholas E D and Thomas W M 1998 *Int. J. Mater. Production Tech.* **13** 45

[168] Estimation enhancement and control of aircraft fatigue performance 1995 *Proc. 24th ICAF Symposium* eds J G Grandage and G S Jost (Warley: EMAS)

[169] Newman J C 1995 *J. Eng. Mater. Tech: Trans. ASME* **117** 433

Chapter 11

Bulk amorphous, nanocrystalline and nanoquasicrystalline aluminium alloys

Akihisa Inoue and Hisamichi Kimura

Introduction

Since the discovery of age-hardening, various kinds of high-strength aluminium-based alloys have been produced over the subsequent eight decades. It is known that these aluminium-based alloys have been developed by use of the following strengthening mechanisms: (1) solid solution strengthening, (2) precipitation strengthening, (3) grain size refinement strengthening, (4) dispersion strengthening, (5) work hardening and (6) fibre reinforcement. However, the use of these conventional strengthening mechanisms leads to an upper limit of tensile strength of 500 to 600 MPa at room temperature. Consequently, in order to develop a new type of aluminium-based alloys with much higher tensile strength, the use of completely different strengthening mechanisms seems to be essential. More recently, we have carried out systematic studies on the development of high-strength aluminium-based alloys by utilization of nonequilibrium phase effects. In particular, great attention has been paid to nonperiodic structure alloys consisting of amorphous and quasicrystalline phases.

In 1988, we succeeded in finding aluminium-based amorphous alloys with high tensile strength exceeding 1200 MPa [1]. Subsequently, we have noticed [2] that the homogeneous dispersion of nanoscale fcc aluminium particles into an amorphous matrix causes a drastic increase in tensile fracture strength to 1560 MPa, which is three times higher than the strength level of conventional high-strength aluminium-based alloys. From the relation between tensile strength and calendar year in the historical progress of the high-strength aluminium-based alloys, a drastic increase in tensile strength is recognized for the recent nonequilibrium alloys including an amorphous phase. Figure 11.1 summarizes the features of the microstructure

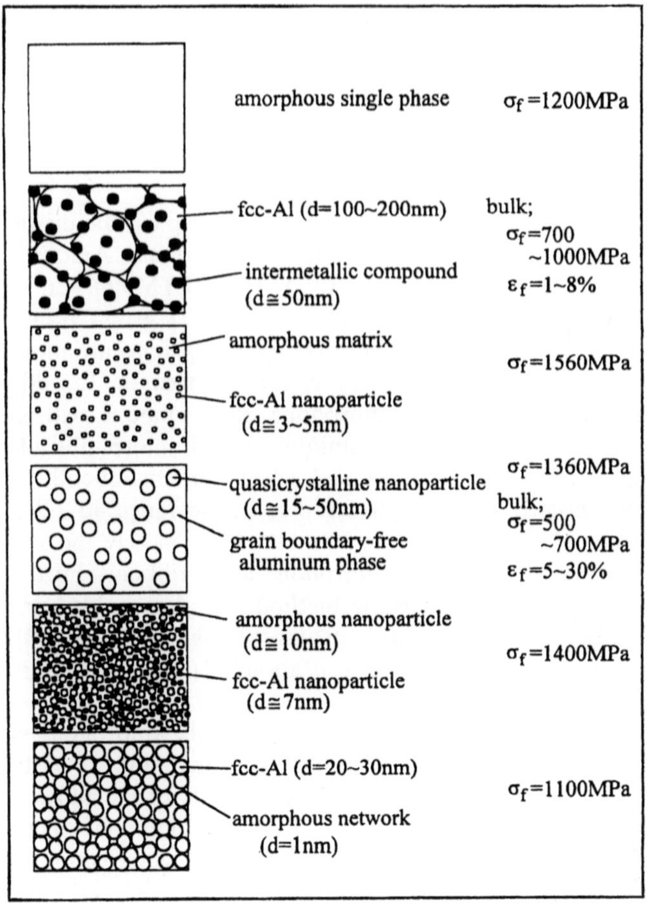

Figure 11.1. Schematic nonequilibrium alloy structures.

and mechanical strength of nonequilibrium aluminium-based alloys developed by our group [3, 4]. The nonequilibrium structures for aluminium-based alloys are classified into the following six types, (1) amorphous single phase, (2) nanostructure of fcc aluminium and intermetallic compounds obtained by crystallization of an amorphous phase, (3) partially crystallized structure of nanoscale fcc aluminium particles embedded in an amorphous matrix, (4) nanoquasicrystalline structure consisting of nanoscale quasi-crystalline particles surrounded by an fcc aluminium phase without grain boundaries, (5) coexistent nanogranular amorphous and fcc aluminium phases, and (6) nanogranular fcc aluminium phase surrounded by an amorphous network. These nonequilibrium aluminium-based alloys exhibit much better mechanical properties compared with conventional crystalline alloys developed up to date. In this chapter, we review our recent results on the synthesis and mechanical properties of the nonequilibrium aluminium-based alloys shown in figure 11.1.

Bulk amorphous phase

The amorphous alloys obtained by melt spinning can be divided into two kinds of metal–metal and metal–metalloid systems. Among these alloy systems, both Al-Ln-TM [1, 5] and Al-ETM-LTM [6] (Ln = lanthanide metal, TM = transition metal, ETM = IV to VI group transition metal, LTM = VII and VIII group transition metal) systems are more important because of the achievement of higher tensile strength. In addition to the melt-spun aluminium-based amorphous alloy ribbons with high tensile strength, we can also produce aluminium-based amorphous alloy wires [7] exhibiting high tensile strength and good bending ductility by a melt-extraction method. The wire diameter is in the range 40–120 μm and the tensile strength lies in the range 900–1100 MPa, nearly the same as those for the equivalent melt-spun amorphous alloy ribbons. Furthermore, the use of a high-pressure die-casting method has also enabled production of aluminium-based bulk alloys with an amorphous surface layer [8]. The thickness of the amorphous surface layer is about 40 μm for a cylindrical alloy of 0.5 mm diameter, and increases to about 150 μm for an alloy of 5 mm diameter. However, we could not obtain a bulk amorphous single phase alloy in aluminium-based alloy systems by any kind of casting method because of the relatively low glass-forming ability of the alloys.

Fortunately, amorphous alloys with high aluminium concentrations of about 85 at% in Al-Ln-TM systems exhibit a glass transition, followed by a supercooled liquid region in the temperature range below the crystallization temperature, though the supercooled liquid region is rather narrow and the temperature interval is less than 30 K [9]. In the supercooled liquid region, significant viscous flow is obtained over an appropriate strain rate range. By use of viscous flow, a bulk amorphous aluminium-based alloy can be produced by extrusion of atomized amorphous powders. By using a gas atomization technique, amorphous alloy powders without any trace of crystalline phase are produced in a particle size fraction range below 25 μm [10]. When the $Al_{85}Ni_{10}Ce_5$ amorphous powder in a size fraction below 25 μm is extruded at the temperatures between 443 and 493 K, at an extrusion velocity of 0.1 mm/s and at extrusion ratios of 2.25 and 2.50, cylindrical bulk alloys with diameters of 10, 15 and 20 mm are produced as exemplified in figure 11.2 [11]. The X-ray diffraction patterns of the bulk alloys consist of a halo peak due to the predominant formation of a single amorphous phase. This is believed to be the first synthesis of aluminium-based bulk alloys with largely a single amorphous phase by the extrusion process. The

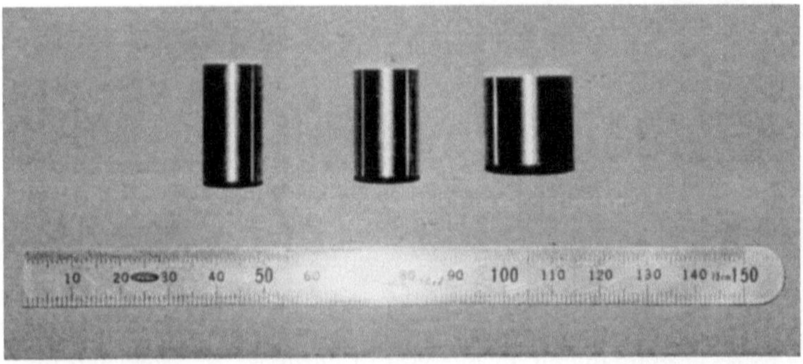

Figure 11.2. Cylindrical bulk amorphous alloys.

recent success in synthesizing aluminium-based consolidated bulk alloys is attributed to more precise control of extrusion conditions, on the basis of further detailed information on the viscous flow behaviour of the supercooled liquid for typical bulk amorphous alloys such as zirconium- and palladium-based systems [12, 13]. Within several years, it is believed that the further extension of the present consolidation techniques will enable us to produce aluminium-based bulk amorphous alloys with high strength and good ductility, which are nearly the same as those for the corresponding melt-spun amorphous alloy ribbons.

Partially crystallized alloys

It is known [14] that the phase transformation from a highly supercooled liquid (or amorphous phase) has the following unique features: (1) homogeneous nucleation, (2) control of high nucleation frequency, (3) control of low growth rate, (4) high solute concentration gradient at the liquid/solid interface, resulting from low atomic diffusivity, (5) formation of metastable phases with new compositions by controlling the redistribution of solute elements, (6) formation of a residual amorphous phase with high solute concentration, (7) defect-free nanocrystalline particles with low residual strain, (8) highly dense packed structure at the liquid/solid interface, (9) nanoscale interparticle spacing effects, and (10) size and shape effects of nanoscale spherical particles. By utilizing these features, new nanoscale mixed phase alloys containing a wide range of phases, including a remaining amorphous phase, are expected to be synthesized even for aluminium-based alloys.

In general, aluminium-rich amorphous alloys with high aluminium concentrations (above 88 at%) crystallize through two stages in which the first-stage exothermic reaction is due to the precipitation of fcc aluminium and the second exothermic peak results from the decomposition of the remaining amorphous phase to intermetallic compounds [14]. When the cooling rate for the aluminium-rich alloys is controlled by varying the rotation speed of the melt spinning wheel, we can obtain a nanoscale mixed structure consisting of fcc aluminium particles with a size of 3–5 nm embedded in an amorphous matrix in the melt-spun ribbons [15]. The volume fraction V_f of the fcc aluminium phase is thus controlled by changing the rotation speed of the melt spinning wheel, and can be evaluated subsequently by the change in the exothermic heat released during crystallization. The tensile fracture strength σ_f increases from about 1100 MPa at $V_f = 0\%$ to 1560 MPa at $V_f = 25\%$, accompanying an increase in hardness H_v from 280 to 400 and Young's modulus E from 63 to 71 GPa, as shown in figure 11.3. The significant decrease in σ_f accompanying further increase in V_f is due to embrittlement of the remaining amorphous phase by the progress of structural relaxation and enrichment of solute elements.

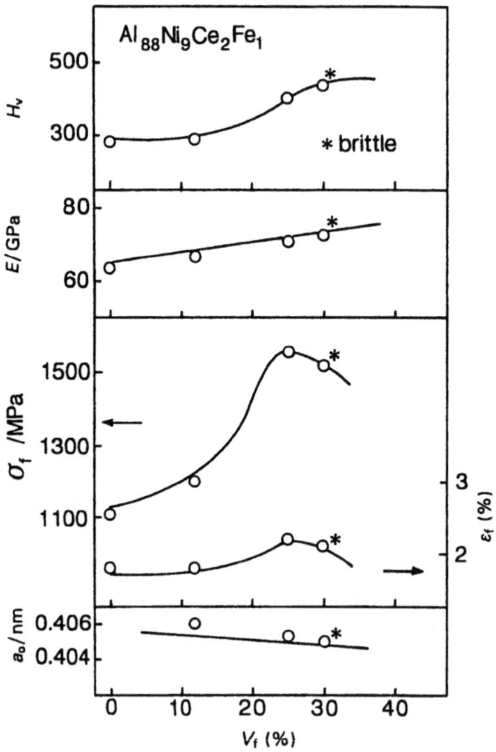

Figure 11.3. Properties of partially crystallized Al-Ni-Ce-Fe alloy as a function of crystal fracture or extent of crystallization.

A similar nanoscale mixed structure can also be formed by more rapid cooling to produce a fully amorphous material followed by annealing in the first-stage exothermic reaction range. The resulting mixed phase alloys exhibit high tensile strength exceeding 1400 MPa, which is about 1.4 times higher than that for the corresponding amorphous single phase alloys [16]. The increase in σ_f caused by the dispersion of nanoscale fcc aluminium particles into the amorphous phase has been thought [14] to result from the combination of the following three effects:

1. defect-free nanoscale fcc aluminium particles, because the aluminium particles are too fine to contain dislocations;
2. interface effects, because the amorphous/aluminium particle interface has a highly dense packed structure without excess vacancies and a corresponding low interfacial energy; and
3. nanoscale effects, because the aluminium particle size is smaller than the width of inhomogeneous shear deformation, and hence the nanoscale aluminium particles can act as an effective barrier against shear deformation of the amorphous matrix.

However, the tensile strength is too high to manufacture a consolidated bulk alloy by warm extrusion at a relatively low temperature range where the nanoscale mixed structure can be maintained.

Bulk nanocrystalline alloys

When the extrusion temperature is raised to the second-stage exothermic reaction range, it is straightforward to produce a fully dense nanocrystalline alloy consisting of fine scale intermetallic compounds with a size of about 50 nm, embedded in an fcc aluminium matrix with a grain size of 100–200 nm [17]. The relation between tensile strength and rotating-beam bending fatigue strength for nanocrystalline and conventional aluminium-based alloys is shown in figure 11.4. The nanocrystalline alloys exhibit good combinations of high tensile strength, above 900 MPa and high fatigue strength, of 300–350 MPa, which are superior to those for conventional aluminium-based alloys [18] as well as for other new aluminium-based alloys [19] developed by rapid solidification and powder metallurgy techniques. Figure 11.5 shows the relation between the yield strength $\sigma_{0.2}$ and the grain size of the fcc aluminium matrix phase for the nanocrystalline aluminium-based alloys, together with data for conventional crystalline aluminium-based alloys. The high tensile strength of the nanocrystalline alloys has been interpreted to result from a combination of the two mechanisms of dispersion strengthening and grain size refinement strengthening. The combination of the two strengthening mechanisms is attributable to the formation of a structure in which the intermetallic compounds are dispersed homogeneously within the aluminium

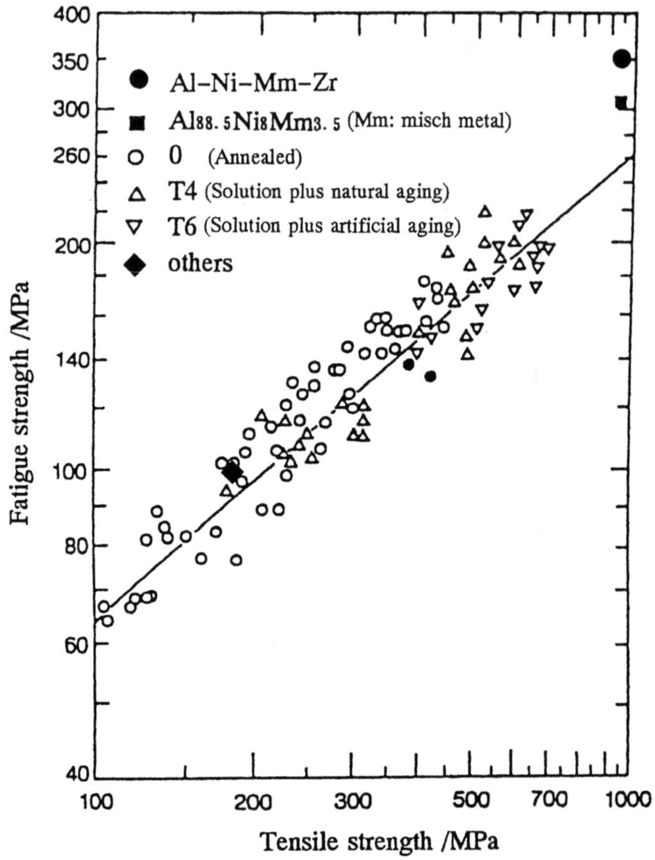

Figure 11.4. Relation between tensile and fatigue strength for nanocrystalline and conventional aluminium alloys.

grains and on the grain boundaries. The dispersion density of the intermetallic compounds is not high enough to suppress all the dislocations, but the dislocations are hindered at the grain boundaries of the fcc aluminium phase. Figure 11.5 shows that the increase in the yield strength by dispersion strengthening and grain size refinement strengthening is approximately 400 and 390 MPa, respectively. When the dispersion strengthening is attributable to Orowan's mechanism [20], the increase in the yield strength $\Delta\sigma$ is given by

$$\Delta\sigma = \frac{2.5Gb}{2\pi(1-\nu)^{1/2}} \frac{\ln(2r/r_0)}{\lambda_s - 2r} \tag{1}$$

where G is the shear modulus of aluminium (27 GPa), b is the Burger's vector of aluminium (2.86×10^{-10} m), ν is the Poisson's ratio of aluminium (0.345), r is the particle radius, r_0 is the core dislocation radius for aluminium

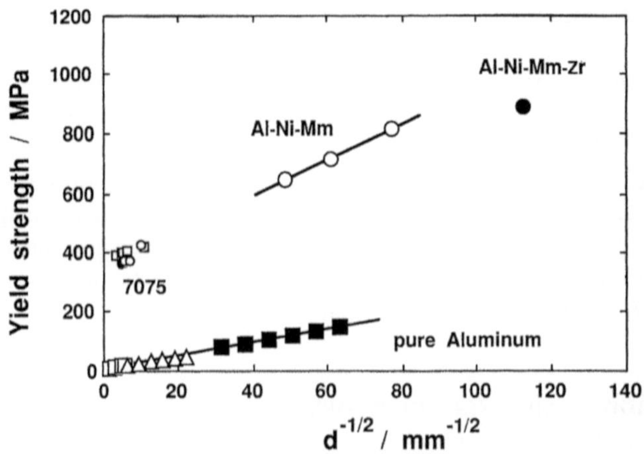

Figure 11.5. Relation between tensile and fatigue strength and grain size of fcc aluminium matrix for nanocrystalline alloys.

(2.86×10^{-10} m), λ_s is $1.25r(2\pi/3f)^{1/2}$ and f is the volume fraction of dispersoids. Values of r and f for the present alloy are measured to be 2.5×10^{-8} m and 0.34 [21], respectively. The value of the resulting increase of yield stress $\Delta\sigma$ is calculated to be 395 MPa.

Furthermore, the increase in yield strength by grain size refinement strengthening is expressed by the Hall–Petch relation [22, 23].

$$\sigma_{0.2} = \sigma_0 + kd^{-n} \tag{2}$$

where σ_0 is the yield strength of a single grain due to all strengthening mechanisms except grain boundary contributions [24], k is a constant related to how effective the grain boundaries are in increasing yield strength, d is the grain or subgrain size, and n is an exponent in the range 0.5 to 1, and assumed to be 0.5 [24]. In the present alloy, σ_0 is due to solid solution strengthening and strain strengthening up to 0.2% elongation and also includes dispersion strengthening. The k value has been reported to be 75 MPa $\mu m^{1/2}$ for commercially pure aluminium [25], 120 MPa $\mu m^{1/2}$ for optimally aged 7075 [26] and 220 MPa $\mu m^{1/2}$ for underaged 7091 [27]. In the present calculation, k is assumed to be 135 MPa $\mu m^{1/2}$. The resulting value of kd^{-n} is 426 MPa for $d = 100$ nm and 302 MPa for $d = 200$ nm. Consequently, the yield strength values $\sigma_{0.2}$ evaluated by the summation of the dispersion strengthening and grain size refinement strengthening are 821 MPa for $d = 100$ nm and 697 MPa for $d = 200$ nm. These calculated yield strength values agree reasonably well with the experimental results shown in figure 11.5, indicating the appropriateness of the strengthening mechanisms.

The nanocrystalline structure has been found [28] to lead to a high-strain rate superplasticity which is evidenced from a high strain-rate sensitivity exponent (*m*-value) of 0.5 as well as a large elongation, reaching about 670% at a high strain rate of $1.0\,s^{-1}$. The nanocrystalline aluminium-based alloys have already been commercialized with the present name GIGAS [29], because they exhibit high tensile strength of about 1 GPa. The GIGAS alloys also have higher specific strength and specific Young's modulus, which are superior to those for conventional aluminium-based and titanium-based alloys. These advantages have already led to some applications, such as rapidly rotating machinery parts requiring high specific strength and high specific modulus, the main construction parts in robots and high elevated temperature die casting materials.

Bulk nanoquasicrystalline alloys

It is known that an icosahedral phase is formed in rapidly solidified Al-TM base alloys containing manganese [30], chromium [31] and vanadium [32, 33] as the TM element. The Al-TM base icosahedral structure has been presumed [34] to consist of Mackay icosahedral clusters containing 55 atoms which are arranged through glue atoms into a three-dimensional quasiperiodic structure. The icosahedral phase can contain dislocations [35, 36]. However, it is very difficult for dislocations to move in the icosahedral phase because movement of the dislocations destroys the quasiperiodic lattice at room and elevated temperatures [35]. We prepared a large single icosahedral $Al_{70}Pd_{10}Mn_{20}$ quasicrystal [37] by the Czochralski method, and measured the mechanical properties as a function of quasicrystal orientation and temperature [38]. The stoichiometric quasicrystal has a high Young's modulus of 200 GPa, a high flexural modulus of 72 GPa, and a high Vickers hardness of about 750 at room temperature, even though the alloy is extremely brittle [38]. The quasicrystal also exhibits high elevated temperature compressive strength of about 650 MPa at 1000 K, accompanying significant elongation [38]. By utilizing the high Young's modulus, high hardness and high elevated temperature strength, an icosahedral base aluminium alloy with high strength can be developed. The key point in the development is the possibility of synthesizing an icosahedral base alloy with good ductility.

It is expected that the high strength and good ductility for icosahedral base alloys will be obtained in a nanostructured material, where the icosahedral phase is coexistent with fcc aluminium, and has a spherical morphology, with small grain sizes and a homogeneous dispersion. The nanostructured state is also expected to exhibit an increase in the stability of the supercooled liquid, against the transition to a crystalline phase. Consequently, significant attention has been paid to the Al-Mn-Ln and Al-Cr-Ln systems, because Al-Mn and Al-Cr alloys are good icosahedral-forming systems, and Al-Ln binary alloys are a good glass-forming system. Figure 11.6 shows the

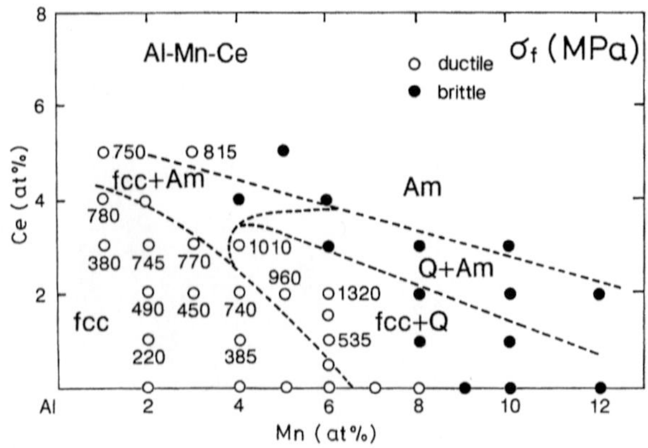

Figure 11.6. Composition dependence of structure, ductility and strength for nanocrystalline Al-Mn-Ce.

compositional dependence of structure, ductility and tensile fracture strength of melt-spun Al-Mn-Ce alloys [39]. The mixed structure consisting of icosahedral plus fcc aluminium phases is formed in the aluminium-rich composition range above 92 at% aluminium and the mixed phase alloys have good bending ductility and high σ_f reaching 1320 MPa. TEM examination revealed that the high-strength $Al_{92}Mn_6Ce_2$ alloy consists of spherical icosahedral particles with a size of about 50 nm surrounded by fcc aluminium with a thickness of 5–10 nm. The particles have random orientations because of primary precipitation of the icosahedral phase, followed by crystallization of the matrix fcc aluminium phase from the remaining liquid. The volume fraction of the icosahedral phase is as high as about 60 to 70%. A similar nanoscale mixed structure is also formed in rapidly solidified Al-Cr-Ce-Co alloys [40]. As exemplified for the $Al_{94.5}Cr_3Ce_1Co_{1.5}$ alloy in figure 11.7, the icosahedral particles are 20–40 nm in size, and are dispersed homogeneously in the aluminium phase without high-angle grain boundaries. This icosahedral alloy again exhibits a high tensile strength of 1340 MPa. Similar high tensile strength values exceeding 1000 MPa combined with high hardness above 400 are obtained for a range of rapidly solidified $Al_{93.5}Cr_3Ce_1Co_{1.5}M_1$ alloys containing various transition metals M [41]. The icosahedral alloys also have good cold deformability, and can be cold rolled up to significant reduction ratios in thickness, above 70% [42]. Even in the heavily deformed alloys, no appreciable cracks are observed. It has been confirmed that the cold-rolled alloys have a much finer mixed structure consisting of smaller icosahedral particles, 5–10 nm in size, which coexist with a nanoscale fcc aluminium phase.

Figure 11.8 illustrates the change in the mixed structures of icosahedral plus fcc aluminium phases with solidification mode. Here, it is important to point out that high tensile strength combined with good ductility is obtained

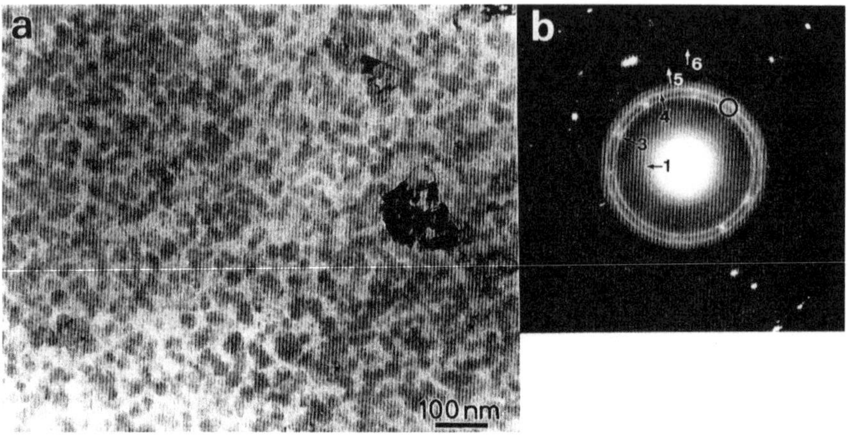

Figure 11.7. Icosahedral particles and diffraction pattern for rapidly solidified Al-Cr-Ce-Co alloy.

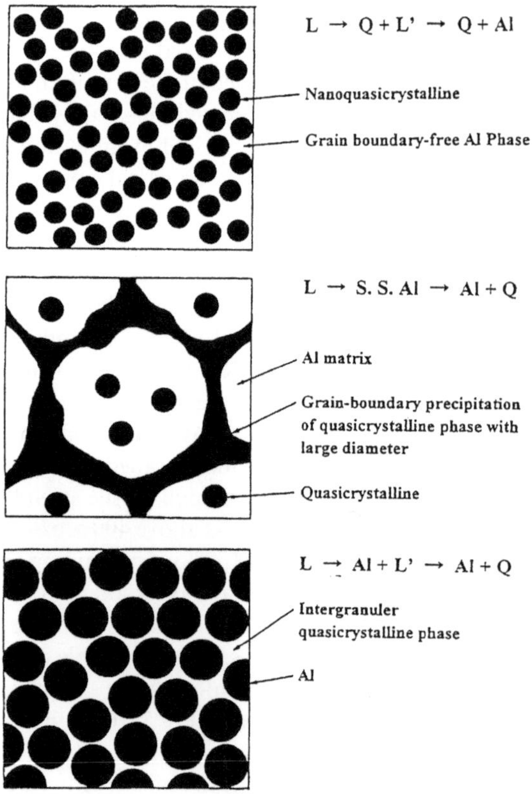

Figure 11.8. Schematic of icosahedral/fcc aluminium structures depending on solidification conditions.

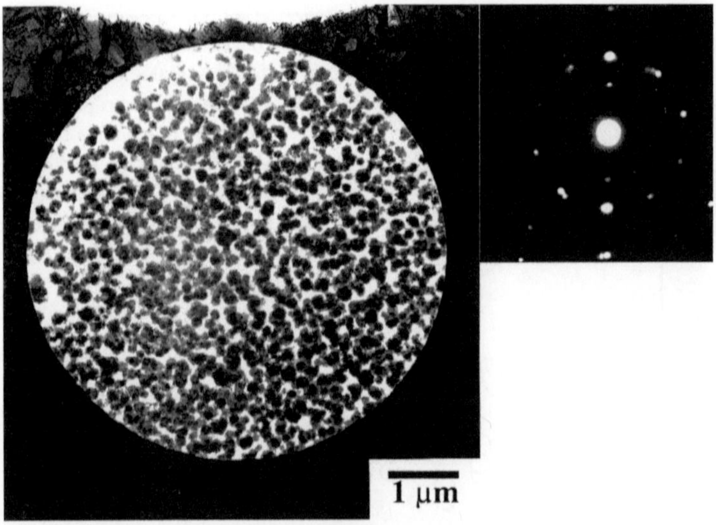

Figure 11.9. Structure of icosahedral Al-Mn-Co alloy.

only for the mixed phase alloys prepared through the unique solidification process in which the nanoquasicrystalline phase precipitates as a primary phase, followed by the solidification of fcc aluminium from the remaining liquid. When the aluminium-based alloys follow other solidification paths, in which an fcc aluminium solid solution is formed followed by icosahedral precipitation within the aluminium grains and along aluminium grain boundaries, or in which primary fcc aluminium particles precipitate, followed by the solidification of an intergranular quasicrystalline phase from the remaining liquid, the resulting alloys do not have high tensile strength and good ductility.

In addition to melt-spun ribbon forms, the same mixed structure of nanoscale icosahedral particles surrounded by fcc aluminium without high-angle grain boundaries is also formed for atomized powders in the Al-Mn-Ln and Al-Cr-Ln-TM systems, as well as in Al-Mn-TM and Al-Cr-TM systems without a lanthanide element, as exemplified for $Al_{93}Mn_5Co_2$ in figure 11.9 [43]. When the atomized powders are extruded in the temperature range below 673 K, which is lower than the decomposition temperature of the icosahedral phase, one can obtain fully dense bulk icosahedral base alloys which have the same mixed structure as the atomized powders. Figure 11.10 shows the relation between tensile strength and plastic elongation for extruded bulk icosahedral alloys [43, 44] in comparison with data for conventional aluminium-based alloys. The bulk icosahedral alloys have high tensile strengths of 500–850 MPa combined with large elongations of 5–25%, which are much superior to conventional aluminium-based alloys.

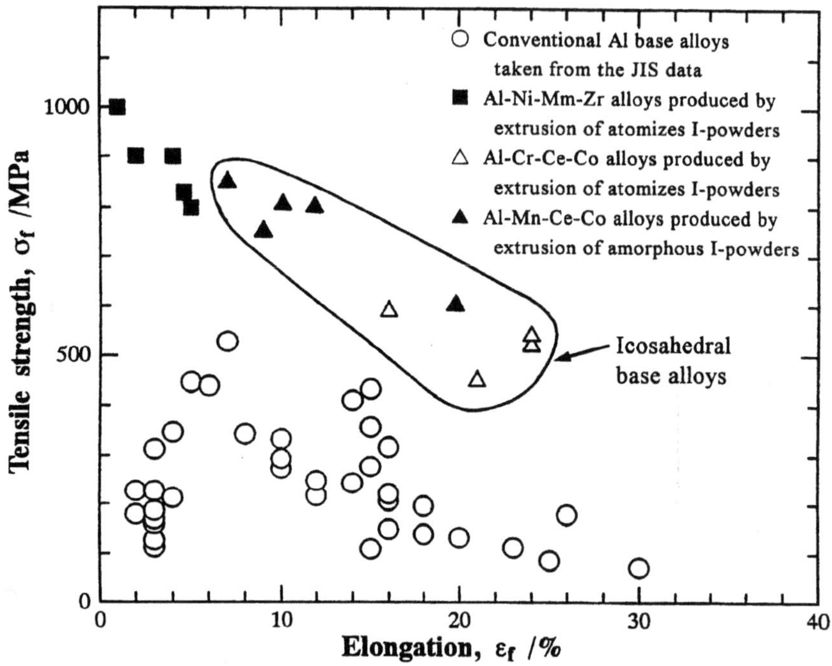

Figure 11.10. Relation between strength and ductility for extruded bulk Al-Mn-Co alloy and others.

The bulk icosahedral base alloys can be divided into two kinds:

1. high strength for Al-Mn(Cr) alloys containing lanthanide elements; and
2. high ductility for Al-Mn(Cr)-TM alloys without lanthanide elements.

The Al-Cr and Al-Mn base alloys containing lanthanide elements also exhibit rather high elevated temperature strengths of 350 MPa at 473 K and 200 MPa at 573 K respectively, in addition to a high Young's modulus of about 100 GPa at room temperature. The bulk icosahedral alloy $Al_{94}Cr_{2.5}Co_{1.5}Mn_1Zr_1$ also exhibits high fatigue strength values of 240 MPa at room temperature and 210 MPa at 423 K, much higher than those (175 and 125 MPa) for conventional Al-Si-Fe alloys, as shown in figure 11.11. On the other hand, bulk icosahedral alloys without lanthanide elements in the $Al_{93-95}(Cr,Mn,Ni)_{5-7}$ and $Al_{93-95}(Cr,Mn,Cu)_{5-7}$ systems exhibit large elongations to failure, reaching 30%, and high impact fracture energies reaching 160 kJ/m^2, combined with high tensile strengths of 500–800 MPa [45].

More recently, we have also found that a mostly single icosahedral phase is formed at a new composition of $Al_{84.2}Fe_{7.0}Cr_{6.3}Ti_{2.5}$ in the melt-spun state, as shown in figure 11.12 [46]. Based on previous data [47] showing that the solute elements Fe, Cr and Ti have very low diffusivities in fcc

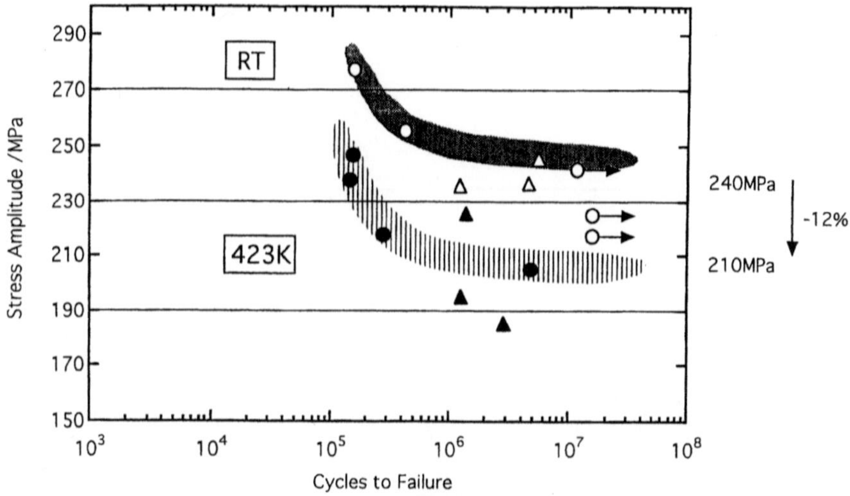

Figure 11.11. Fatigue strength (S-N curves) for bulk icosahedral Al-Cr-Co-Mn-Zr alloy.

aluminium, the icosahedral alloys are expected to sustain high strength to elevated temperatures. Even for an aluminium-rich $Al_{93}Fe_3Cr_2Ti_2$ alloy, a similar mixed structure consisting mainly of icosahedral particles and fcc aluminium is formed in atomized powders with a particle size fraction range below 125 μm [46]. The resulting bulk alloys prepared by extrusion

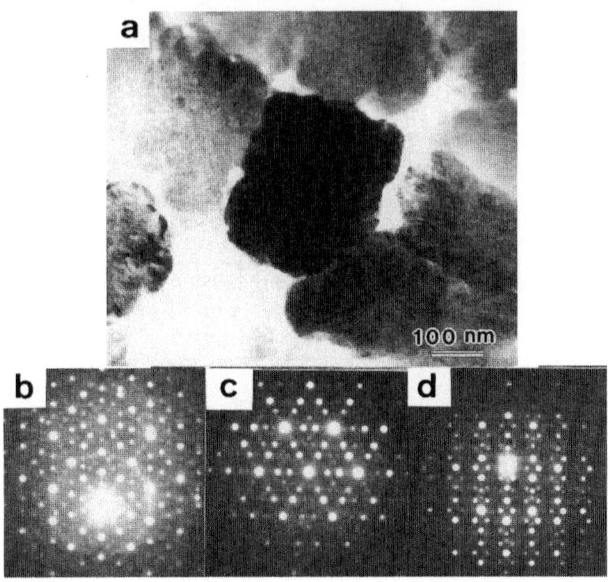

Figure 11.12. Structure and diffraction patterns from icosahedral Al-Fe-Cr-Ti.

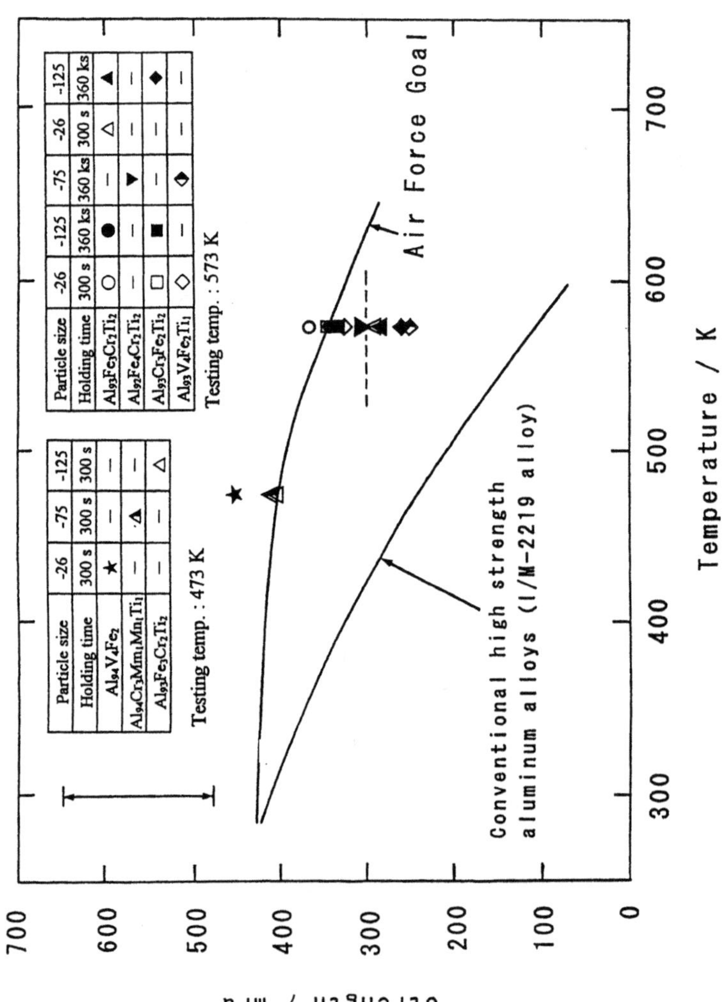

Figure 11.13. Elevated temperature strength of bulk icosahedral alloys.

Table 11.1. Alloy system, microstructure and mechanical properties for the quasicrystalline base bulk aluminium alloys.

Type	Alloy system	Structure	Mechanical properties
High-strength alloy	Al-Cr-Ce-M Al-Mn-Ce	Al + Q	$\sigma_f \fallingdotseq 600 \sim 800$ MPa $\varepsilon_p \fallingdotseq 5 \sim 10\%$
High-ductility alloy	Al-Mn-Cu-M Al-Cr-Cu-M	Al + Q	$\sigma_f \fallingdotseq 500 \sim 600$ MPa $\varepsilon_p \fallingdotseq 12 \sim 30\%$
High-elevated temperature strength alloy	Al-Fe-Cr-Ti	$Al + Q + Al_{23}Ti_9$	$\sigma_f \fallingdotseq 350$ MPa at 573 K

of the atomized powders also consist mainly of icosahedral plus fcc aluminium phases. Figure 11.13 shows the elevated temperature strength of the bulk icosahedral base alloys, in comparison with US Air Force goal and conventional high-strength aluminium alloys [46]. The new Al-Fe-Cr-Ti alloys exhibit excellent elevated temperature strength of 400–460 MPa at 473 K and 350–360 MPa at 573 K, which exceed the US Air Force goal. It has further been confirmed that the high elevated temperature strength of extruded $Al_{93}Cr_3Fe_2Ti_2$ is maintained even after annealing for 1000 h at 573 K. The annealed alloy also maintains a fine mixed structure consisting of fine icosahedral particles with sizes less than 400 nm and fine fcc aluminium grains with sizes of about 500 nm.

Table 11.1 summarizes the features of alloy systems, microstructures and mechanical properties for these quasicrystalline based aluminium alloys. The icosahedral aluminium alloys can be classified overall into

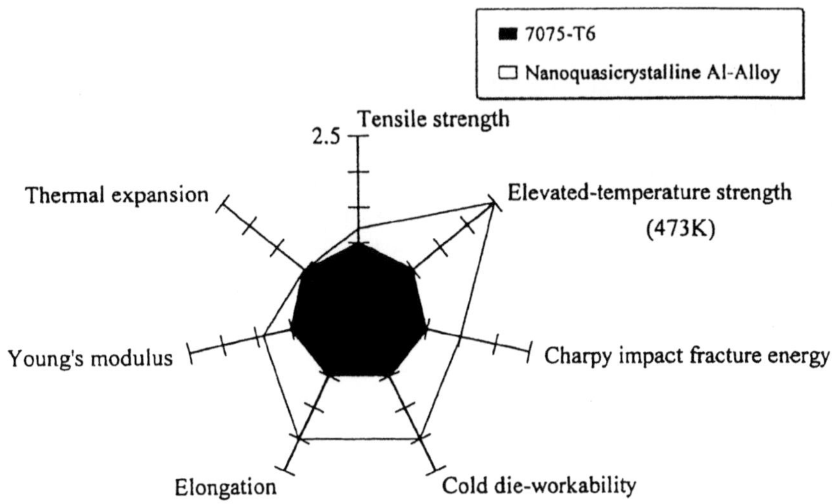

Figure 11.14. Property envelope for nanoquasicrystalline alloy, compared with 7075.

three groups as follows: (1) high strength Al-Mn-Ln and Al-Cr-Ln-TM alloys; (2) high ductility Al-Mn-Cu-TM and Al-Cr-Cu-TM alloys; and (3) high-elevated temperature strength Al-Fe-Cr-Ti alloys. When the mechanical properties are compared with those for 7075-T6, the nanoquasicrystalline aluminium alloys exhibit much better combinations of almost all properties of tensile strength, elevated temperature strength, Charpy impact fracture energy, cold workability, elongation and Young's modulus, but not thermal expansion, as illustrated in figure 11.14. These properties are attractive for future aerospace applications, as aluminium-based alloys with high tensile strength, high ductility and high elevated temperature strength.

Nanogranular amorphous alloys

The data described in the last section were limited to Al-TM alloys such as Al-Mn, Al-Cr and Al-Fe based systems. We subsequently examined the structure and mechanical properties of melt-spun Al-V and Al-Ti based alloys containing solute elements belonging to groups IV and V in the periodic table. The $Al_{94}V_4Fe_2$ alloy consists of nanogranular amorphous grains with a size of about 10 nm coexisting with fcc aluminium grains with a size of about 7 nm [48]. A similar nanogranular amorphous structure is also formed in melt-spun $Al_{95}Ti_3Fe_2$ [49]. These nanogranular mixed structure alloys in melt-spun ribbon form have good bending ductility and exhibit high tensile strengths of 1380 MPa for $Al_{94}V_4Fe_2$ and 1200 MPa for $Al_{95}Ti_3Fe_2$. Table 11.2 summarizes the structures of melt-spun Al-Ln-TM and Al-ETM-LTM alloys containing high aluminium contents of 93 to 95 at%. There is a systematic change in the structures, from aluminium + intermetallic compound → aluminium + icosahedral quasicrystal → aluminium + amorphous → fully amorphous, with decreasing group number of the transition metal element. Such a systematic change in the rapidly quenched structure is presumably because the decrease in the group number of the transition metal element causes an increase in the ability to achieve a stabilized supercooled liquid by increasing the atomic size ratio and the negative heat of mixing. The structures of melt-spun aluminium-based binary and ternary alloys are summarized in figure 11.15. Easy

Table 11.2. Structures of melt-spun Al-Ln-TM and Al-ETM-LTM alloys containing high aluminium contents of 93 to 95 at%.

	Lanthanide	ETM			LTM
Element	Ln	Ti	V	Cr-Mn	Fe-Co-Ni-Cu
Structure	Am	Al + Am	Al + Am/Al + Q	Al + Q	Al + Comp.

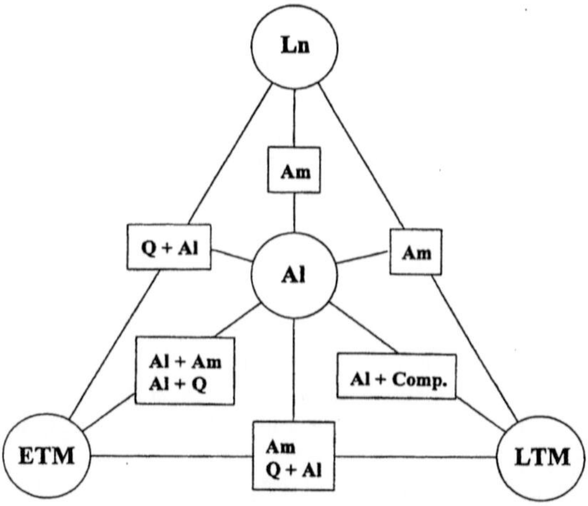

Figure 11.15. Structures of melt-spun aluminium-based binary and tertiary alloys.

formation of nonperiodic amorphous and nanoquasicrystalline phases is found in aluminium-rich composition ranges above about 92 at% aluminium in Al-Ln and Al-TM binary, and Al-Ln-ETM, Al-Ln-LTM and Al-ETM-LTM ternary systems by melt spinning. From the easy formation of nonperiodic structures and the achievement of good mechanical properties for a number of these aluminium-based alloys, it is expected that the nonperiodic aluminium-based alloys may be developed as new high-strength engineering materials for aerospace and other applications.

Summary

We have succeeded in developing a range of new types of high-strength aluminium-based alloys containing nanoscale nonperiodic phases, by controlling the atomic structure, composition and stability of the supercooled liquid. The new synthesis methods involve controlling the stability, composition and nonperiodic structure of the supercooled region, and will lead to the future development of other types of new aluminium-based materials with high functional properties.

References

[1] Inoue A, Ohtera K, Tsai A P and Masumoto T 1988 *Jpn. J. Appl. Phys.* **27** L280
[2] Kim Y H, Inoue A and Masumoto T 1990 *Mater. Trans. JIM* **31** 747
[3] Inoue A 1994 *Mater. Sci. Eng. A* **179/180** 57

[4] Inoue A and Kimura H M 1997 *Mater. Sci. Forum* **235–238** 873
[5] He Y, Poon S J and Shiflet G J 1988 *Science* **241** 1640
[6] Tsai A P, Inoue A and Masumoto T 1988 *Metall. Trans. A* **19** 1369
[7] Inoue A, Amiya K, Yoshii I, Kimura H M and Masumoto T 1988 *Jpn. J. Appl. Phys.* **27** L1579
[8] Inoue A, Onoue K and Masumoto T 1994 *Mater. Trans. JIM* **35** 808
[9] Inoue A, Ohtera K, Tsai A P, Kimura H M and Masumoto T 1988 *Jpn. J. Appl. Phys.* **27** L1579
[10] Inoue A, Kita K, Ohtera K and Masumoto T J 1988 *Mater. Sci. Lett.* **7** 1287
[11] Inoue A, Kawamura Y and Sasamori K to be submitted
[12] Inoue A 1995 *Mater. Trans. JIM* **36** 866
[13] Inoue A, Zhang T and Takeuchi A 1998 *Mater. Sci. Forum* **269–272** 855
[14] Inoue A, Kimura H M and Kita K 1997 *New Horizons in Quasicrystals* ed A I Goldman, D J Sordelet, P A Thiel and J M Dubois (Singapore: World Scientific) p 256
[15] Kim Y H, Inoue A and Masumoto T 1991 *Mater. Trans. JIM* **32** 599
[16] Inoue A, Nakazato K, Kawamura Y, Tsai A P and Masumoto T 1994 *Mater. Trans. JIM* **35** 95
[17] Ohtera K, Inoue A, Terabayashi T, Nagahama H and Masumoto T 1992 *Mater. Trans. JIM* **33** 775
[18] *Metals Databook* 1983 ed Japan Inst. Metals, Maruzen, Tokyo
[19] Vasudevan A K and Doherty R O 1989 *Aluminium Alloys* (London: Academic Press)
[20] Martin J W 1980 *Micromechanism in Particle Hardened Alloys* (Cambridge: Cambridge University Press) p 40
[21] England R O, Pickens J R, Kumar K S and Langan T J 1988 *Dispersion Strengthened Aluminium Alloys* ed Y W Kim and W M Griffith (Minerals, Metal and Materials Society) p 371
[22] Hall R O 1951 *Proc. Phys. Soc. London* **B64** 747
[23] Petch N J 1953 *J. Iron Steel Inst.* **174** 25
[24] Bennett V R, Nix W D and Tettleman A S 1973 *The Principles of Engineering Materials* (Englewood Cliffs, NJ: Prentice-Hall) p 226
[25] Decker R F 1973 *Metall. Trans.* **4** 2495
[26] Wert J A 1980 *Strength of Metal Alloys* ed R C Gifkins (Oxford: Pergamon Press) p 339
[27] Kim Y W and Griffith W M 1984 *PM Aerospace Materials* (Shrewdburg: MPR) p 1
[28] Higashi K, Mukai T, Tanimura S, Inoue A, Masumoto T, Kita K, Ohtera K and Nagahora J 1992 *Scr. Metall.* 26 191
[29] YKK Catalog 1995
[30] Shechtman D, Blech L A, Gratias D and Cahn J W 1984 *Phys. Rev. Lett.* **53** 1951
[31] Inoue A, Kimura H M and Masumoto T J 1987 *Mater. Sci.* **22** 1758
[32] Rao K V, Fildler J and Chen H S 1986 *Europhys. Lett.* **1** 647
[33] Inoue A, Arnberg L, Lehtinen B, Oguchi M and Masumoto T 1986 *Metall. Trans. A* **17** 1657
[34] Elser Y and Henley C L 1985 *Phys. Rev. Lett.* **55** 2883
[35] Takeuchi S 1992 *Tetsu-to-Hagane* **78** 1517
[36] Voisin E and Pasturel A 1987 *Phil. Mag. Lett.* **55** 123
[37] Yokoyama Y, Tsai A P, Inoue A and Masumoto T 1991 *Mater. Trans. JIM* **32** 1
[38] Yokoyama Y, Inoue A and Masumoto T 1993 *Mater. Trans. JIM* **34** 135

[39] Inoue A, Watanabe M, Kimura H M, Takahashi F, Nagata A and Masumoto T 1992 *Mater. Trans. JIM* **33** 723
[40] Inoue A, Kimura H M, Sasamori K and Masumoto T 1994 *Mater. Trans. JIM* **35** 85
[41] Inoue A, Kimura H M, Sasamori K and Masumoto T 1995 *Mater. Trans. JIM* **36** 6
[42] Inoue A, Kimura H M, Watanabe M and Kawabata A 1997 *Mater. Trans. JIM* **38** 756
[43] Inoue A and Kimura H M 1997 *Mater. Sci. Forum* **235–238** 873
[44] Inoue A, Kimura H M and Sasamori K 1997 *Chemistry and Physics of Nanostructures and Related Non-Equilibrium Materials* ed E Ma, B Fultz, R Shull, J Morral and P Nash (Minerals, Metals and Materials Society) p 201
[45] Inoue A, Kimura H M, Sasamori K and Kita K 1998 *Aluminium Alloys* **3** 1841
[46] Inoue A, Kimura H M and Sasamori K 1998 *Advanced Materials* **4** 91
[47] Vasudevan A K and Doherty R O 1989 *Aluminium Alloys* (London: Academic Press)
[48] A Inoue, H M Kimura and T Masumoto 1996 *Nanostruct. Mater.* **7** 363
[49] Kimura H M, Sasamori K and Inoue A 1996 *Mater. Trans. JIM* **37** 1722

Chapter 12

High toughness metal matrix composites

Toshiro Kobayashi

Introduction

Metal matrix composites (MMCs) are expected to be next generation materials, especially for the aerospace industry. However, their low toughness must be overcome.

Many researchers have reported that when a main crack propagates in a discontinuously-reinforced MMC, numerous microcracks are initiated ahead of the main crack caused by reinforcement fracture and/or reinforcement-matrix interfacial debonding. This suggests that crack propagation with accompanying microcracking has a significant influence on the overall fracture of the MMC. Such crack growth mechanisms can be described in terms of fracture mechanics, and computer simulations of crack growth in MMCs have been performed [1]. Such simulations show that some segregated SiC whiskers may improve the toughness of SiCw/6061 Al alloy MMCs.

For MMCs in which age-hardenable aluminium alloys are used as the matrix, very thin precipitate free zones (PFZs) are formed around the reinforcements, and the precipitate free zones grow with ageing treatment [2]. Based on the results of detailed analysis of precipitation and mechanical behaviour, a new retrogression and re-ageing (RRA) heat-treatment procedure has been developed for MMCs, capable of improving the precipitation structure of the age-hardenable matrix, in order to optimize the mechanical properties of the composite [3]. This chapter describes computer simulation of crack growth in a particulate MMC, and the effect of retrogression and re-ageing heat treatment.

Computer simulation

To avoid difficulties which arise when materials with anisotropy or heterogeneity are analysed by fracture mechanics, it is convenient to limit the computer simulation to isotropic materials such as particulate MMCs or MMCs in which the orientations of short fibres can be assumed to be random. The size of the reinforcement is also assumed to be small in comparison with the crack-tip stress field so as to produce homogenous and isotropic behaviour at both the mesoscopic and macroscopic levels. Similarly, materials with an anisotropic matrix grain structure as a result of secondary processing are also out of the scope of the computer simulations.

Because of restrictions on total computation time, a two-dimensional plane strain model is adopted. Various characteristics of the reinforcement, such as shape, size, distribution, volume fraction, and orientation angle in the case of fibres, are input beforehand. Of all the characteristics except distribution, strength and orientation of the reinforcement, averaged values are assigned equally to every reinforcement particle or fibre. The

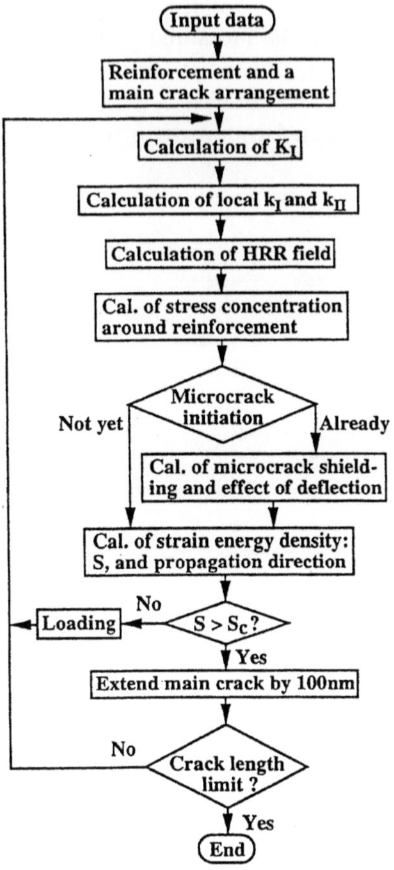

Figure 12.1. Flow chart of the simulation program.

distribution of reinforcement strengths is taken into account by using a three-parameter Weibull distribution. Inhomogeneous spatial distribution of the reinforcement is always found in real MMCs, and the corresponding quantitative analysis is not simple and straightforward. To overcome this problem, the frequency distribution of the number of the reinforcements within minute unit meshes are expressed by random number generation, following one of three discrete distribution functions, selected according to the desired distribution mode. After the construction of the material model, the position of the main crack is also determined using random number generation [1].

Generally, stress distributions in the zone governed by the J integral are solved as a Hutchinson–Rice–Rosengren (HRR) field in the elastic–plastic condition of the power-law hardening material [4]. On the other hand, the stress singularity at the tip of the growing crack in the ductile material is solved as a Rice–Drugan–Sham (RDS) field with lower intensity caused by non-proportional loading [5]. The crack-tip singularity is identified by comparing the measured distribution of the crack opening displacement (COD) along the crack surface with theoretically-predicted values based on these two analyses. The crack opening displacement measurement was performed by in-situ observation in fracture toughness

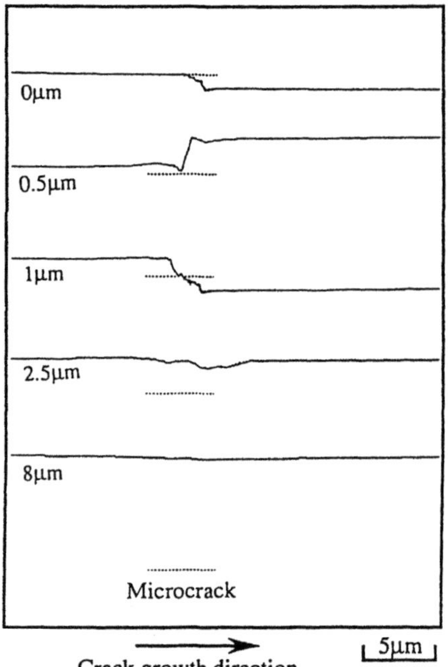

Figure 12.2. Crack path morphologies when a crack passes by a single microcrack.

testing of a peak-aged 22%SiCw/6061 composite which acted as a model material for the simulation. The in-situ experiments showed that the crack-tip stress field was dominated by the HRR singularity for both stationary and growing cracks in the composite.

Several other approaches have been reported previously for crack-tip stress fields in damaged materials with microcracks and microvoids. However, if the distribution of the damage is uniform, the HRR stress singularity for the damaged material is found to be similar to that for sound material to a surprising degree [6]. With this in mind, and also taking into consideration the difficulty in determining damage parameters, the ordinary HRR singularity for sound material is used to calculate the stress distribution even after the initiation of the microcracks. Many other assumptions are required for the fracture mechanical analysis, and they have been reported elsewhere [1].

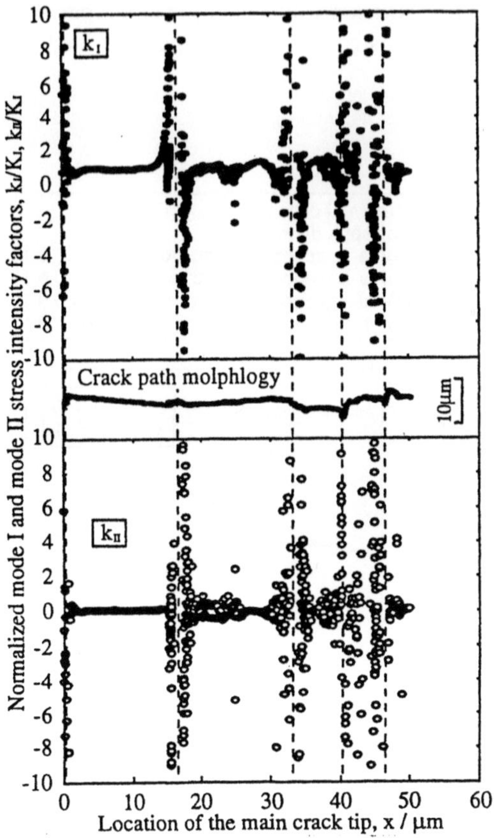

Figure 12.3. Variation of K_I and K_{II} along the crack path.

Figure 12.1 shows the flow chart of the simulation program. When a microcrack exists ahead of the main crack, the crack will be deflected as shown in figure 12.2. When the main crack passes through the microcrack (dotted line), anti-shielding and then shielding effects appear as shown in figure 12.3.

High toughness metal matrix composite

Segregated SiC whiskers made by spray-drying were embedded into a 6061 alloy matrix. Figure 12.4 shows the result. Appropriate agglomeration of the reinforcement gives the maximum roughness of the crack profile, and therefore the optimum propagation toughness. The normal fracture toughness K_{IC} corresponds to initiation toughness, and is almost constant. In other words, segregated whiskers act effectively to improve the toughness, especially with respect to crack propagation. Microcracking appears first in the segregated SiC whiskers, and then the crack passes through the aluminium matrix. The microstructure of the agglomerated SiCw/6061 MMC is shown in figure 12.5.

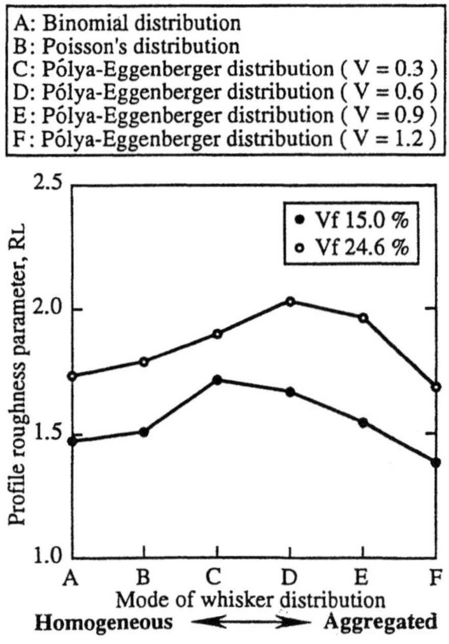

Figure 12.4. Effect of the spatial pattern of the reinforcement distribution on the crack propagation resistance.

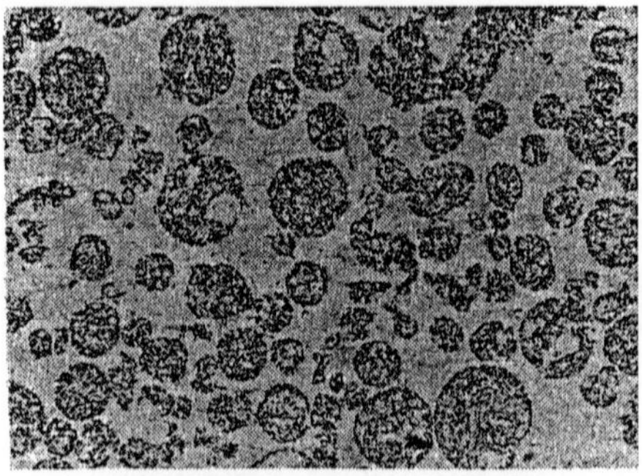

(b) MT-No.2 (d = 44.4 μm)

Figure 12.5. Typical microstructure of an agglomerated metal matrix composite.

Improvement of metal matrix composite toughness by retrogression and re-ageing

Recently, it has been reported that thin precipitate free zones are formed around SiC reinforcements in 6061 Al alloy [2]. In such a case, solute atoms (i.e. magnesium) in the matrix segregate in the vicinity of the interface. The effects of such inhomogeneous matrix microstructure on the mechanical properties can be analysed by means of non-linear finite element calculations. The precipitate free zone promotes significant local plastic deformation around the corner of the whisker, and the strength is predicted to decrease with increasing thickness of the precipitate free zone, as shown in figure 12.6, which gives the distribution of the effective plastic strain. In spite of the microscopic thickness of the precipitate free zone (44 nm), large scale yielding is extended substantially beyond the zone itself, into the age-hardened matrix interface region, from the early stage of loading.

To strengthen the MMC, therefore, it is important to eliminate the precipitate free zone. This can be achieved by utilizing the segregation of magnesium content at the interface and applying a retrogression and re-ageing (RRA) treatment [2]. Solvus curves for GP(II) zones and the intermediate phase β' in the pseudo-binary phase diagram of Al-Mg_2Si are shown in figure 12.7 and indicate that both solvus curves have rather large slopes against Mg_2Si concentration. The values of the solvus temperatures thus vary with distance from the reinforcement because of solute segregation in the MMC. Heating up to temperature T_A above the pre-ageing temperature causes local re-dissolution of the GP(II) zones in regions far from the reinforcement, because T_A exceeds the solvus temperature for lower Mg_2Si concentration. However, in the vicinity of the reinforcement, formation of the GP(II) zones and precipitation of the intermediate phase proceed because of the high local solvus temperature caused by the solute segregation near

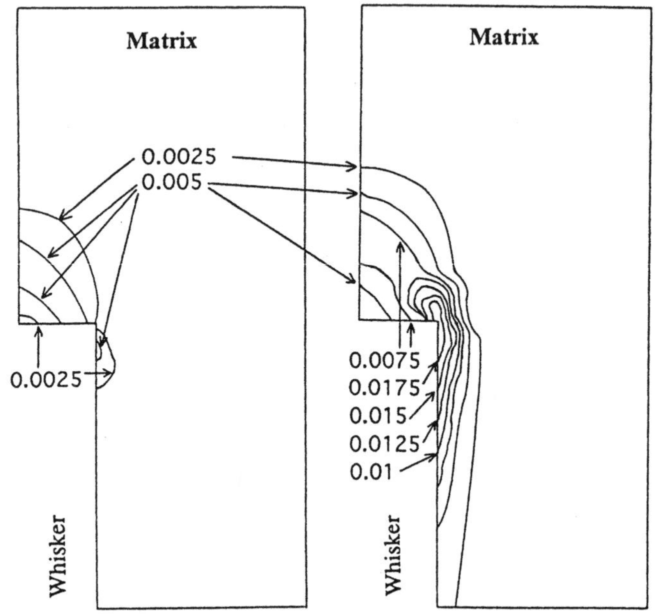

Figure 12.6. Effects of the existence of a precipitate free zone on the distribution of effective plastic strain in the matrix. Results of the finite element modelling using a quarter model of the cell containing a single whisker.

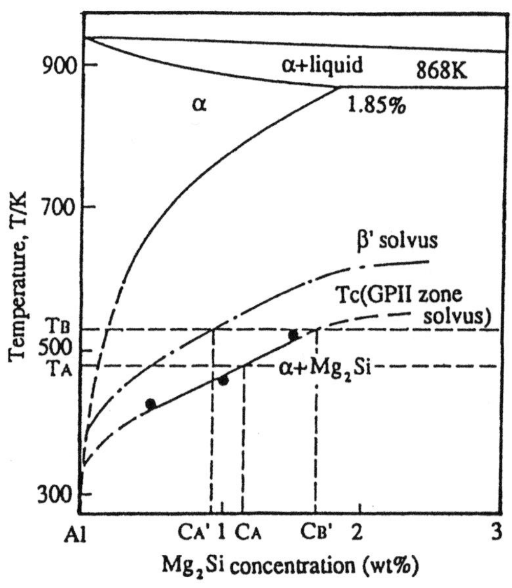

Figure 12.7. Pseudo-binary phase diagram for Al-Mg-Si alloys, including β' and GP zone solvus temperatures.

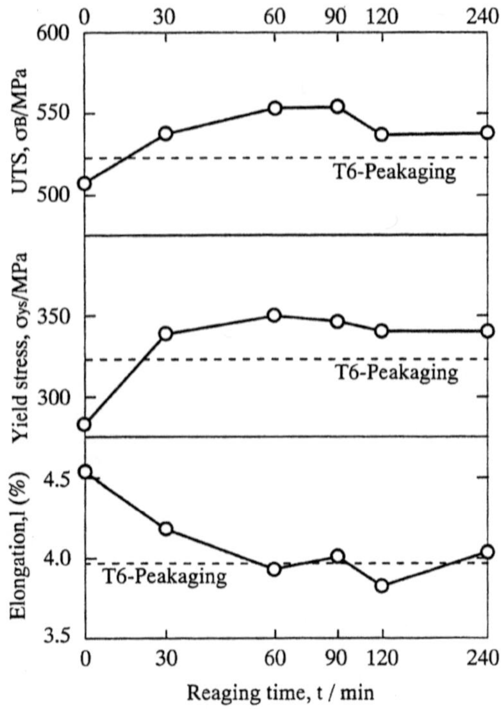

Figure 12.8. Variations in tensile strength, 0.2% proof stress and elongation during re-ageing at 450 K in the 6061/SiCw composite. Measurements were after pre-ageing at 450 K for 7.2 ks and following retrogression at 493 K for 120 s.

the interface. After a retrogression treatment of this type, adjustment of the mechanical properties to enhance strength requires the addition of a subsequent re-ageing treatment, as shown in figure 12.8.

Summary

Fracture mechanics simulation of crack growth in MMCs shows methods of achieving high toughness and resistance to crack propagation. A combination of controlling the reinforcement distribution and eliminating precipitate free zones by retrogression and re-ageing can be successfully applied.

References

[1] Toda H and Kobayashi T 1997 *Met. Mat. Trans. A* **28** 2149
[2] Kobayashi T and Toda H 1996 *Proc. ICAA5 Mat. Sci. Forum* **217-222** 1127

[3] Kobayashi T and Toda H 1997 *Proc. ICAA5 Mat. Sci. Forum* **242** 193
[4] Rice J R and Rosengren G F 1968 *J. Mech. Phys. Solids* **16** 1
[5] Rice J R, Drugan W J and Sham T L 1980 *Fracture Mechanics ASTM STP* **700** 189
[6] Wang J and Chow C L 1992 *Int. J. Frac.* **54** 165

Chapter 13

Matrix and fibre systems in polymer matrix composites

Hazel Assender

Introduction

The development of novel polymer matrix composite materials for use in aerospace technologies is driven on a number of fronts including improved reliability, damage self-sensing materials, high temperature systems, improved toughness and impact resistance, and improved processing routes. Often it is the development of novel or modified matrix or fibre systems which allows such improvements. With a view to the next decade of development in polymer matrix composites for aerospace applications, a number of development areas will be discussed in this chapter, both those in which clear progress is already under way and some rather more speculative areas which are, as yet, the 'glint in the eye' of researchers.

In the past decade, significant use has been made of polymer matrix composites as structural materials for airframes in both military and civilian applications. Their anticipated wider application has driven much research and development work in this class of materials in recent years. There are, however, still significant areas of potential and necessary improvement which will continue to guide development of polymer matrix composites into the new millennium. Concerns about the performance and cost effectiveness of polymer matrix composites have already led to doubts as to the extent of growth of these materials in the aerospace industry. The next decade will be crucial to the development of polymer matrix composites if they are going to continue as an attractive materials choice for a wide range of aerospace components.

Materials properties

To improve the underlying materials performance of a polymer matrix composite, one can explore improvements to the fibres and the matrix resins as well as the interfaces between the fibres and the matrix. Fibres require improved toughness, compressive strength and ease of processing. The announcement by the Defence and Evaluation Research Agency (DERA), Structural Materials Centre in the UK in 1996 [1] of the production of the first continuous hollow carbon fibre showed the potential for use of such materials as a fibre reinforcement in polymer matrix composites. The fibres are produced by oxidation of hollow PAN fibres. The hollow core provides, apart from the obvious weight saving, the improved compressive properties one would anticipate from a cylindrical structure. Microbuckling of the fibres on compression is reduced. The hollow core would also allow for the development of a multifunctional or 'smart' fibre by, for example, inclusion of a second phase material in the core, or by activating the inner surface of the carbon in some way. The development of hollow carbon fibres for the filtering of toxins using activated carbon, for example, is well established [2–4].

If one might extend of the concept of a hollow carbon fibre to nano-structured materials then carbon nanotubes provide the opportunity for ultra-fine scale reinforcement of polymer matrices. Nanoscale composite materials is a field of keen interest currently throughout the materials science community due to potential advantages in mechanical properties that can be achieved. The advent of carbon nanotubes based upon cylinders of graphitic sheet end-capped with half a 'bucky-ball' (C_{60}) has led to wide speculation as to their application for a variety of materials functions, one of which is their use as reinforcing fibres in polymer composites. The interest in such materials for structural systems arises from the exceptionally high modulus that had been predicted (up to 5 TPa) [5] using molecular modelling and that has recently been measured as 1.8 TPa [6]. The measurement involved the observation, by transmission electron microscopy, of the vibrations of a single fibre, held at one end. As the ambient temperature was increased, so did the magnitude of the vibrations, and this information could be directly related to the modulus of the fibres. The very high modulus of the carbon nanotubes is attractive for their use as a reinforcing fibre in combination with high resilience of the fibres under bending and buckling [7]. A single nanotube can be bent with the tip of a scanning probe microscope to very high deformations without imparting permanent damage upon the fibre. Elementary trials of carbon nanotubes embedded in a cross-linked polymer matrix to examine the stress transfer between the fibre and the matrix have been made [8, 9]. In both cases, a very good interfacial bond strength was observed between the fibre and the matrix. On application of load, the failure was found to be only once the tubes fractured. In the case of a multi-walled nanotube, the inner tube fractures and the structure telescopes out of the outer sheath which is well bonded to the matrix resin. For materials systems based upon polymer matrices reinforced with carbon nanotubes to come into widespread use, the problems associated with large-scale manufacture (and its commensurate cost) and routes for processing such composites would have to be overcome. Nevertheless, carbon nanotube reinforcement holds the promise of high modulus and tough composite systems.

Turning now to matrix resins for fibre reinforced polymers, there are a number of active research fronts to explore, including improvement of toughness, high temperature and 'hot–wet' performance, and oxidative stability.

The toughness of typical matrix resins such as epoxies has been improved by the incorporation of a second phase. Resins were initially toughened with rubbers such as carboxy-terminated butadiene-acrylonitrile, but this had the effect of decreasing the modulus and the temperature at which the epoxy may be used, due to the low glass transition temperature of the rubber. To increase the temperature stability, siloxanes and even inorganic fillers such as silica have been used as toughening agents, although the effect of silica on fracture toughness is highly dependent upon the particle size and shape of the filler [10]. The move to thermoplastic additives incorporates material with a higher glass transition temperature than rubber and with higher inherent toughness than inorganic fillers, while at the same time allowing for a fine dispersion of the two phases by the process of spinoidal decomposition during cure. This phase separation process can lead to a number of phase morphologies which result from the interplay between the phase separation and the reaction kinetics. Figure 13.1 illustrates three typical phase morphologies that are observed with increasing volume fraction of thermoplastic [11]. The mechanical properties of the resulting material are highly dependent upon this morphology which itself depends upon the composition and reaction conditions of the blend. A co-continuous structure of the two phases shows the highest toughening contribution from the two phase structure. Some careful studies have been made of this phase separation and curing process. Using a combination of modelling and neutron scattering techniques, a phase diagram may be constructed [12] showing the spinodal decomposition and nucleation-and-growth regions at various stages of cure. As cure progresses, the phase boundaries move to lower temperatures, i.e. phase separation occurs as the molecular weight of the curing species increases.

To control the microstructures that occur on curing one needs an understanding of the kinetics, in addition to understanding the thermodynamics of phase separation. Using a combination of light scattering and torsional braid analysis techniques, Kim *et al* [13] have constructed a time–temperature transformation (TTT) curve for the curing and phase separation process, recording the onset of phase separation, the gelation point, the stage when the microstructural geometry is fixed, the end of phase separation and finally the vitrification point when the glass transition temperature of the cured resin falls below the cure temperature. From this kind of analysis, one can measure

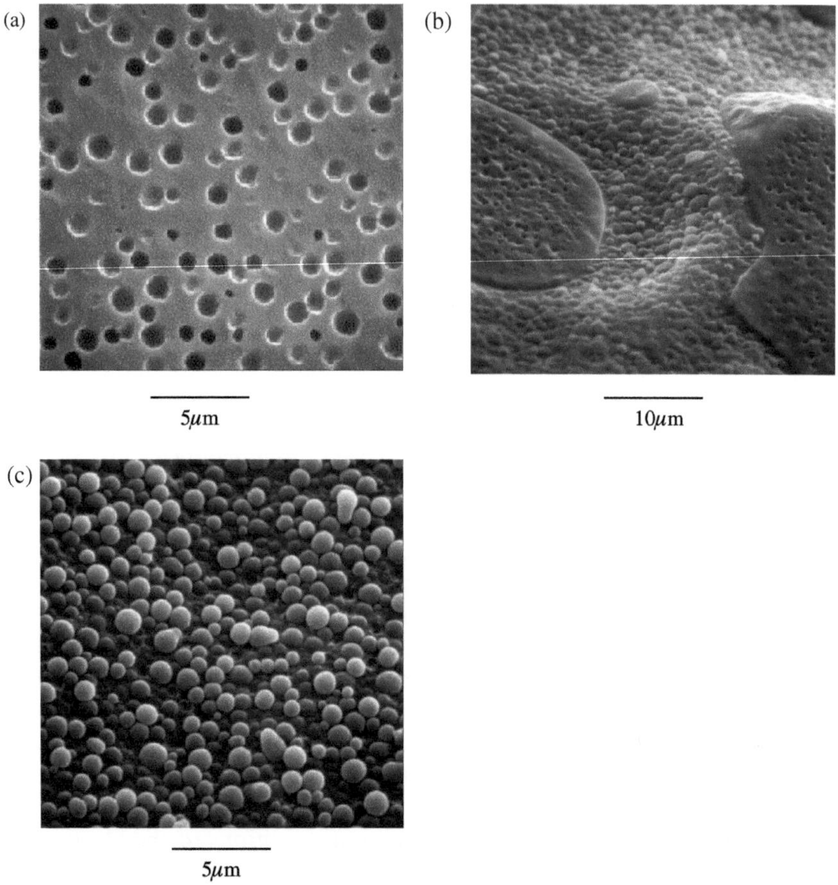

Figure 13.1. Micrograph of the morphologies of thermoplastic toughened epoxy. Epoxy resin blended with (*a*) 10%, (*b*) 20% and (*c*) 30% thermoplastic. The morphology moves from thermoplastic inclusions through co-continuous to a phase inverted morphology as the proportion of thermoplastics is increased. The mechanical properties of the resins appear to be linked to these morphologies [11].

and control the microstructure available for the matrix resin system, and hence modify the toughness of the resin, however, the toughness that is observed in the bulk resin is not necessarily transferred to the composite due to modification of the phase morphology and restriction of the crack-tip plastic zone by the reinforcing fibres. It is noteworthy that the size of the morphology that typically occurs in a co-continuous structure, is comparable with the distance between the fibres in a typical fibre–matrix composite material. Thus it is not unexpected that the presence of fibres embedded in the matrix will modify the toughening effect due to the morphology. Modification of the phase morphology by a variety of fibre

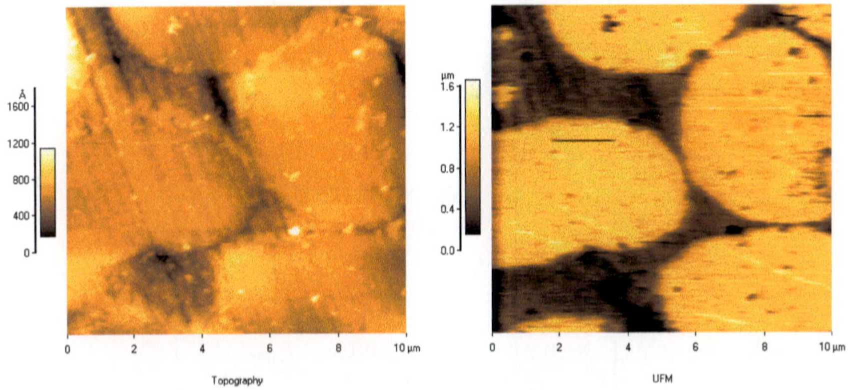

Figure 13.2. Atomic force microscope (AFM) and ultrasonic force microscope (UFM) images of a carbon-fibre reinforced epoxy viewed parallel to the fibres The stiffer carbon fibre phase is clearly distinguished and the interface and fibre morphology may be examined.

types has already been reported [14]. The thermoplastic toughening of matrix resins is described in more detail in chapter 14.

For a wider exploitation of such materials, a better understanding is required of the spinodal decomposition and curing of the resin and the interaction of the toughened matrix with the reinforcing fibres. Evidently, microstructural studies are crucial to such understanding, and one method which may help in the characterization of these small scale structures and the morphological changes during cure is ultrasonic force microscopy [15] and related scanning probe microscopy techniques. Figure 13.2 shows how this has already been used to distinguish the phase morphology in toughened acrylics and in fibre reinforced epoxies [16].

In addition to toughening, matrix resins need to withstand higher temperatures, including wet and oxidative environments. Hydrophobic matrix resins are discussed in chapter 15. The development of resins with very high glass transition temperature which are stable to degradation at these high temperatures necessitates the incorporation of heteroatoms such as silicon, boron, nitrogen or phosphorus into the polymer backbone. Phthalonitrile [17] and benzoxazine [18] based resins have both been tested as matrix resins in glass and carbon fibre reinforced composites. Other potential resin systems are still under development such as polyphosphazenes, siloxane–acetylene polymers and silicon–oxygen based polymers. Examples of candidate structures are shown in figure 13.3. In order for these materials to be exploited as matrix resins the processing routes and the fibre–matrix interfacial characteristics must be controlled, and such materials will tend to be brittle, so toughening mechanisms will need to be introduced.

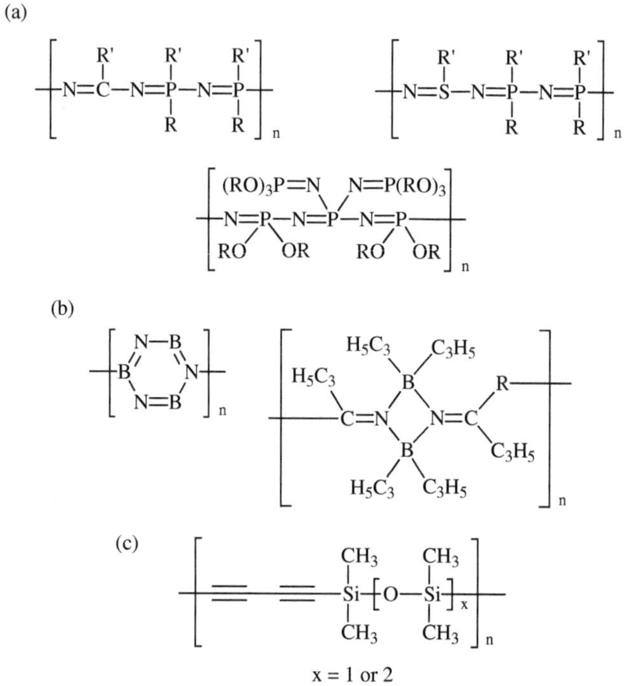

Figure 13.3. Potential polymers for use in ultra-high temperature matrix systems: (*a*) modified polyphosphazenes (*b*) boron-based compounds and (*c*) siloxane–acetylene.

Lifetime performance

A broad issue in materials science is that of lifetime performance including reliability of manufacture and testing methods, materials lifing, repair methods and, with the advent of smart materials, self-sensing and self-repairing materials. This issue is one that will play an increasing role in materials selection and application in the aerospace industry in the next decade, and it is one that is particularly pertinent for polymer matrix composite systems in which modelling for design and manufacture of components is not well advanced, non-destructive testing is challenging, and consumer confidence is low.

Composite materials are particularly prone to internal damage which is difficult to detect and which affects the structural performance of the material and its lifetime. One way in which internal damage may be monitored in carbon fibre reinforced polymers is by exploitation of the low electrical resistance of the carbon fibre. Chung [19] has shown that the resistance of the composite changes during the fatigue life of material, and has been able to relate this directly to fibre breakage. This property enables the production of an intrinsically self-monitoring system in order to predict a component life, or perhaps to locate composite damage, enabling component repair. An elementary self-repairing system has been demonstrated by workers at the University of Illinois based upon hollow glass fibres [20]. Glass fibres were filled in pairs with a two-component epoxy system. Alternatively, a cyanoacrylate containing glass fibre was paired with a metal wire. On cracking of the matrix, the glass tube breaks, releasing the adhesive which locally repairs the cracks. It was shown that on further loading the new cracks did not form at the previous crack sites, and thus the original cracks had been repaired to give a strong matrix once more.

Processing routes

The processing of polymer matrix composites has already been mentioned in the above discussions, and advances such as reactive resin transfer moulding and pultrusion have been developed as production techniques. A major influence here is the drive to reduce production costs and this is discussed further in chapters 6 and 16. However, there are other areas related to the processing of these materials which require consideration. These include repair of composites, non-destructive testing for control of manufacture, and joining techniques (improved mechanical fastenings and adhesive technology).

Summary

Polymer matrix composites have a wide current and potential application in the aerospace industry and developments in such materials systems continue apace. There are a number of research fronts which are making good progress to solve some of the key issues such as temperature resistance, toughening, lifing and processing, and others that are farther from the market place such as carbon nanotube reinforced composites, exotic chemistries, and sophisticated self-repairing systems. This chapter can by no means give an exhaustive account of all developments relevant to the aerospace industry but has provided a discussion of several key areas of current and future interest.

References

[1] 'Hollow Carbon-Fiber' news item 1996 *Materials World* **4** 316

[2] Koresh J K and Sofer A 1983 Molecular-sieve carbon permselective membrane. 1. Presentation of a new device for gas–membrane separation *Separation Sci. Tech.* **18** 723

[3] Henmi M and Yoshioka T 1993 Hollow fiber membrane based on polyacrylonitrile with a high degree of polymerization and its properties *J. Membrane Sci.* **85** 129

[4] Yang M-C and Yu D-G Catalytic oxidation of sulfur dioxide on polyacrylonitrile-based active hollow carbon fiber 1996 *J. Appl. Polymer Sci.* **62** 2287

[5] Sinnott S B, Shenderova O A, White C T and Brenner D W 1998 Mechanical properties of nanotubule fibers and composites determined from theoretical calculations and simulations *Carbon* **36** 1

[6] Treacy M M J, Ebbesen T W and Gibson J M 1996 Exceptionally high Young's modulus observed for individual carbon nanotubes *Nature* **381** 678

[7] Falvo M R, Clary G J, Taylor II R M, Chi V, Brooks Jr F P, Washburn S and Superfine R 1997 Bending and buckling of carbon nanotubes under large strain *Nature* **389** 582

[8] Wagner H D, Lourie O, Feldman Y and Tenne R 1998 Stress-induced fragmentation of multiwall carbon nanotubes in a polymer matrix *Appl. Phys. Lett.* **72** 188

[9] Lourie O, Cox D M and Wagner H D 1998 Buckling and collapse of embedded carbon nanotubes *Phys. Rev. Lett.* **81** 1638

[10] Nakamura Y, Yamaguchi M, Okubo M and Matsumoto T 1992 Effects of particle-size on mechanical and impact properties of epoxy resin filled with spherical silica *J. Appl. Polymer Sci.* **45** 1281

[11] Chaplin A, Davies T J, Jones D A, Shaw S J and Tudgey G F Novel hydrophobic tough and high temperature matrix resins for polymer composites *Plastics Rutter and Composites* 1999 **28** 191

[12] Elliniadis S, Higgins J S, Clarke N, McLeish T C B, Choudhery R A and Jenkins S D 1997 Phase diagram prediction for thermoset/thermoplastic polymer blends *Polymer* **38** 4855

[13] Kim B S, Chiba T and Inoue T 1993 A new time–temperature-transformation cure diagram for thermoset/thermoplastic polymer blends *Polymer* **34** 2809

[14] Varley R J and Hodgkin J H 1997 Effect of reinforcing fibres on the morphology of a toughened epoxy/amine system *Polymer* **38** 1005

[15] Kolosov O and Yamanaka K 1993 Nonlinear detection of ultrasonic vibrations in an atomic force microscope *Jap. J. Appl. Phys. part 2 Lett.* **32** 1095

[16] Dinelli F, Assender H E, Takeda N, Briggs G A D and Kolosov O V 1999 Elastic mapping of heterogeneous nano-structures with ultrasonic force microscopy UFM *Surf. Interface Anal.* **27** 562

[17] Sastri S B, Armistead J P and Keller E M 1996 Phthalonitrile–carbon fibres composites *Polymer Comp.* **17** 816

[18] Shen S B and Ishada H 1996 Development and characterization of high-performance polybenzoxazine composites *Polymer Comp.* **17** 710

[19] Chung D D L 1998 Self-monitoring structural materials *Mat. Sci. Eng.* **R22** 57

[20] Dry C 1999 Procedures for self-repair of polymer matrix composite materials *Composite Struct.* **35** 263

Chapter 14

Toughened thermoset resin matrix composites

Hajime Kishi and Nobuyuki Odagiri

Introduction

Carbon fibre reinforced composite materials have become familiar as structural materials, especially for weight-critical aeroplane parts, using their high strength, high modulus and light weight. Thermosetting resins such as epoxy resins have been used for the matrix resins of composites rather than thermoplastic resins, due to their better processability in prepregs, their tackiness and drapability. However, the low fracture toughness (compressive strength after impact energy (C_{AI})) of thermoset resin matrix composites compared with metals is one of the factors which limits their application as primary structural materials. Therefore, improvement of fracture toughness has been a hot topic of the composites industry.

Elastomer toughening of epoxy resins has been researched by many people. Pearson and Yee at the University of Michigan clearly explained the toughening mechanism [1]. The toughening effect depends very much on the shear deformation capacity of the matrix resin itself which includes elastomer particles. They indicated the importance of the crosslink density of the matrix epoxy resins. Comparatively lightly crosslinked epoxy was easily toughened by elastomer addition, but highly crosslinked epoxy was not. As the resin matrix is crosslinked more loosely, however, the compressive strength under hot–wet environment (C_{HW}) of the composite laminates, which is also an important property for structural materials for aeroplanes, tended to decrease in general.

With this background, thermoplastic resin blended thermoset resins for composites have been researched since the early 1980s in order to achieve both high impact resistance and high hot–wet performance. The first approach to toughening is thermoset/thermoplastic polymer alloys [3, 4]. McGrath [2] indicated that functional group terminated polysulphone was useful as the toughening modifier of epoxy resins. Polysulphone is one of the so-called engineering thermoplastic resins which have high heat resistance. A micro-phase separated morphology is formed during the curing process of the resin blend. The toughness of the cured resins depends on the phase separated morphology.

The second approach to enhance the compressive strength after impact of composite laminates is selective toughening of the interlayer zone, which is the weakest portion of the laminates. In the middle of the 1980s the concept of a thermoplastic resin film interleaf layer was proposed by the American Cyanamid Company (now Cytec-Fiberite) [5]. This was effective in improving the compressive strength after impact, but it also had disadvantages: there is an upper limit in fibre volume fraction due to the presence of a discrete homogeneous film in the interlayer zone, and the pre-preg requires special care in lay-up procedure due to the one-sided tack of the interleaved pre-preg. Following this, Toray Industries developed a heterogeneous interlayer toughening technology using a thermoplastic resin particle dispersion [6].

This chapter reviews the two elemental toughening technologies mentioned above: thermoset/thermoplastic polymer alloys and thermo-plastic-particulate interlayer toughening, mainly based on research activities by Toray Industries [7].

Thermoset/thermoplastic polymer alloys

As an example of such polymer alloying, epoxy resin of diglycidyl ether of resorcinol cured with 4,4′-diaminodiphenylsulphone (DDS) has been toughened with three types of amino-terminated polyimide oligomers (PI-A, PI-Si and PI-F), synthesized as toughening modifiers. The chemical structures of the oligomers and the formulated molecular weights (FMW) are shown in figure 14.1. The resin and the polyimide oligomer were mixed and heated to 130°C. After the oligomer was completely dissolved, the mixture was cooled to 80°C and the chemical equivalent amount of 4,4′-diaminodiphenyl-sulphone added to the epoxy resin. The oligomer content was chosen to be 30 wt% in the whole resin composition. The formulated epoxy resins were cured at 120°C for three hours followed by 180°C for two hours.

The fracture toughness (strain energy release rate G_{IC}) of the resulting resins was measured by the double torsion method and the flexural elastic modulus E measured by three-point bending. The fracture surfaces of the resins after double torsion test were observed using scanning electron microscopy (SEM). Scanning electron micrographs of the micro-phase separated morphologies of these resins are presented here to illustrate the characteristic structures. The component phases were identified by electron probe x-ray analysis (EPMA).

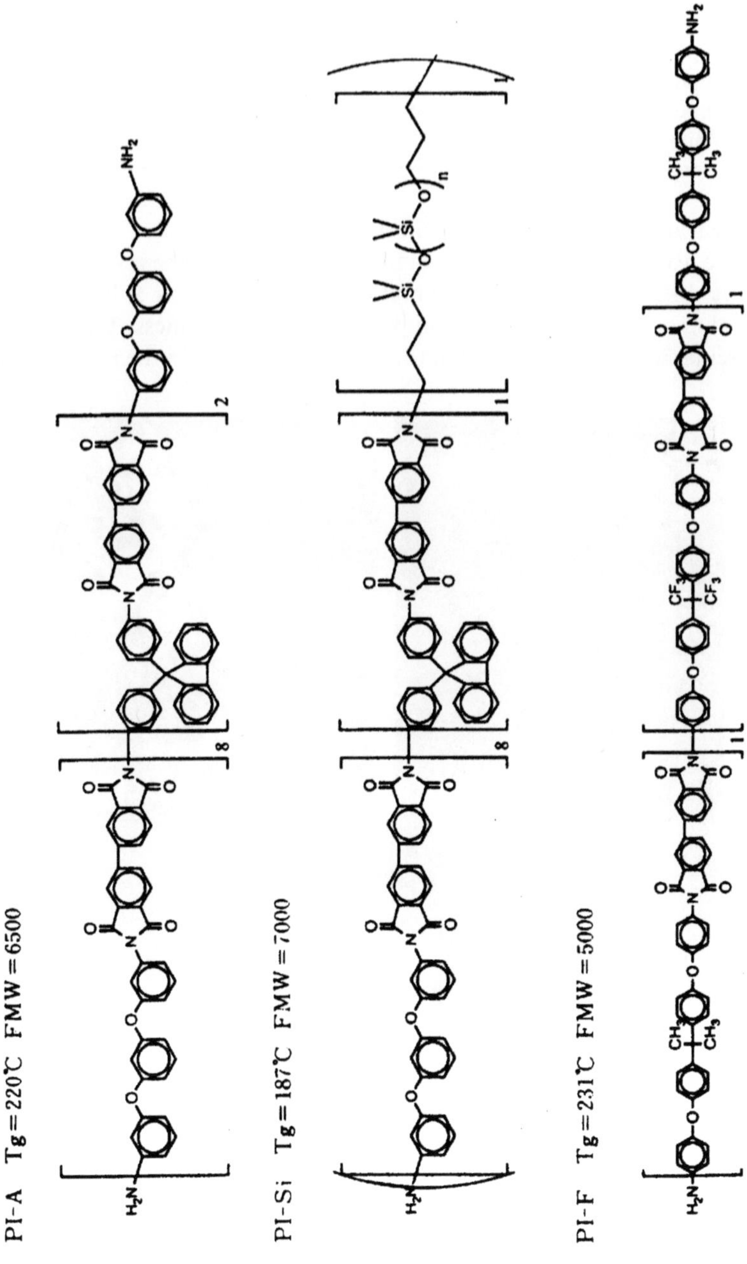

Figure 14.1. Chemical structures of imide oligomers.

Table 14.1. Mechanical properties of cured resins.

Resin system	G_{IC} (kJ m^{-2})	Elastic modulus (MPa)
Epoxy/PI-A (30 wt%)	0.30	3620
Epoxy/PI-Si (30 wt%)	1.22	3580
Epoxy/PI-F (30 wt%)	1.34	3430
Pure epoxy	0.15	3920

The fracture toughness G_{IC} and flexural elastic modulus E of the cured resins are listed in table 14.1 and the fracture surfaces of the resins are shown in figure 14.2. The epoxy resin modified with PI-A has a homogeneous morphology, with only a small improvement in toughness. On the other hand, remarkable improvement in toughness is achieved in the resin modified

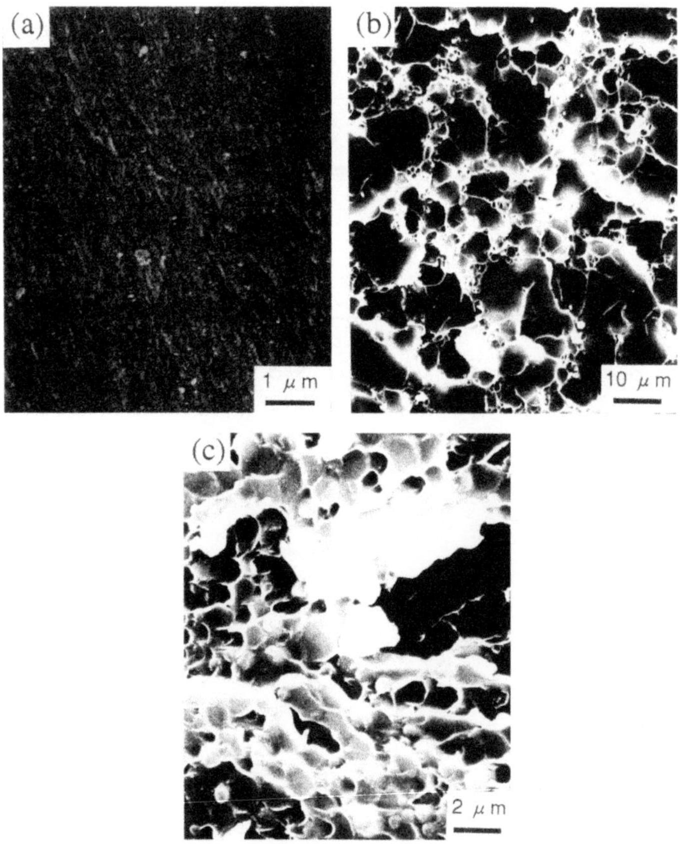

Figure 14.2. Scanning electron micrographs of the fracture surfaces of cured resins: (*a*) epoxy/PI-A (30 wt%), (*b*) epoxy/PI-Si (30 wt%), (*c*) epoxy/PI-F (30 wt%).

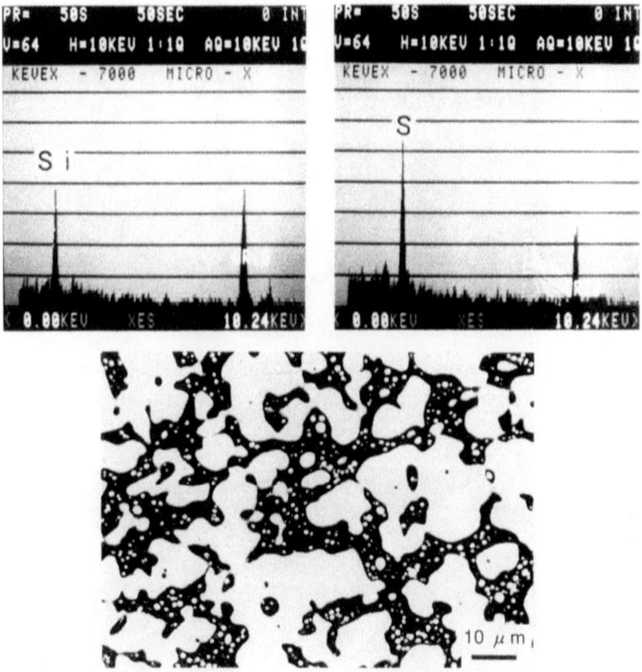

Figure 14.3. Scanning electron micrograph (backscattered electron mode) of epoxy/PI-Si cured resin: (1) darker phase: imide oligomer-rich phase which contains silicon, (2) brighter phase: epoxy-rich phase which contains sulphur.

with PI-Si and the resin modified with PI-F. The cured resin modified with PI-Si has a co-continuous phase separated morphology, with both phases having dispersed domains consisting of the other component. Figure 14.3 gives an example of this morphology. EPMA analysis revealed that the darker phase is rich in thermoplastic oligomers and the brighter phase is the epoxy-rich phase containing sulphur originating from the DDS. The resin modified with PI-F has a phase separated morphology, shown in figure 14.4, with spherical domains of epoxy-rich phase dispersed in the thermoplastic oligomer-rich continuous phase.

Comparison between the resin modified with PI-Si and the resin modified with PI-F shows a difference in morphology. However, the thermoplastic oligomer-rich phase forms a continuous phase in both cured resins resulting in excellent toughness, even though the oligomer is the minor component (30 wt%). Two energy absorbing mechanisms were found and characterized from micrographs such as those in figures 14.5 and 14.6. These mechanisms would account for the high fracture toughness of the cured resins:

1. ductile drawing of the thermoplastic oligomer-rich continuous phase;

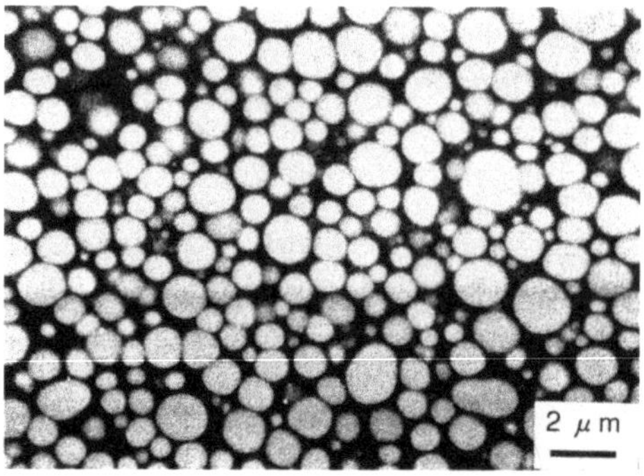

Figure 14.4. Scanning electron micrograph (backscattered electron mode) of epoxy/PI-F cured resin: (1) darker phase: imide oligomer-rich phase, (2) brighter phase: epoxy-rich phase.

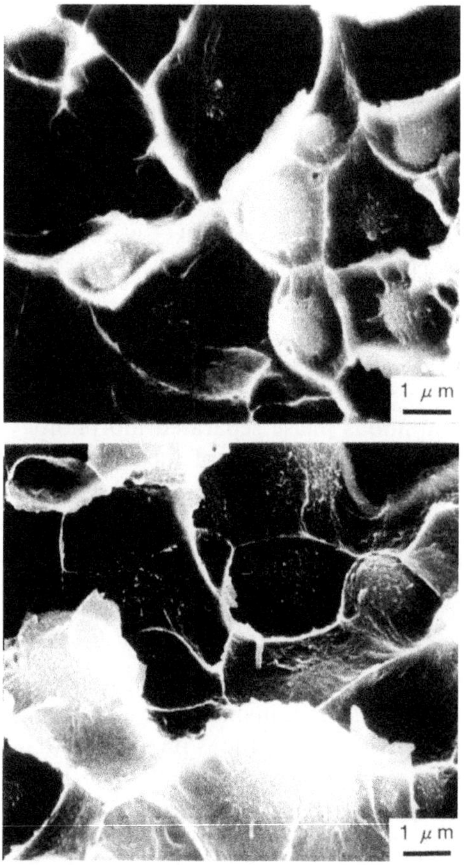

Figure 14.5. Scanning electron micrographs of the fracture surfaces of cured resins: (epoxy/PI-Si (30 wt%) blend) showing ductile drawing of the thermoplastic continuous phase.

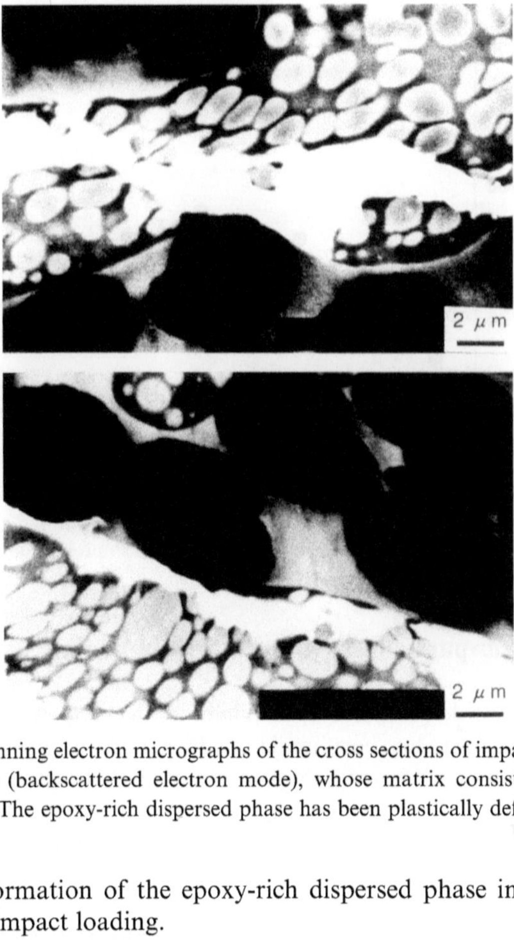

Figure 14.6. Scanning electron micrographs of the cross sections of impact loaded carbon fibre composites (backscattered electron mode), whose matrix consists of epoxy/PI-Si (30 wt%) blend. The epoxy-rich dispersed phase has been plastically deformed.

2. plastic deformation of the epoxy-rich dispersed phase in the interlayer zone after impact loading.

The first mechanism has been reported in the case of an epoxy/polyetherimide alloy system by Bucknall [3]. The second mechanism indicates that even the crosslinked resins can deform plastically under a shear stress condition. Good interfacial adhesion between the two phases is observed, which originates from the reaction between epoxy and amino-terminated imide oligomers.

Thermoset/thermoplastic polymer alloy technology improves resin toughness with a small reduction of hot–wet performance, but translation of the resin toughness to the composite toughness is marginal. Despite improvements in resin toughness, the compressive strength after impact of the composites gradually reaches a plateau, as illustrated in figure 14.7. One explanation of this effect is the constraint of the plastic deformation zone caused by the existence of reinforcing fibres [8]. To enhance the impact resistance of laminates, an interlayer toughening approach may be chosen.

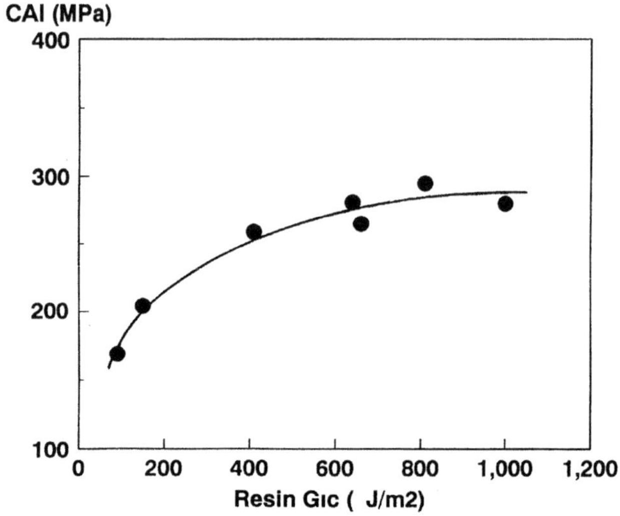

Figure 14.7. Correlation of resin toughness (G_{IC}) with compressive strength after impact (C_{AI}).

Thermoplastic-particulate interlayer toughening

To illustrate the role of interlayer toughening, the properties of an interlayer toughened composite has been made in which Torayca T800H carbon fibres are used as the reinforcement. Conventional highly crosslinked epoxy resin, which consists mainly of tetraglycidyl diaminodiphenylmethane (TGDDM)/diaminophenylsulphone, was used as the matrix resin for a pre-preg. Three types of unidirectional pre-preg tape with a fibre areal weight 145 g m^{-2} were prepared. The resin content of the pre-preg was 35 wt%. Two pre-preg samples with different thermoplastic polymer particles and one control pre-preg sample without the polymer particles were prepared. The polymer particles used were a modified amorphous polyamide with high T_g and a crystalline polyamide with a low T_g. The particles were placed on both surfaces of the pre-preg. The pre-preg was cut and layed up according to two different stacking configurations: $[+45/0/-45/90]_{4s}$ for compressive strength after impact tests (C_{AI}) and $[0]_6$ for hot–wet compression tests (C_{HW}). Cure of the stacked pre-preg was performed in an autoclave at 177°C for 120 min with 61 MPa nitrogen pressure. C_{AI} testing was conducted with an impact energy of 67 J cm^{-1} by dropping a weight on to a test panel. The damaged area was measured by ultrasonic C-scan inspection, followed by compressive strength measurements. Hot–wet compression test were performed on the test specimens at 82°C after immersion in water for 14 days at 71°C.

Crack propagation due to stress concentration was induced in the interlayer zone during impact. Compressive strength after impact was found to increase with a decrease in the damage area upon interlayer toughening. Thus, the delamination can be suppressed by selective toughening of the interlayer zone.

In this interlayer toughening technology, the performance of the thermosetting resin and the thermoplastic resin were separately utilized as follows: the thermosetting resin was utilized as a base matrix resin to impregnate between carbon fibres throughout the intraply, while the thermoplastic resin was selectively distributed as fine particles onto the surface of the pre-preg, in order to form resin with a high concentration of thermoplastic particles in the resin interlayer in the laminates. This thermoset/particulate-thermoplastic hybrid approach, figure 14.8, results in a pre-preg system that

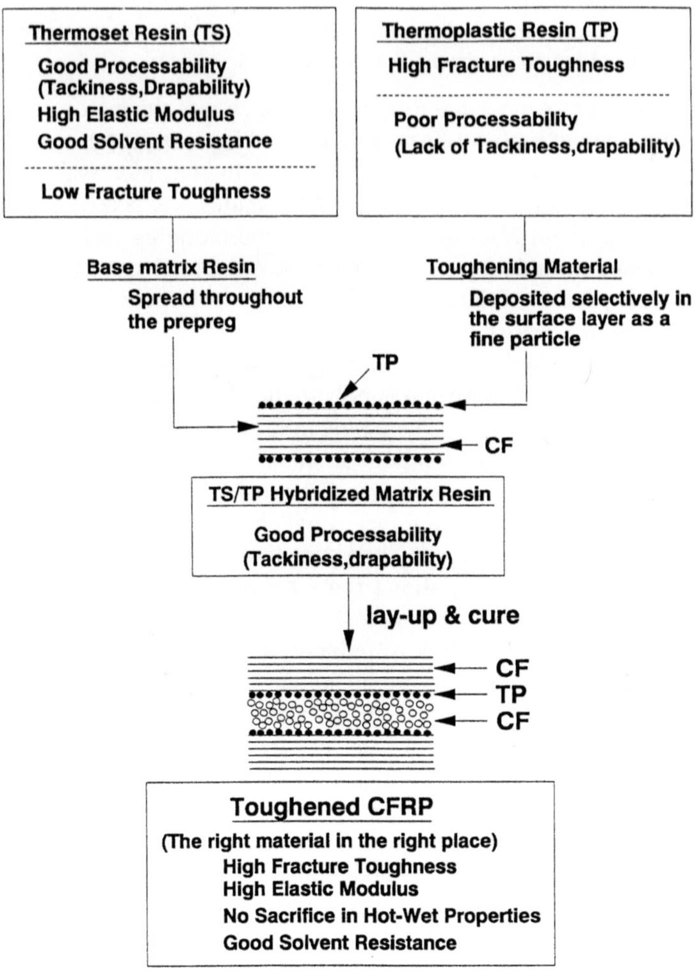

Figure 14.8. Concept of thermoplastic-particulate interlayer toughened pre-preg.

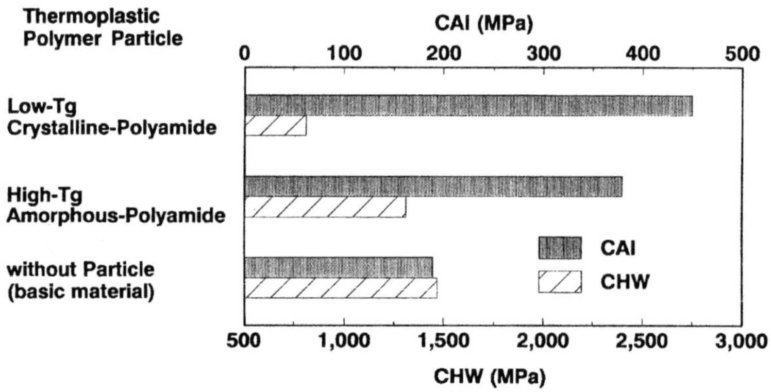

Figure 14.9, Effects of thermoplastic particulate materials on compressive strength after impact (C_{AI}) and hot–wet compressive strength (C_{HW}).

possesses good tackiness and drapability, which was the major advantage of the existing epoxy matrix pre-preg systems. After pre-preg lay-up and cure, thermoplastic constituents are concentrated in the interlayer zone and a resulting high fracture toughness of the laminates is expected by decreasing the damage area on fracture. The high elastic modulus and high T_g of the intraply resin should be reflected by high compressive strength and high hot–wet properties of laminates. In addition to these advantages good solvent resistance is also expected.

Two types of polyamide, a low T_g crystalline polyamide ($T_g = 55°C$) and high T_g modified amorphous polyamide ($T_g = 160°C$), were utilized as the thermoplastic polymers. These polyamides had a lower water absorption than cured epoxy resin. A comparison of compressive strength after impact (C_{AI}) and hot–wet compressive (C_{HW}) strength of the laminates using the thermoplastic-particulate interlayer toughening technology with the non-modified control laminates, is shown in figure 14.9. The composite with low T_g polyamide particles shows remarkably high C_{AI} values, while C_{HW} values are relatively low. The composite with the high T_g polyamide particles shows improved C_{AI} with less reduction in C_{HW} properties. Cracking especially in mode II (shear mode) under impact loading is observed to propagate through the thermoset/thermoplastic heterogeneous interlayer zone. A large amount of the impact energy was found to be absorbed by both ductile deformation and ultimate failure of the thermoplastic particles, as illustrated by the micrograph in figure 14.10.

Fracture toughness and T_g of the thermoplastic polymer particles are found to reflect the toughness and hot–wet properties of the composite. The observations indicate that it is possible to meet various material requirements by selecting the chemical structure of the thermoplastic polymers. Well balanced properties required for aircraft primary structures can be achieved by introduction of high T_g modified-amorphous polyamide particles.

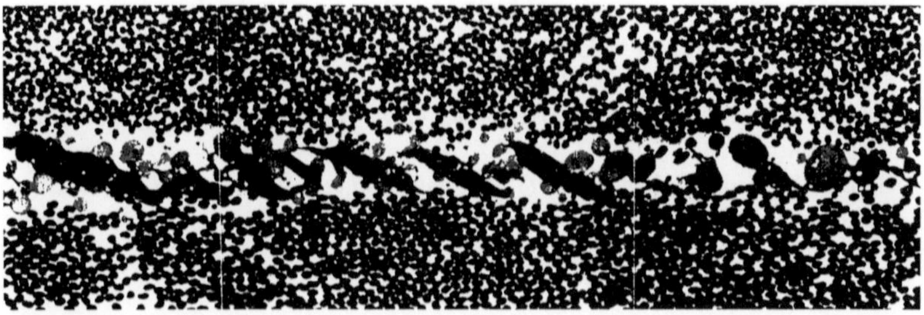

Figure 14.10. Cross section of impact loaded composite using particulate interlayer toughening.

Torayca T800H/3900-2 toughened pre-preg system [6] has been chosen as a primary structural material for the Boeing 777.

Forward view

As one of the key issues in the development of composites, the translation from the resin properties to the composite properties is an important issue to be intensively researched. If the interface adhesion is poor, the toughness of the resin is not fully utilized in the composites. Other properties, even the tensile strength of laminates which is generally a fibre-dominant property, should also be affected by the resin properties and interface adhesion. The role of the interface on mechanical properties has been discussed conceptually, but not yet quantitatively. More research on the interface issues is expected.

Reductions in the cost of both materials and fabrication processes will be a critical issue. Increased performance seems to become less important in order to propagate the application of composites. Instead, affordability is now more important. Revolutionary low-cost fabrication processes and materials (fibres, preforms and resins) which enable processing with less reduction of the performance of the resulting composites is today's hot topic in the composite industry.

References

[1] Pearson R A and Yee A F 1989 *J. Materials Sci.* **24** 2571

[2] Cecere J A Sender J S and McGrath J E 1987 *International SAMPE Symposium* **32** 1276

[3] Bucknall C B and Gilbert A H 1989 *Polymer* **30** 213

[4] McGrail P T, Sefton M S, Almen G R and Wilkinson S P 1989 European Patent Application 0311349

[5] Masters J E, Courter J L and Evans R E 1986 *International SAMPE Symposium* **31** 844
[6] Odagiri N, Kishi H and Nakae T 1991 *Proc. American Society for Composite 6th Tech. Conf.* **43** 7–9 October
[7] Odagiri N and Kishi H 1992 *Proceedings of the American Chemical Society Polymer Preprints* **33** 384
[8] Bradley W L and Cohen R N 1985 *Delamination and Debonding of Materials ASTM STP* 876 389–410

Chapter 15

Hydrophobic epoxies for polymer matrix composites

Steve Shaw, Graham Tudgey, Adam Chaplin, Jim Davies and Alex Cain

Introduction

Several factors have prevented the most efficient use of composites in aerospace applications. Most notable among these concerns the damaging effect of atmospheric moisture on composite performance, where the influence of water on the matrix phase of the composite and thus on key matrix-dominated properties has been the cause of most concern. A reduction in the water absorption characteristics of the matrix resin would probably lead to improved hot–wet properties, and could allow the design of thinner, lighter structures.

For demanding applications, particularly those requiring an elevated temperature capability, epoxy resins having high functionalities are generally required. Most notable among these is the epoxy resin based on tetraglycidyl 4,4′-diaminodiphenylmethane (TGDDM), which is currently the system favoured for aerospace structural applications. Investigations conducted on TGDDM resins cured with 4,4′-diaminodiphenylsulphone (DDS) have shown water uptake values for this system of the order of 5%, suggesting a reduction in glass transition temperature T_g of up to 100°C if a highly moist atmosphere were to be encountered [1]. Matrix-dominated properties which would inevitably suffer under these circumstances include compression behaviour and interlaminar shear strength (ILSS).

Initial investigations at DERA Farnborough involving the synthesis of halogen-substituted TGDDM resins [2, 3] showed that this approach to hydrophobic enhancement can reduce the water uptake of TGDDM resins by as much as 40%, while showing relatively minor changes in T_g for the chlorinated and brominated systems. This chapter will discuss the hydrophobic

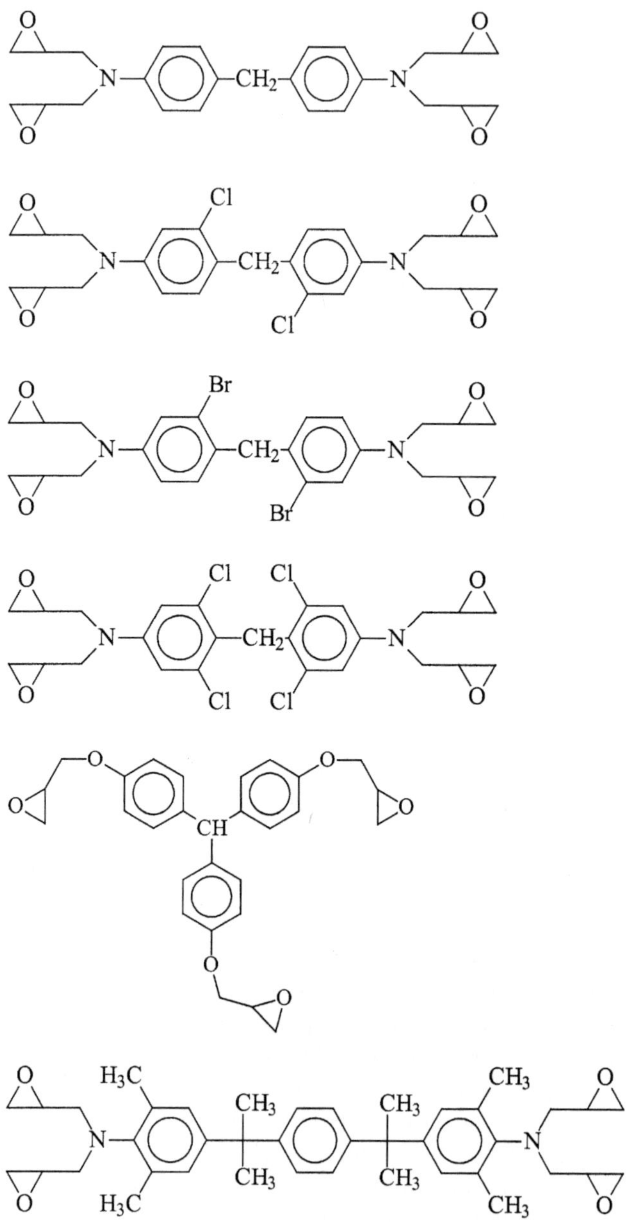

A

B

C

D

E

F

Figure 15.1. The epoxy resins with and without hydrophobic substituent groups: **A**, a commercial epoxy (TGDDM); **B**, epoxy (Cl_2-TGDDM); **C**, epoxy (Br_2-TGDDM); **D**, epoxy (Cl_4-TGDDM); **E**, epoxy: a high temperature commercial epoxy based on triglycidyl ether of tris(hydroxy)methane; **F**, epoxy: a commercial epoxy containing hydrocarbon substituents and offering low water absorption.

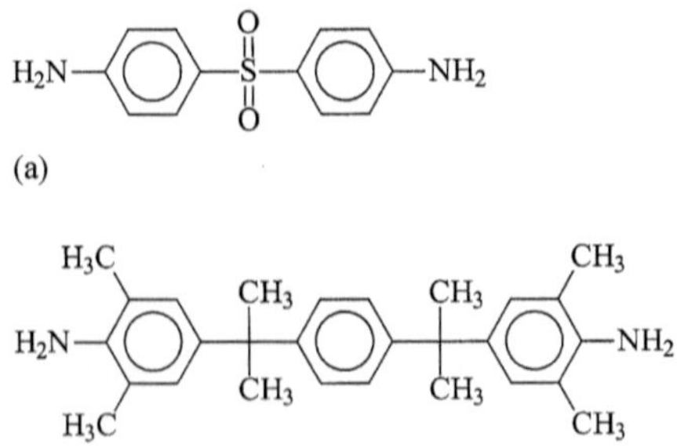

Figure 15.2. The structure of the curing agents employed: (*a*) 4,4′-diaminodiphenylsulphone; (*b*) diamine (X).

substituent groups in matrix resins and their effect on composite water absorption, T_g and interlaminar shear strength. The epoxy resins evaluated as composite matrices are given in figure 15.1. The epoxy systems (A–E) were all cured with 4,4′-diaminodiphenylsulphone (DDS), the structure of which is shown in figure 15.2(*a*). The epoxy systems A, B and F were also cured with X a commercial diamine containing hydrocarbon substituents and offering low water absorption. The structure of X is shown in figure 15.2(*b*).

Various bulk properties of these polymers have been assessed, most notably modulus, fracture toughness/energy, T_g and water uptake behaviour. A range of composite properties including interlaminar shear strength and compression strength after impact have also been investigated.

Bulk matrix properties

Table 15.1 shows the values obtained for T_g, density, flexural modulus and fracture toughness and energy of the various epoxy systems. The flexural modulus of the Cl_2-TGDDM/DDS is significantly higher (+18%) than the base-line TGDDM/DDS system, and has similar values for T_g, fracture toughness K_{IC} and fracture energy G_{IC} with only a minor increase in density. This increase in flexural modulus offers a potential advantage in composite matrix properties where compression properties are often related to matrix modulus.

Specimens were aged hygrothermally at 70°C/83% relative humidity (RH) to high levels of saturation. Table 15.2 shows the reduced levels of moisture uptake due to the various halogen substituents. Thus, the Cl_2-TGDDM/DDS system offers the best combination of reduced moisture absorption, high T_g and high flexural modulus, with the potential for improved performance of hot–wet composite properties.

Table 15.1. Measured values of T_g, density, flexural modulus and fracture toughness and energy for the selected epoxies.

Epoxy	Curing agent	T_g (°C)	Density ($g\,cm^{-3}$)	Flexural modulus (GPa)	Fracture toughness K_{IC} ($MN\,m^{-3/2}$)	Fracture energy G_{IC} ($J\,m^{-2}$)
TGDDM (A)	DDS	260	1.26	4.06	0.47	48
Cl_2-TGDDM (B)	DDS	260	1.35	4.78	0.47	41
Br_2-TGDDM (C)	DDS	235	1.46	–	0.47	41
Cl_4-TGDDM (D)	DDS	235	1.39	–		
Epoxy (E)	DDS	314	1.25	3.60		
TGDDM (A)	(X)	242	1.14	3.62		
Cl_2-TGGDM (B)	(X)	234	1.19	4.13		
Epoxy (F)	(X)	243	1.06	–		

Table 15.2. Moisture absorption levels (% weight increase) attained by 0.5 mm thick epoxy specimens after 96 days at 70°C/83% RH.

Epoxy	Curing agent	%H_2O 96 days at 70°C/83% RH
TGDDM (A)	DDS	4.39
Cl_2-TGDDM (B)	DDS	3.34
Br_2-TGDDM (C)	DDS	2.87
Cl_4-TGDDM (D)	DDS	2.68
Epoxy (E)	DDS	4.45
TGDDM (A)	(X)	3.61
Cl_2-TGGDM (B)	(X)	2.61
Epoxy (F)	(X)	2.80

Initial unidirectional carbon fibre composite properties

High quality unidirectional carbon fibre composite test specimens have been prepared from the eight matrix systems comprising the various epoxy resin/curing agent combinations listed in table 15.2. The initial composite properties of the eight matrix systems tested at 120°C (a standard elevated test temperature for aircraft materials) and 150°C (a target temperature for high performance applications) are shown in table 15.3 and figure 15.3. The halogen substituted TGDDM epoxies cured with DDS all show slightly higher interlaminar shear strength values than TGDDM. These improvements are shown clearly by the black hatched areas on the bar charts in

Table 15.3. Mean ILSS (MPa) and T_g (°C) of composites produced from the epoxy matrices.

Epoxy	Curing agent	T_g (°C) (log E'')	ILSS at 120°C (MPa)	ILSS at 150°C (MPa)
TGDDM (A)	DDS	263	89 (2.6%)	80 (1.6%)
Cl_2-TGDDM (B)	DDS	265	93 (3.2%)	85 (1.4%)
Br_2-TGDDM (C)	DDS	246	96 (1.5%)	82 (2.2%)
Cl_4-TGDDM (D)	DDS	232	90 (4.7%)	80 (3.2%)
EPOXY (E)	DDS	329	71 (4.7%)	67 (3.4%)
TGDDM (A)	X	234	72 (4.5%)	60 (6.6%)
Cl_2-TGDDM (B)	X	238	78 (4.4%)	66 (3.2%)
EPOXY (F)	X	249	66 (3.9%)	59 (5.7%)

Values in parentheses are the percentage coefficients of variation for mean ILSS.

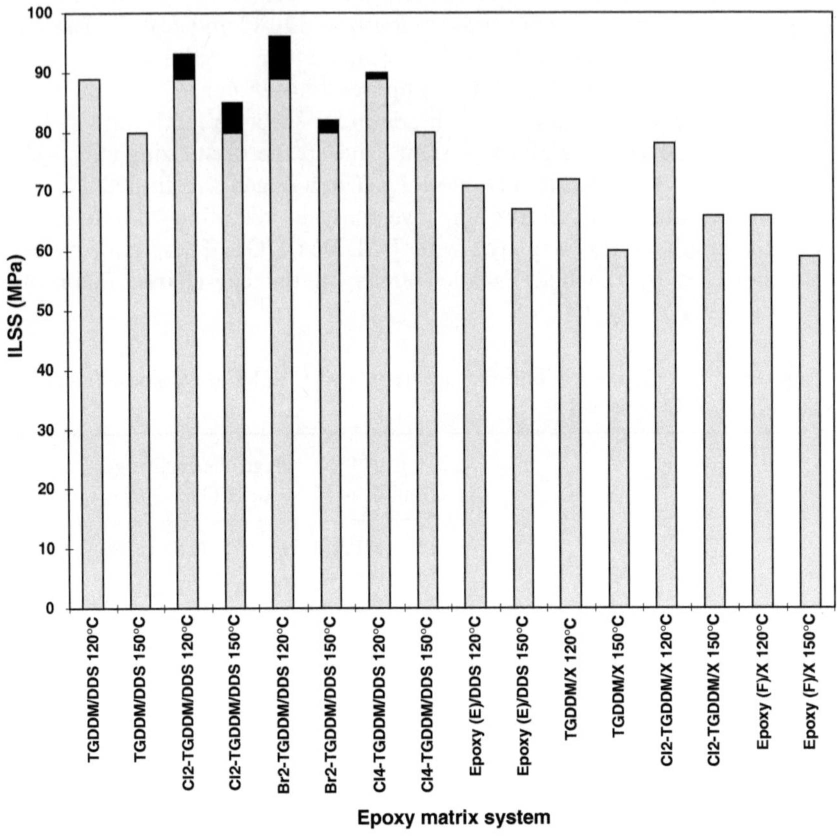

Figure 15.3. Mean interlaminar shear strength (MPa) of unaged composites tested at 120°C and 150°C.

figure 15.3. Br_2-TGDDM had the highest value (96 MPa) at 120°C, while Cl_2-TGDDM had the highest value (85 MPa) at 150°C reflecting its higher T_g. There was also a general trend that the composites with higher T_g show a greater relative retention of interlaminar shear strength at 150°C when compared with their mean interlaminar shear strength at 120°C. The Cl_2-TGDDM/DDS system has a slightly higher T_g than TGDDM/DDS. The use of the curing agent X with TGDDM and Cl_2-TGDDM considerably lowers the interlaminar shear strength values compared with those where DDS is used as the curing agent.

The effect of hygrothermal ageing on composite properties

Composite samples prepared from the various matrix resin/curing agent combinations were hygrothermally aged for approximately 150 days at 70°C/83% RH to a high level of saturation at which point they were removed and tested for interlaminar shear strength at 120°C and 150°C. The mean interlaminar shear strength results obtained compared with all unaged samples are shown in table 15.4 and figures 15.4–15.6.

As indicated, a severe fall in interlaminar shear strength both at 120°C and 150°C occurred for all the systems, due to the plasticizing effect of the water and the resultant reduction in T_g. The halogen-substituted TGDDM systems produced significant improvements in retention of interlaminar shear strength values compared with TGDDM/DDS. These improvements are indicated by the black shaded areas on the bar charts. Figure 15.4

Table 15.4. The effect of hygrothermal ageing at 70°C/83% RH on composite mean ILSS at 120°C and 150°C.

			Unaged composite		Hygrothermally aged at 70°C/83% RH			
			ILSS (MPa)		wt gain	ILSS (MPa)		Wet T_g (log(E′))
Energy	Curing agent	Carbon fibre	120°C	150°C	(%)	120°C	150°C	(°C)
TGDDM (A)	DDS	XAS	89	80	1.37	54	44	177
Cl_2-TGDDM (B)	DDS	XAS	94	85	1.15	60	48	199
Br_2-TGDDM (C)	DDS	XAS	96	82	1.08	58	45	183
Cl_4-TGDDM (D)	DDS	XAS	90	80	0.97	58	46	181
Epoxy (E)	DDS	XAS	71	67	1.30	41	36	150
TGDDM (A)	(X)	XAS	72	60	0.87	55	44	190
Cl_2-TGDDM (B)	(X)	XAS	78	66	0.71	59	47	195
Epoxy (F)	(X)	XAS	66	59	0.68	39	35	–

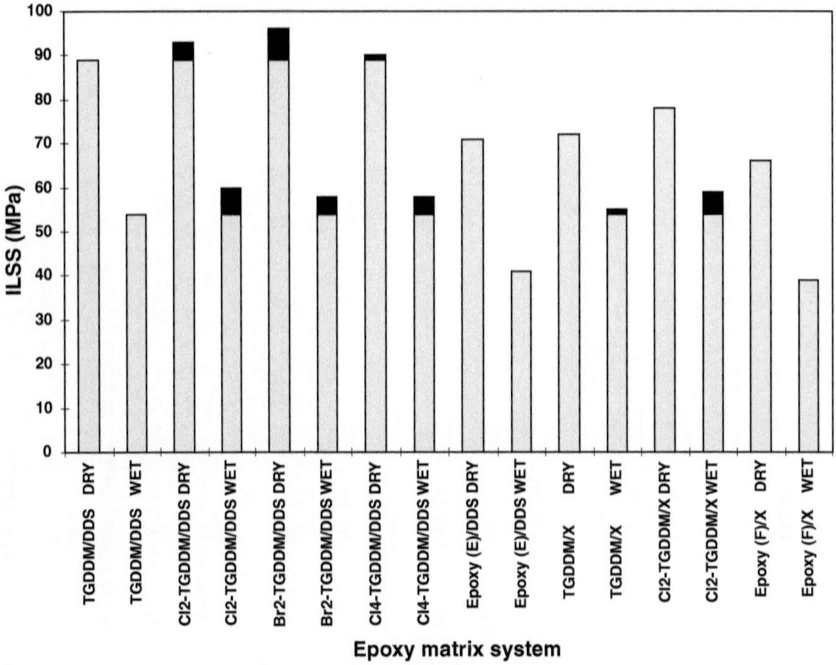

Figure 15.4. The effect of hygrothermal ageing on mean interlaminar shear strength at 120°C.

facilitates the comparison of both dry and wet interlaminar shear strength values at 120°C and 150°C for the halogen substituted TGDDM composites and clearly shows the advantage conferred by the Cl_2-TGDDM/DDS matrix system.

Comparative evaluation of the commercially pre-pregged composite properties of Cl_2-TGDDM with a proprietary TGDDM composite

As a result of the improved hot–wet composite properties offered by the Cl_2-TGDDM/DDS matrix system, a major epoxy composite manufacturer prepared carbon fibre pre-preg (Toray T800H carbon fibre) using the matrix system (hereafter referred to as Cl_2-TGDDM) with the same formulation of toughening agent as that of their standard toughened system (hereafter referred to as TGDDM). The Cl_2-TGDDM system was prepared successfully as a thermoplastic toughened pre-preg and easily processed in a similar manner to the proprietary toughened TGDDM pre-preg. The results of tests at 20°C are shown in table 15.5.

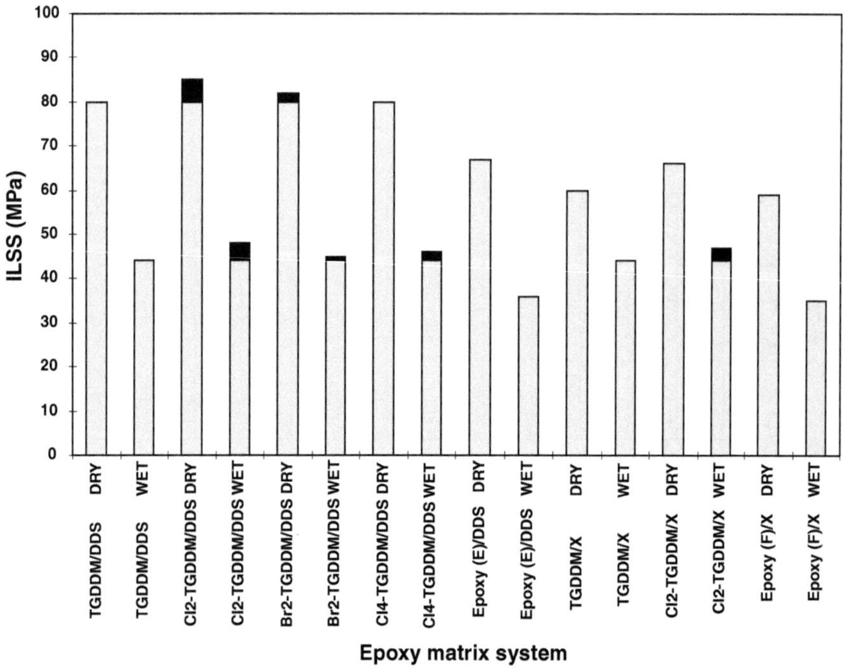

Figure 15.5. The effect of hygrothermal ageing on mean interlaminar shear strength at 150°C.

The results of the tests performed at 20°C demonstrate that the composite properties of the Cl_2-TGDDM system are comparable or better than the proprietary TGDDM system with the standard 180°C cure and showed further improvements with an additional post-cure at 200°C. In addition to the comparable T_g, the Cl_2-TGDDM system displays outstanding improvements (up to approximately 100%) in fracture energy G_{IC} and G_{IIC} (shown in figure 15.7 by the black shaded areas) and a 10% improvement in compression after impact (C_{IA}) combined with a significantly reduced area of impact damage.

The effect of hygrothermal ageing

Composite samples prepared from the standard proprietary toughened TGDDM system and the toughened Cl_2-TGDDM matrix system were all aged hygrothermally for 204 days at 70°C/83% RH. At this point, a high level of saturation was attained and the specimens were removed and tested for interlaminar shear strength and open-hole notch compression

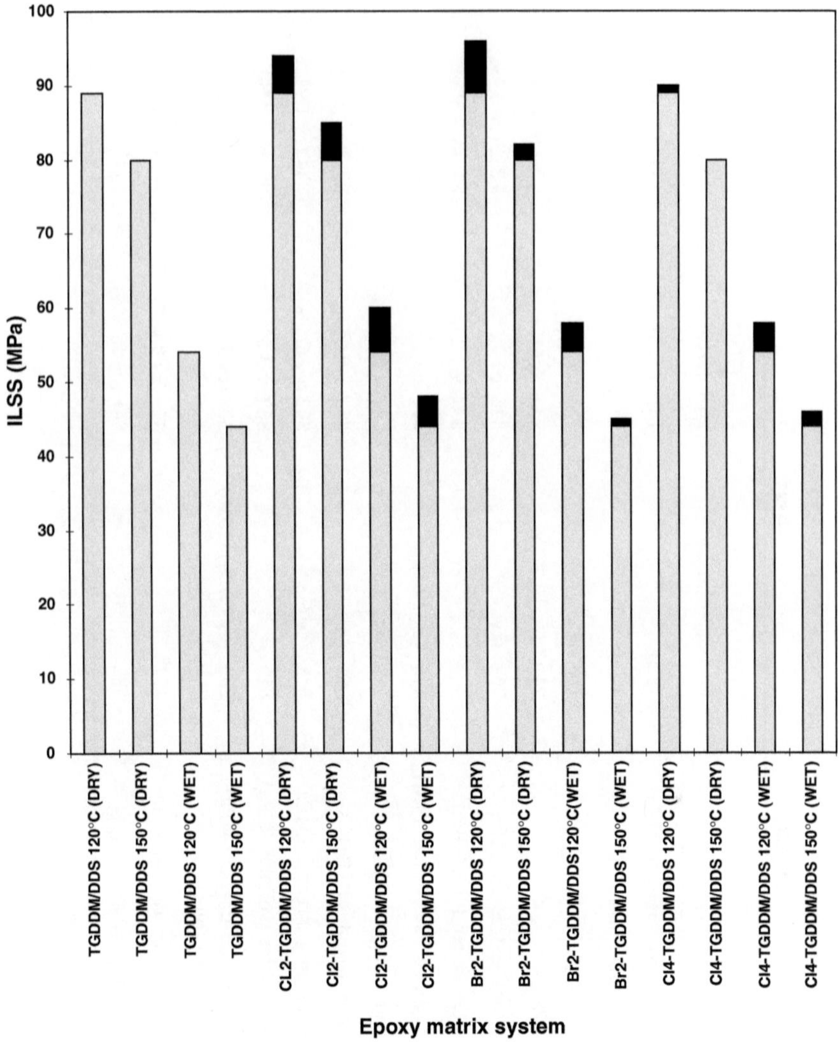

Figure 15.6. The effect of hygrothermal ageing on mean interlaminar shear strength at 120°C and 150°C and improvements over TGDDM/DDS shown by halogen substituted TGDDM/DDS.

strengths at room temperature (20°C) and 120°C (a standard elevated test temperature for aircraft materials). Table 15.6 shows that the Cl_2-TGDDM matrix system has moisture absorption approximately 20% (180°C post-cure) and 15% (200°C post-cure) lower in comparison with the standard proprietary TGDDM system. These reductions in moisture absorption results in a slightly higher wet T_g of the Cl_2-TGDDM composites.

Table 15.5. Composite properties of the proprietary TGDDM pre-preg system compared with Cl_2-TGDDM pre-preg system at room temperature (20°C).

Fibre orientation	Mechanical test	TGDDM post-cured at 180°C	Cl_2-TGDDM post-cured at 180°C	Cl_2-TGDDM post-cured 200°C
35-ply/uni-directional	UD/Compressive strength (MPa)	1520[a]	1410 (cv 19.8%)[c]	1540 (cv 9.4%)
8-ply/uni-directional	UD/tensile strength (MPa)	2700[a]	2421 (cv 7.1%)	2840 (cv 1.6%)
8-ply/uni-directional	UD/tensile modulus (GPa)	168[a]	155 (cv 6.5%)	163 (cv 5.5%)
16-ply/multi-directional	ILSS (MPa)	74.1[b] (cv 10.2%)	82.6 (7.6%)	77.2 (6.3%)
16-ply/multi-directional	Notched tensile strength (MPa)	762[b] (cv 3.3%)	826 (cv 7.6%)	1262 (cv 6.3%)
16-ply/multi-directional	Notched tensile modulus (GPa)	–	83 (cv 1.2%)	84 (cv 4.8%)
16-ply/multi-directional	Notched compressive strength (MPa)	478[b] (cv 3.6%)	495 (cv 3.8%)	501 (cv 5.2%)
16-ply/multi-directional	Plain tensile strength (MPa)	1413[b] (cv 2.6%)	1412 (cv 2.3%)	1563 (cv 7.6%)
16-ply/multi-directional	Plain tensile modulus (GPa)	85[a]	87 (cv 2.3%)	89 (cv 4.5%)
16-ply/multi-directional	Plain compressive strength (MPa)	853[b] (cv 6.3%)	802 (cv 6.5%)	834 (cv 1.9%)
16-ply/multi-directional	Plain compressive modulus (GPa)	81[a]	86 (cv 3.5%)	89 (cv 3.4%)
16-ply/multi-directional	Compression-after-impact (MPa)/ delaminated area (mm)	293[b] (cv 6.1%) (1350)[b]	328 (cv 2.6%) (814)	320 (cv 3.3%) (724)
24-ply/uni-directional	Mode I fracture energy G_{IC} ($J\ m^{-2}$)	172[b]	325 (cv 9.4%)	343 (cv 4.7%)
24-ply/uni-directional	Mode II fracture energy G_{IIC} ($J\ m^{-2}$)	582[b]	841 (cv 18.3%)	723 (cv 11.6%)

[a] Data provided by manufacturer.
[b] DERA data.
[c] cv, Coefficient of variation.

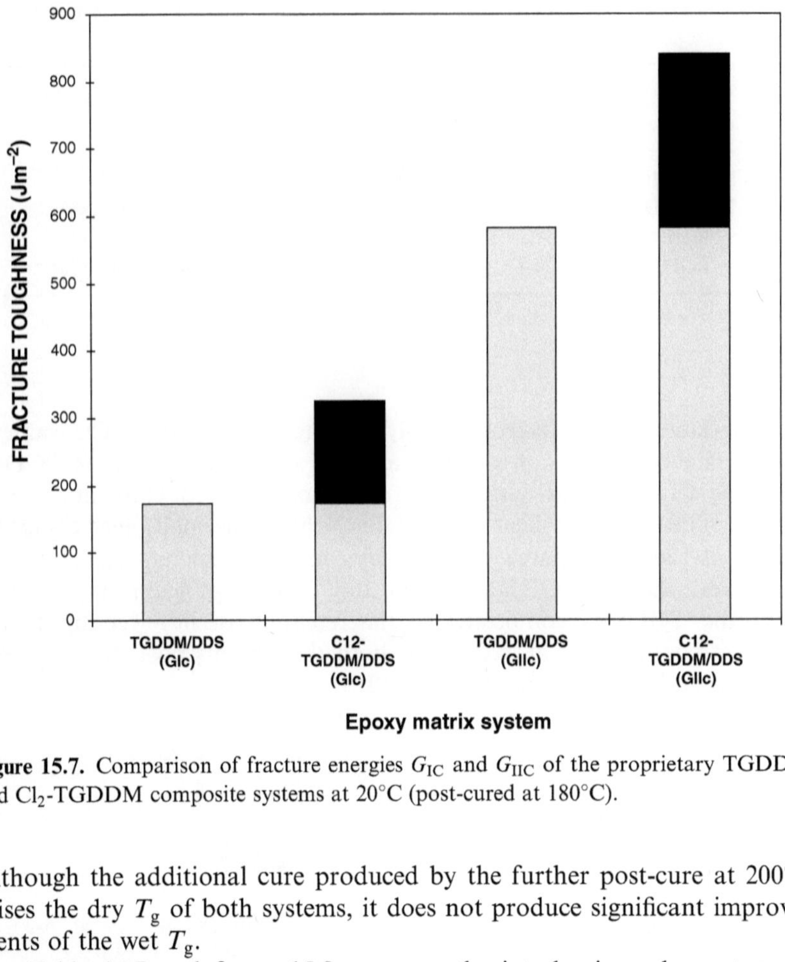

Figure 15.7. Comparison of fracture energies G_{IC} and G_{IIC} of the proprietary TGDDM and Cl_2-TGDDM composite systems at 20°C (post-cured at 180°C).

Although the additional cure produced by the further post-cure at 200°C raises the dry T_g of both systems, it does not produce significant improvements of the wet T_g.

Table 15.7 and figure 15.8 compare the interlaminar shear strength values of both dry composite specimens in their dry unaged state at 20°C

Table 15.6. Comparison of composite moisture uptake and wet T_g of the proprietary TGDDM and the Cl_2-TGDDM composite systems.

	Dry T_g °C		Wet T_g °C		
Composite matrix	E''	E'	E''	E'	Average water uptake (%)
F924C (2 h, 180°C)	220	229	176	166	1.45
Cl_2-TGDDM (2 h, 180°C)	219	228	180	167	1.18
F924C (2 h, 200°C)	225	234	177	164	1.47
Cl_2-TGDDM (2 h 200, °C)	224	233	181	169	1.26

Table 15.7. Comparison of dry and hot–wet composite ILSS of the proprietary TGDDM and the Cl_2-TGDDM composite systems.

Composite matrix	Dry ILSS at 20°C (MPa)	Wet ILSS at 120°C (MPa)
F924C (2 h, 180°C)	74.1 (cv 10.2%)[a]	42.1 (cv 3.0%)
Cl_2-TGDDM (2 h, 180°C)	82.6 (cv 7.6%)	49.8 (cv 2.8%)
F924C (2 h, 200°C)	70.2 (cv 8.1%)	41.6 (cv 2.5%)
Cl_2-TGDDM (2 h, 200°C)	77.2 (cv 6.3%)	46.7 (cv 2.5%)

[a]cv, Coefficient of variation.

and specimens aged hygrothermally then tested at 120°C. The values of percentage coefficients of variation are also shown. For both the TGDDM and the Cl_2-TGDDM systems there is a marked reduction in the mean values of interlaminar shear strength for the hygrothermally aged composites tested at 120°C compared with the dry, unaged specimens tested at 20°C. However, the Cl_2-TGDDM composites show significant improvements over the TGDDM composites in the retention of hot–wet interlaminar shear strength values at 120°C, for both the systems post-cured at 180°C (+18%) and post-cured at 200°C (+12%).

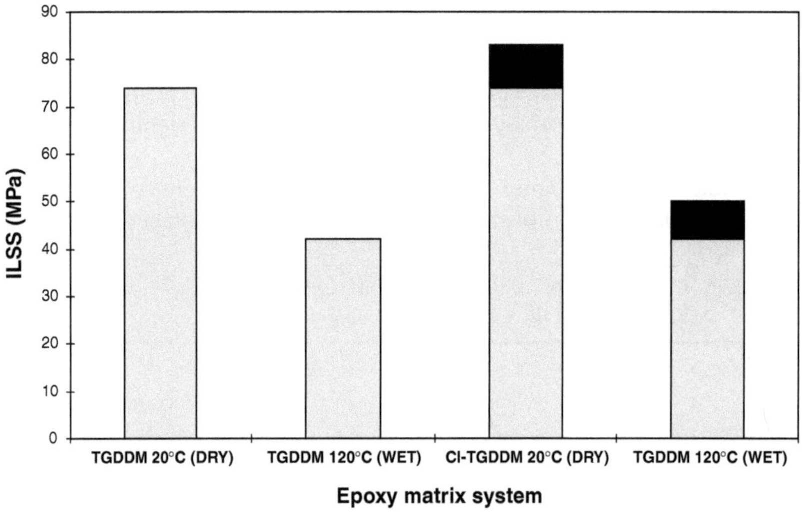

Figure 15.8. Comparison of dry (20°C) and hot–wet (120°C) composite interlaminar shear strength of the proprietary TGDDM and the Cl_2-TGDDM composite systems (post-cured at 180°C).

Table 15.8. Mean notch-compression strengths of dry unaged composite specimens at 20°C and composite specimens hygrothermally aged for 204 days at 70°C/83% RH and tested at 120°C.

Composite matrix	Notched compression strength at room temperature (dry) (MPa)	Notched compression strength at 120°C (wet) (MPa)
F924C (2 h, 180°C)	478 (cv 3.6%)[a]	355 (cv 4.7%)
Cl_2-TGDDM (2 h, 180°C)	495 (cv 3.8%)	447 (cv 9.7%)
F924C (2 h, 200°C)	483 (cv 4.5%)	367 (cv 5.2%)
Cl_2-TGDDM (2 h, 200°C)	501 (cv 5.2%)	435 (cv 6.5%)

[a]cv, Coefficient of variation.

Table 15.8 and figure 15.9 compare the open-hole notch compression strength values of both composite specimens in their unaged state at 20°C and specimens aged hygrothermally to saturation levels and tested at 120°C. The Cl_2-TGDDM composites show significant improvements over the TGDDM for both the composites post-cured at 180°C (mean value 26% higher) and post-cured at 200°C (mean value 19% higher).

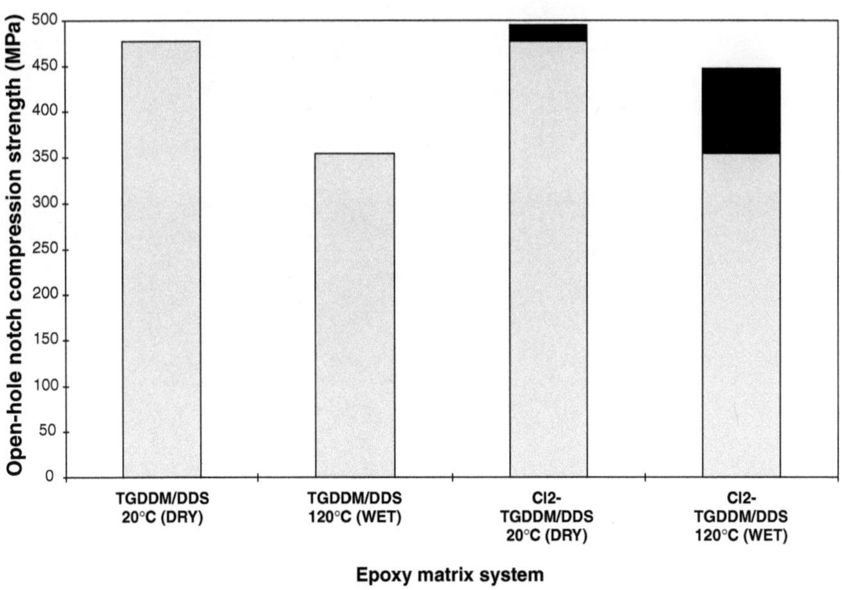

Figure 15.9. Comparison of dry (20°C) and hot–wet (120°C) open-hole notch compression strengths of the proprietary TGDDM and the Cl_2-TGDDM composite systems (post-cured at 180°C).

Summary

The hydrophobically enhanced epoxy resin systems discussed above demonstrate superior properties over current resin materials. Significant improvements in the retention of hot–wet carbon fibre composite properties were obtained with the thermoplastic-toughened Cl_2-TGDDM matrix compared with the standard proprietary thermoplastic-toughened TGDDM. Their potential for commercial exploitation is currently being assessed and it is likely that polymers of this type will achieve commercial utilization within the next ten years.

References

[1] Wright W W 1981 *Composites* **12** 201
[2] Johncock P and Tudgey G F 1983 *Brit. Polym. J.* **15** 14
[3] Johncock P and Tudgey G F 1986 *Brit. Polym. J.* **18** 292

Chapter 16

Technical and economic considerations influencing the role of advanced polymer composites in airframe applications

Michael Bader

Introduction

Advanced polymer composites using carbon, boron and aramid fibres first became available, albeit only in development quantities, some 30 years ago. Immediately their potential for aerospace applications was appreciated [1], but it is only over the past 10 years or so that appreciable quantities have been incorporated into production airframes. During the intervening period world production of reinforcing fibres has increased, the properties of the fibres have been improved and the price of fibres has fallen dramatically. Over the same period the requirements for aerospace materials for both military and civil applications have shifted from being performance driven to being cost-effectiveness driven, and materials must be evaluated for their impact on both acquisition and life-cycle costs. It is notoriously difficult to estimate materials and manufacturing costs accurately, and to make meaningful comparisons between competing systems. This problem is addressed by the introduction of a cost–performance index [2]. This provides a means for numerical comparison based on an assigned value for performance enhancement. Whilst it does not purport to provide a full assessment it may prove useful for preliminary design and trade studies.

Performance

Polymer matrix composites were first proposed for use in airframes on the basis of the possibility of their enhanced structural efficiency when compared

Table 16.1. Properties of composite laminates fabricated from epoxy resin pre-preg.

Material code	Fibre	Format	Young's modulus E_1 (Pa)	Shear modulus G_{12} (Pa)	Allowable strain ε_a	Allowable stress σ_a (Pa)	Max. shear stress τ_a (Pa)	Density ρ ($kg\,m^{-3}$)	Feed stock cost C_{FS} ($\$\,kg^{-1}$)
GQ1	E-glass	Q1	2.20×10^{10}	8.80×10^{09}	0.005	1.10×10^{08}	4.40×10^{07}	2000	65
GA3	E-glass	A3	3.00×10^{10}	8.80×10^{09}	0.005	1.50×10^{08}	4.40×10^{07}	2000	65
AQ1	Aramid	Q1	2.80×10^{10}	1.14×10^{10}	0.004	1.12×10^{08}	4.56×10^{07}	1382	95
AA3	Aramid	A3	4.30×10^{10}	1.14×10^{10}	0.005	2.15×10^{08}	5.70×10^{07}	1382	95
HQ1	HS carbon	Q1	5.50×10^{10}	2.20×10^{10}	0.004	2.20×10^{08}	8.80×10^{07}	1580	100
HA3	HS carbon	A3	8.20×10^{10}	2.20×10^{10}	0.005	4.10×10^{08}	1.10×10^{08}	1580	100
IQ1	IM carbon	Q1	7.00×10^{10}	2.70×10^{10}	0.004	2.80×10^{08}	1.08×10^{08}	1580	220
IA3	IM carbon	A3	1.02×10^{11}	2.70×10^{10}	0.005	5.10×10^{08}	1.35×10^{08}	1580	220
2024	Al alloy	sheet	7.20×10^{10}	2.70×10^{10}		4.00×10^{08}	2.50×10^{08}	2700	25

Table 16.2. Performance indices for various loadings, normalized with respect to Al alloy 2024.

Material code	Stiffness plane tension (E/ρ)	Stiffness plate bend ($E^{1/3}/\rho$)	Stiffness plane shear (G_{12}/ρ)	Comp. buckle ($E^{1/2}/\rho$)	Strength plane tension (σ/ρ)	Flex. strength ($\sigma^{2/3}/\rho$)	In-plane shear strength (τ/ρ)
GQ1	0.41	0.91	0.44	0.75	0.37	0.57	0.24
GA3	0.56	1.01	0.44	0.87	0.51	0.70	0.24
AQ1	0.76	1.43	0.82	1.22[a]	0.55	0.84	0.36
AA3	1.17	1.65	0.82	1.51[a]	1.05	1.29	0.45
HQ1	1.31	1.56	1.39	1.49	0.94	1.15	0.60
HA3	1.95	1.78	1.39	1.82	1.75	1.74	0.75
IQ1	1.66	1.69	1.71	1.68	1.20	1.35	0.74
IA3	2.42	1.92	1.71	2.03	2.18	2.01	0.92

[a]The low compressive strength of aramid fibres limits their usefulness in this loading mode.

with traditional aluminium alloy fabrications. This is easily demonstrated in table 16.1 which gives the properties of a typical set of laminates compared with a standard aluminium alloy, and in table 16.2 on which are plotted performance indices for typical composites and an aluminium alloy, for various static loading conditions. Selected data from table 16.2 are also shown graphically in figures 16.1 and 16.2.

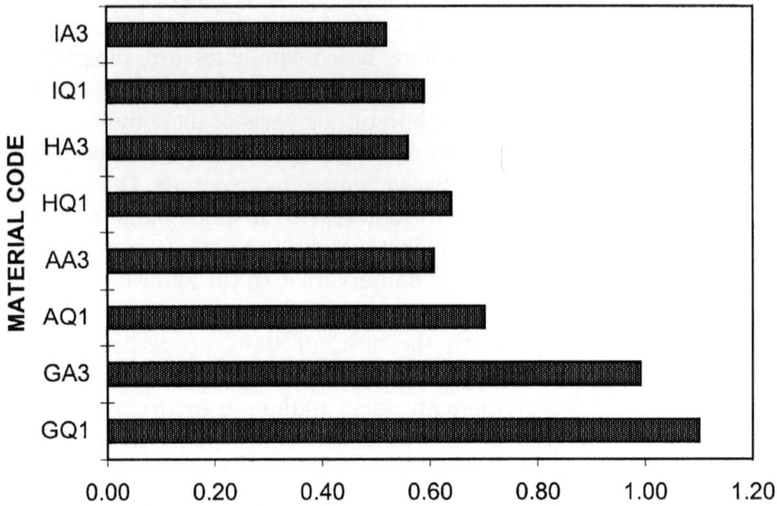

Figure 16.1. Mass for equal bending stiffness—normalized to aluminium alloy 2024.

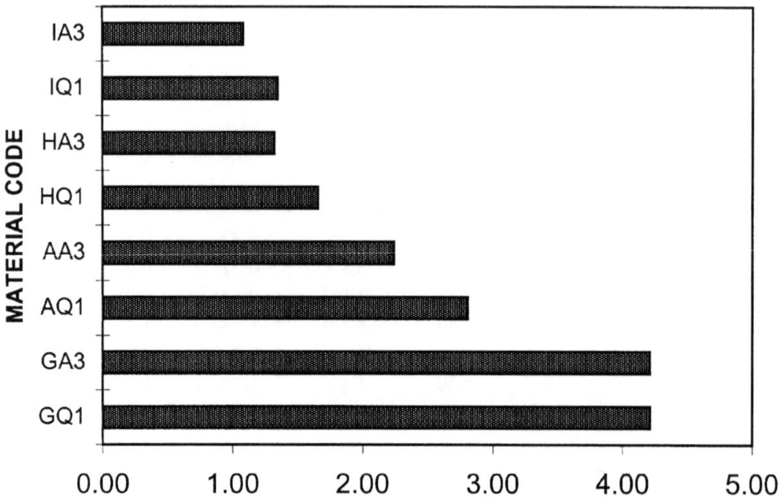

Figure 16.2. Mass for equal shear strength—normalized to aluminium alloy 2024.

The values for the composites are based on two laminate configurations, a quasi-isotropic (Q1) [0°, 90°, ±45°] and one with a greater proportion of zero degree fibres (A3) [0_2°, ±45°]$_s$. Four different fibres have been selected: E-glass is a base line fibre, the HS carbon is the most widely utilized fibre in aerospace, the intermediate modulus (IM) carbon is the standard premium fibre. Aramid is favoured for impact critical structures or where non-conducting fibres are required. The performance indices are calculated according to the method recommended by Ashby [3]; higher values indicate enhanced performance. The masses of structures of equal performance are in inverse proportion to these indices. Thus, with the exception of in-plane shear strength, all the carbon-fibre based laminates are superior to the aluminium alloy (i.e. index <1), but the glass-fibre laminates are inferior. Aramid is less attractive than carbon on the basis of these indices.

Many aircraft components are designed for stiffness, which allows for straightforward comparison between competing materials. Other aspects of performance are more difficult to represent by a single parameter. Fatigue is a major consideration, but fortunately the fatigue performance of composites, at least at the very conservative strain allowables which are typical of current practice, is generally better than that of aluminium alloys, so that comparison on the basis of static properties is probably adequate. Likewise, durability under typical corrosive environments is also generally better, although there are some additional environmental hazards with composites. In general the two most critical performance aspects of currently available polymer composites are the hot–wet and compression-after-impact conditions. Both of these conditions may be represented by knocked down design strain or stress allowable.

Cost

The cost of a manufactured composite component is the sum of a number of contributory items. The principal categories are:

- raw materials (fibres and resins)
- conversion costs (weaving, pre-pregging, etc.)
- tooling
- plant—capital and operation (presses, autoclaves, curing-ovens, etc.)
- consumables
- direct labour.

Indirect costs (i.e. overheads) must also be added, but no attempt is made to estimate these costs in the present work as the levels are strongly influenced by local financial practices. Materials utilization, i.e. buy-to-fly ratio, is also a major consideration.

In the aerospace industry the use of pre-preg coupled with autoclave processing has become standard practice. However, both woven and non-crimp multi-layered fabrics, processed by resin transfer moulding (RTM) or resin film infiltration (RFI), are now also being considered. The achievable fibre fraction V_f and hence mechanical properties are somewhat lower when fabrics are used, but this disadvantage is off-set by lower feedstock and processing costs and also by greater productivity. Typical costs for these materials are given in table 16.3. Note that the areal density of the pre-preg is such that each layer consists of only 110–188 $\mathrm{g\,m^{-2}}$ of fibre, whereas for fabrics the values are three to four times higher. Eight layers of pre-preg are required for each mm of laminate, compared with two for the fabrics. Fabrics are, of course, available in many weights, but the weights selected represent a reasonable compromise between adequate drapability and fast build rate. The weights given are for the fibre content only, as this controls

Table 16.3. Typical feedstock costs.

	Fibre cost	Std pre-preg $t = 0.125$ mm		Woven cloth $t = 0.5$ mm		N-C fabric 4-ply $t = 0.5$ mm	
Fibre	(\$/kg)	(g/m^2)[a]	(\$/kg)[b]	(g/m^2)[a]	(\$/kg)[b]	(g/m^2)[a]	(\$/kg)[b]
E-glass	3.75	188	65	625	24	690	38
Aramid	25	110	95	370	43	405	63
HS-carbon	27	135	100	450	44	495	65
IM-carbon	120	135	220	450	108	495	147

[a] Mass of fibre only.
[b] Includes resin cost.

Table 16.4. Cost estimates for plates of equal in-plane stiffness.

					Costs				
Fibre	Format	Number of plies	Process	Mass (kg)	Feedstock ($)	Lay-up debulk ($)	Tooling ($)	Process ($)	Total ($)
E-glass	Pre-preg. A3	90	A/C	22.30	1449	550	120	200	2319
	Woven cloth	32	RTM	30.00	720	194	60	20	994
	Non-crimp fabric	26	RFI	25.20	957	163	50	60	1230
Aramid	Pre-preg. A3	62	A/C	10.80	1026	380	120	200	1726
	Woven cloth	21	RTM	14.30	615	135	60	20	830
	Non-crimp fabric	18	RFI	12.40	781	114	50	60	1005
HS carbon	Pre-preg. A3	32	A/C	6.40	640	194	120	200	1154
	Woven cloth	11	RTM	8.40	370	70	60	20	520
	Non-crimp fabric	10	RFI	7.70	500	66	50	60	676
IM carbon	Pre-preg. A3	26	A/C	5.20	1122	163	120	200	1605
	Woven cloth	9	RTM	6.80	734	63	60	20	877
	Non-crimp fabric	8	RFI	6.20	911	48	50	60	1069

the mechanical properties. Pre-preg laminates achieve a V_f of about 0.6, woven cloth 0.5 and the non-crimp fabric about 0.55.

The costs of laminates produced from each of the four fibres in three formats with different process routes have been estimated. The data are given in table 16.4 and shown graphically in figure 16.3. All the laminates are of the specified fibre in an appropriate hot-cured epoxy resin. These estimates are for plates of equal performance, defined as in-plane stiffness. The standard laminate is 32 plies of the HS carbon. It is 4 mm thick and weighs 6.4 $kg\,m^{-2}$. The masses of the other plates have been obtained by using the performance indices given in table 16.3. The nearest integral number of plies to achieve the required performance has been used, but balance and symmetry have not been considered.

Lay-up and debulking costs are estimated on the basis that one operative can lay a 1 m^2 ply in two minutes. and the cost of labour is \$100 per hour per operative. Other labour charges are also estimated at this rate. Tooling costs are based on commercial estimates for a simple curved stiffened panel of 1 × 1 m plan form. The costs are spread on the basis of a full year's operations. A year is defined as 250 days, each of 8 working hours. For autoclave processing composite tooling is used. In the case of resin transfer moulding the tooling is steel, heated by circulating oil and set in a hydraulic press. Similar tooling is used for resin film infiltration.

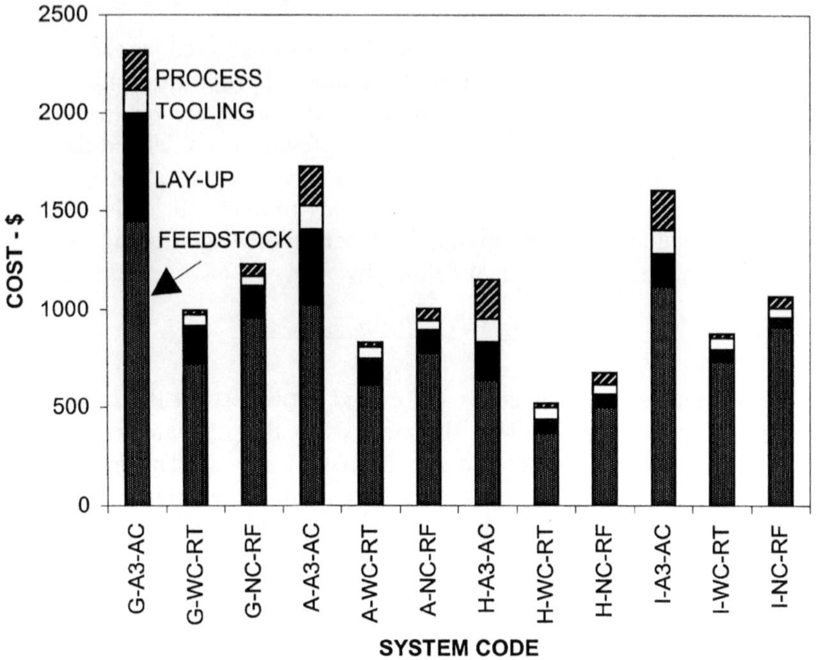

Figure 16.3. Cost for equal in-plane stiffness.

The estimates of process cost are more speculative but, since these are in no case greater than 10% of the total costs, this is unlikely to be critical. Plant costs are calculated on the basis of a ten-year write-off period at an interest rate of 10%. Thus, the annual cost of a plant initially valued at $1 000 000 is $155 000. Operating costs, labour and consumables are included in the process cost. The autoclave is 4 m diameter and 10 m long, and is considered to have a capacity of eight moulds. It operates one cycle per working day. Thus, a single set of tooling can produce 250 components per year. The resin transfer moulding and resin film infiltration are faster and each tool is capable of three parts per day, 750 per year. It is readily apparent from these estimates that the dominant costs are those of feedstock and lay-up. The pre-preg/autoclave route is relatively costly. The least costly is the HS carbon/woven cloth processed by resin transfer moulding. The IM carbon/woven cloth is some 65% more costly but offers a weight saving of 24%. It is perhaps surprising that the cheap or E-glass is not only much heavier, but also more costly, regardless of format. This is mainly due to the extra materials and processing costs of the increased number of plies required to meet the performance criterion. The aramid would not be chosen either on the basis of structural efficiency or of cost.

Cost–performance effectiveness

In aerospace engineering, it is always advantageous to reduce the mass of a structure or component. Selection of the most appropriate combination of material and process route may be effected objectively by the use of a cost–performance index. In this context performance is defined as a minimum mass for a specified mechanical or non-mechanical property, e.g. stiffness, strength or life in an aggressive environment. It is also necessary to assign a value for weight saving, v_p, $ per kg of weight saving. A relative cost–performance index is then defined by:

$$I_{PC} = \frac{(m_0 - m_i)v_p - (c_i - c_0)}{c_0} \qquad (1)$$

where m_0 and m_i are the masses and c_0 and c_i the costs of the standard and compared systems respectively. In most cases the standard system is the cheapest of those under consideration so that $c_i > c_0$. The index is then simply the value of the mass reduction, minus the additional cost, all normalized with respect to the cost of the standard system. The index is positive for systems which are more, and negative for systems which are less cost–performance effective than the standard system. The standard system always has an index of zero. The index is the fractional difference in value per $, in comparison with the standard system. The index has been calculated for values of v_p ranging from $100 to $1000 per kg and is shown in table 16.5.

Table 16.5. Cost-performance indices for selected laminate systems.

Fibre	Format	Number of plies	Process	Mass (kg)	Cost ($)	Relative cost–performance index I_{PC}: cf HCC/WR @ v_p ($/kg) indicated			
						100	250	500	1000
E-glass	Pre-preg. A3	90	A/C	22.30	2319	−6.13	−10.14	−16.83	−30.19
	Woven cloth	32	RTM	30.00	994	−5.07	−11.30	−21.68	−42.45
	Non-crimp fabric	26	RFI	25.20	1230	−4.60	−9.44	−17.52	−33.67
Aramid	Pre-preg. A3	62	A/C	10.80	1726	−2.78	−3.47	−4.63	−6.93
	Woven cloth	21	RTM	14.30	830	−1.73	−3.43	−6.27	−11.94
	Non-crimp fabric	18	RFI	12.40	1005	−1.70	−2.86	−4.78	−8.63
HS carbon	Pre-preg. A3	32	A/C	6.40	1154	−0.83	−0.26	0.70	2.63
	Woven cloth[a]	11	RTM	8.40	520	**0.00**	0.00	0.00	0.00
	Non-crimp fabric	10	RFI	7.70	676	−0.17	0.04	0.37	1.05
IM carbon	Pre-preg. A3	26	A/C	5.20	1605	−1.47	−0.55	0.99	**4.07**
	Woven cloth	9	RTM	6.80	877	−0.38	**0.08**	0.85	2.39
	Non-crimp fabric	8	RFI	6.20	1069	−0.63	0.00	**1.06**	3.18

[a] Reference system.

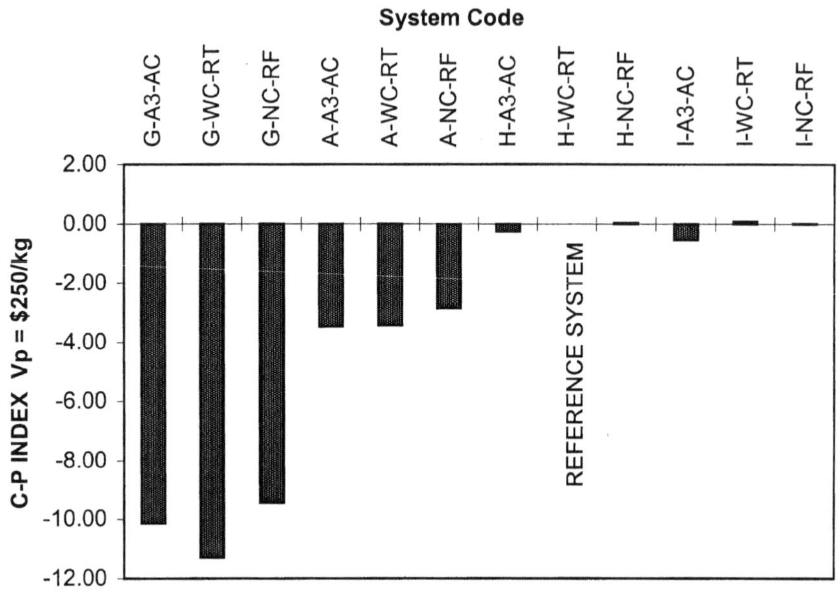

Figure 16.4. Cost performance indices ($v_p = 250\,\$/\text{kg}^{-1}$).

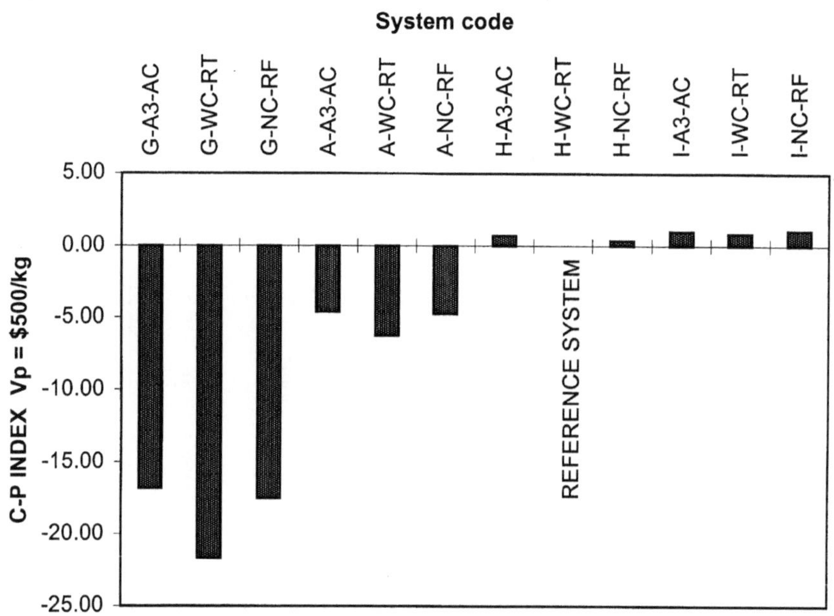

Figure 16.5. Cost performance indices ($v_p = 500\,\$/\text{kg}^{-1}$).

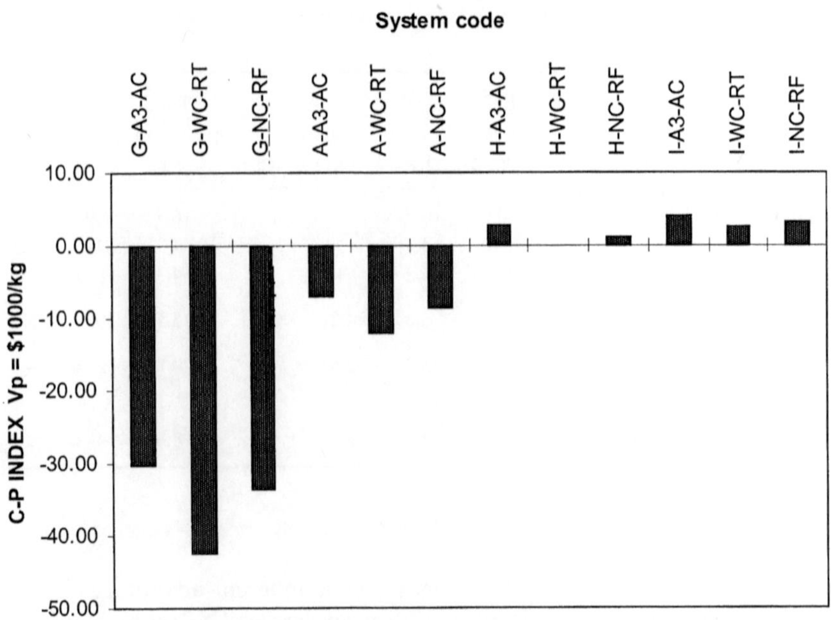

Figure 16.6. Cost performance indices ($v_p = 1000\,\$/\text{kg}^{-1}$).

The most cost-performance effective system for each value of v_p is emboldened. The data are also shown graphically in figures 16.4–16.6. It is clear that when v_p is less than $250 per kg, the HS carbon/woven cloth system is the best, at $250/kg the IM carbon/woven cloth is marginally better, at $500 per kg the IM non-crimp fabric and at $1000 per kg and above the IM pre-preg becomes the system of choice. It is suggested that the values of v_p which have been explored are realistic for civil airframe manufacture. Resin transfer moulding and resin film infiltration appear to be attractive alternatives to pre-preg and consideration should be given to use of higher specification fibres in formats suited to these processes.

Reducing manufacturing costs

What is the scope for further reduction in the cost of manufactured laminates? It is unlikely that the price of the raw fibres will be significantly reduced. However, large scale adoption of premium fibres in place of the commodity HS carbon (e.g. T300) could lead to a modest reduction. All hand lay-up processes are labour intensive, and pre-preg especially so because of the very low build rates. Conversion costs for pre-preg are also very high. Switching to woven and non-crimp fabrics reduces conversion

Table 16.6. Comparison of preliminary processing alternatives for carbon fibre reinforced laminates.

Process	Format	Build rate[a] (kg/h)	Labour ($/kg)	Capital ($/kg)	M/c op ($/kg)	Total lay-up ($/kg)	Feed-stock ($/kg)	Total ($/kg)
Hand lay-up	Pre-preg	6.0	16.7			16.7	100.0	117
	Woven cloth	23.0	4.3			4.3	44.0	48
	NC fabric	23.0	4.3			4.3	65.0	69
Tape lay M/C	Pre-preg	18.0	5.6	4.3	5.6	15.5	100.0	115
Tow place M/C	Impregnated tow	3.8	13.2	2.0	5.6	20.8	27.0	48
Pultrusion	Woven cloth	48.0	2.1	2.1	4.2	8.4	44.0	52

[a]Per operator or per machine.

costs and increases build rates. This with the inherent advantages of resin transfer moulding and resin film infiltration can lead to more economical processing. Another possibility is replacement of hand lay-up by automated lay-up processes. Three methods have been assessed: tape lay-up, tow-placement and pultrusion. Tape lay-up was considered using 50 mm wide pre-preg tape laid at a rate of $1\ \mathrm{m\,s^{-1}}$ and assuming a 50% effective utilization of the equipment, i.e. $4\ \mathrm{h\,day^{-1}}$. For tow-placement, carbon tow is laid at $5\ \mathrm{m^{-1}}$, again with a 50% utilization factor. These rates may be somewhat optimistic. Pultrusion is of limited utility but could be used to produce a simple section, e.g. a 1 m wide, 4 mm thick plate with integral T stiffeners. The feedstock would be woven cloth, precut into suitable widths. A pultrusion rate of $15\ \mathrm{m\,h^{-1}}$ has been assumed; this is fairly conservative. The results of this assessment are given in table 16.6. The values estimated for capital and machine operating costs are highly speculative but fortunately do not have a critical influence on the overall results. The important factors of build rate, labour and feedstock cost are considered to be reasonably realistic. The most significant observation is that whilst automated tape-laying gives a faster build rate than hand-laid pre-preg, the cost is unlikely to be less.

The build rate for tow placement is low (unless multiple tows are laid simultaneously), but cost might be much lower due to the reduced cost of the tow feedstock. Hand-laid fabrics give high build rates and low total costs. Pultrusion has the possibility of high output at costs comparable with resin transfer moulding and resin film infiltration. However, it would be viable only if a standard section could be effectively utilized in the structure—perhaps a wing plank.

Summary

Composite laminates will continue to be materials of choice for many components of airframe structures, where structural efficiency is the principal criterion. There may also be significant cost advantages arising from parts consolidation when composite laminates are used. This brings the advantage of a reduced parts count and also reduced assembly and inspection costs. Carbon fibre reinforced laminates processed from pre-preg in the autoclave have become the favoured route in aerospace. They are expensive to produce but have been able to compete with aluminium alloys on price as well as performance in many applications. However, if costs could be significantly reduced the range of applicability could be vastly increased. A methodology has been proposed to assist with preliminary selection. It has been shown that, although feedstock costs are dominant, it is often less costly overall to use higher performance feedstock, such as T800 type carbon fibre in place of the more commonly specified T300/HTA type, or even E-glass. It is also shown that processes such as resin transfer moulding and resin film infiltration can offer significant cost and productivity advantages over the autoclave route.

References

[1] Watt W, Phillips L N and Johnson W 1966 *The Engineer* **221** 815

[2] Bader M G 1997 'Strategies for the design of cost-effective composite laminates' in *Technology and Materials Needs for the Year 2000* ed M A Erath, H Meyer-Piening, J Kenny, J Hognat and A Green (Niederglatt: SAMPE Europe) pp 255–266

[3] Ashby M F 1992 *Materials Selection in Mechanical Design* (Oxford: Pergamon Press)

SECTION 3

HIGH TEMPERATURE MATERIALS

Since the invention of the gas turbine aeroengine, operating temperatures have risen steadily. Gas temperatures in some areas of the modern aeroengine now exceed the melting point of component alloys, with component surface temperatures approaching 75–80% of the melting point. The upward trend in operating temperature is set to continue, driven by the need to increase thrust, reduce fuel consumption and reduce emissions. Concurrently, aeroengine manufacturers are under increasing pressure to reduce costs and increase profit margins, placing constraints on the implementation of new materials and technology. Nonetheless, successful aeroengine performance requires the development of improved directionally solidified and single crystal technologies, thermal barrier coatings and improved understanding of failure modes including creep and fatigue.

As further improvements in performance inevitably become harder and harder to realize, incremental progress in basic understanding of the factors controlling materials performance will always be needed if current materials are to be exploited fully, and new materials and technology are to be developed. Successful new developments in high temperature aeroengine materials will be those which provide an integrated solution to performance needs, addressing not only economic and environmental aspects such as lower fuel consumption and emissions at lower cost, but also allowing improved predictions of component life and reliability.

This section discusses many of the critical issues and performance aspects of high temperature titanium and nickel based materials, as well as thermal barrier coatings which will be increasingly employed in the highest temperature regions of the aeroengine. Chapter 17 describes titanium aluminide based alloys which allow the use of titanium based materials at higher operating temperatures, mitigating the need to use much denser nickel alloys in a variety of components. The importance of developing corresponding new processing technologies for titanium aluminide alloys is highlighted, as well as the role of minority alloying elements on alloy performance. Chapter 18 again focuses on materials and processing strategies to extend the operating temperature of weight saving titanium alloys, in this case by reinforcing the matrix with monofilament silicon carbon fibres to form a metal matrix composite, and facilitating dramatic weight savings in rotating compressor components. Chapter 19 describes a numerical model to explain the important rafting phenomenon observed during high temperature creep of nickel superalloys containing a high volume fraction of strengthening γ'-Ni_3Al precipitates. The model shows good agreement with experimental measurements of rafting behaviour, but also provides an insight into the underlying thermodynamics driving the development of a rafted microstructure. Chapter 20 reviews the mechanisms of creep in single crystal nickel based superalloys widely used for turbine blades. Creep phenomena are described by a unified model for high temperature creep deformation, allowing both microscopic crystallographic changes and macroscopic shape changes to be predicted. Chapter 21 focuses on the role of minority alloying elements in controlling creep behaviour in nickel based alloys. Finally, chapter 22 reviews thermal barrier coating materials and processing, describing the principles of thermal barrier coating operation and the benefit to aeroengine performance, manufacturing techniques, failure modes and likely future developments.

Chapter 17

TiAl-based alloys for aeroengine applications

Mike Loretto

Introduction

The influence of process route on the properties and on the structures of some TiAl-based alloys is reviewed. The thermal stability of these structures is also considered together with the factors that are responsible for the changes in properties in differently processed material. The results reported here confirm that extrusion at high temperatures can give higher strengths than forging, and that for lamellar alloys, the strength increment is mainly due to the very fine lamellar spacing. The current limitations of the production of TiAl alloys using cold-wall furnaces for melting the alloys, either for the production of powders or for the production of castings, is highlighted. Although the process route has a more significant effect on properties than composition, the thermal stability of some processed alloys places a limit on their subsequent operating temperature. Further experimental work is required, aimed at understanding the changes that occur during cooling and/or holding samples at high temperatures, in order to optimize the process route and to develop cost-effective heat treatments of the wide range of TiAl-based alloys that have been developed. The implications of these factors on the potential applications of TiAl-based alloys, particularly in aeroengines, is discussed.

TiAl-based alloys with a fully lamellar structure generally have a better balance of properties than alloys with a duplex or equiaxed structure [1]. Work has therefore concentrated mostly on controlling the lamellar spacing and grain size in order to improve ultimate tensile strength and ductility at room temperature, as well as the creep properties. Significant improvements in properties can be obtained by extruding samples in the α phase field and then cooling reasonably rapidly after working [2]. This treatment can result in very finely spaced lamellae (the order of tens of microns) and a fine grain size. Isothermal forging, the most investigated alternative processing route for high temperature working, followed by heat treating within the α phase field can also result in a finely spaced, fully lamellar structure, if the cooling rate is fast enough. However, strengths are significantly lower than those found for extruded alloys.

There are obvious disadvantages in process routes that rely upon relatively rapid cooling. First, the structure is likely to be unstable when it is reheated, since non-equilibrium phase distributions and morphologies may be introduced; and second, the structure found in thinner regions may not be present in thicker regions of the product. In the case of cast alloys, the microstructure is generally lamellar with a grain size determined mainly by the alloy composition, and the lamellar spacing determined by the local cooling rate. Subsequent hot isostatic pressing (HIPing) and/or heat treatment may or may not change the lamellar structure of the as-cast alloy to equiaxed: again the final microstructure depends partly on the alloy composition and partly on the hot isostatic pressing conditions.

In addition to the choice of processing route, and something that is inexorably linked to processing, is the ability to change alloy structure and properties by appropriate alloying. A great deal of research has been aimed at optimizing the compositions of TiAl-based alloys, and a large number of alloy compositions have been patented. Broadly, there are three types of patents in TiAl alloy development. First, there are patents concerning the addition of boron to alloys to refine both the as-cast structure and the structure after thermo-mechanical processing. Second, there are patents that cover alloys with additions of elements such as niobium, tungsten, silicon, carbon etc. where the individual additions are typically in the range of 1 or 2 at% or much lower, as is the case for carbon additions. Finally, there are alloys which have much larger individual alloying additions, totalling about 8 at% which, it should be noted, generally increase the cost of the alloy. In all cases the aluminium contents in the patented alloys lie between 43 and 48 at%. It should be noted that the strength of TiAl-based alloys increases with decrease of aluminium content (because the volume fraction of the relatively strong α_2 phase increases) and it is necessary to consider this when comparing the apparent strengthening effect of alloying additions.

In this chapter several different TiAl-based alloys produced using different methods of processing will be considered briefly. The factors that appear to be significant in determining the structure and properties of these variously processed alloys will be discussed. The thermal stabilities of the variously processed alloys are considered in terms of their microstructure and the influence of exposure at high temperatures on properties, since the in-service stability of the variously processed alloys is clearly important.

Since the uppermost factors which will determine whether or not TiAl-based alloys will be used in aircraft engines are safety and reliability, it is essential that the manufacturing route for components meets these requirements. The cost of components is, however, an important factor in controlling the introduction into engines of components manufactured using a new alloy or a new processing route, and to a great extent the requirements of safety and reliability drive the processing route (and to some extent the alloy composition) towards a higher cost. Having considered the influence of alloy composition and processing route on the properties of TiAl-based alloys, the significance of these factors on the cost of components manufactured to appropriate standards and their significance in determining the timescale for major applications of TiAl-based alloys in aeroengines and in land-based power plant is discussed.

TiAl alloys

The alloys used in the work reported here were all plasma melted using titanium sponge and a range of master alloys as feedstock. Either 1 kg buttons or ingots up to 1 m in length and 150 mm in diameter were produced. In the case of the buttons they were typically inverted and remelted five times in order to minimize heterogeneity. The ingots which were to be used for forging or extrusion were all double-melted—an essential procedure for cold hearth melting furnaces with a single small hearth [3].

Sections from the ingots were used as the feedstock for the production of powders and castings. These sections were remelted in a bottom-pouring cold-wall induction furnace and either atomized using argon or, for the production of castings, poured directly into unheated ceramic moulds [4]. The powders were sieved to produce powders within specified size ranges between 45 and 250 μm diameter. These were hot isostatically pressed (HIPed) at temperatures between 1000 and 1370°C for 4 h using pressures of between 100 and 200 MPa [5]. The buttons and ingots were generally processed either by isothermal forging at 1180°C which was followed by high temperature heat treatment and cooling at different rates, or by high temperature extrusion at a nominal temperature of 1380°C. The oxygen contents in all alloys and powders used in the investigation of properties was in the range 500 to 800 ppm.

Transmission electron microscopy was carried out using either a JEOL 4000FX or a Philips CM20 both operating at 200 kV. Standard methods were used to produce the thinned samples. Scanning electron microscopy was carried out using a JEOL 840. The tensile tests were carried out using a 200 kN Zwick 1484 mechanical testing machine at a strain rate of $10^{-4}\,s^{-1}$ at room temperature. The specimens were ground to a gauge diameter of 4 mm and a gauge length of 12 mm. Creep tests were done using a Mayes creep testing rig with an arm ratio of 15 : 1 under constant stress conditions. Strain was measured by a differential capacitor transducer method. Measurements were done at 700°C in air at stresses of either 70 or 200 MPa and strain was measured up to times of 1000 h. The creep samples were 4 mm in diameter with a gauge length of 25.4 mm. High temperature heat treatments were carried out in air and, before either carrying out mechanical tests or assessing the changes in microstructure caused by these treatments, the surface layer affected by oxidation was removed.

TiAl alloy properties

Room temperature strength

The reference alloys used in these investigations were Ti-48(or 47)Al-2Mn-2Nb and Ti-48.6Al-2Cr-2Nb-1B and the range of properties which can be obtained from these alloys by different processing routes is shown in table 17.1. The room temperature strengths of the ingot-route materials were strongly influenced by the thermo-mechanical processing route with, as expected, the strongest alloy being the boron-containing sample extruded at 1380°C. The ductility of the alloys was always 1–2% or worse. The structures of the cast and extruded alloys were fully lamellar but the powder-route

Table 17.1. The range of typical properties obtained for two TiAl-based alloys using different processing routes.

Alloy	Treatment	Yield (MPa)	Elong (%)	UTS (MPa)	Primary creep	Secondary creep
Ti-47Al-2Mn-2Nb	Cast ingot		0.1	244	0.1%/26 h	58×10^{-10}
					70 MPa at 700°C	
	Cast + HIP	326	0.6	369		
	Forged ingot + 48 h 1200°C. AC	351	<0.2	430		
	Remelted + investment cast + HIP	420	0.8			
	Remelted + atomized[a] + HIP + HT	439	1.9	562		
Ti-48Al-2Cr-2Nb-1B	As cast	349	1.0	426	0.4%/265 h	12×10^{-10}
					200 MPa at 700°C	
	As cast + 1380°C 1 h SC	309	1.1	404	0.4%/510 h	13×10^{-10}
					200 MPa at 700°C	
	As cast + 1380°C 1 h + AC	440	1.3	557	–	–
	Extruded at 1380°C	671	1.5	761	–	–

[a]The powder sizes used for these samples lay between 70 and 150 μm and the HIPing conditions were 1100°C at 150 MPa for 4 h.

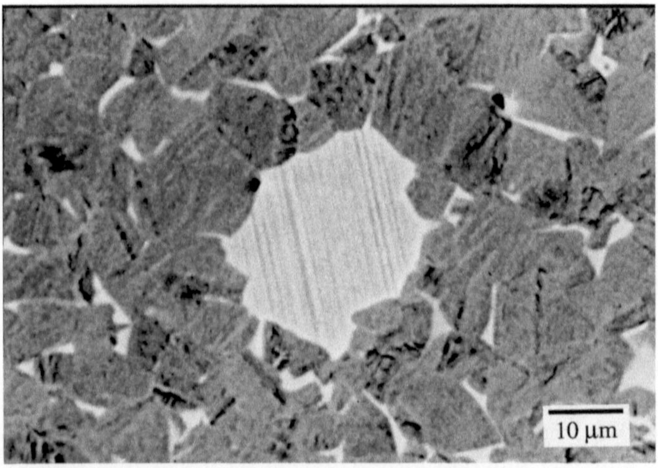

Figure 17.1. Scanning electron micrograph of a HIPed sample of atomized Ti-48Al-2Mn-2Nb showing a region with a lamellar structure, despite the fact that the sample was HIPed well below the α transus. EDX shows this region to be relatively rich in Ti.

sample and the investment-cast material were duplex or equiaxed because they were heat treated in the two phase $(\alpha + \beta)$ field after HIPing.

The room temperature strength of the HIPed powder-route material could be occasionally comparable with that of ingot route material (see table 17.1) but was commonly relatively poor and fracture usually occurred either just before or just after yield. Examination of samples HIPed within the two phase region revealed regions which were high in titanium and which showed a lamellar structure, as shown in figure 17.1, because at the HIPing temperature these regions were in the single phase region. Examination of fracture surfaces shows that these regions were clearly the origin of fracture. It has been shown [6] that the fracture sites originate from incipient dendrites in the mushy zone adjacent to the solid at the cold-wall, which occasionally pass through the nozzle during atomization. These dendrites will be rich in titanium (because of the partition coefficient between the liquid and the solid) and clearly, unless high temperature HIPing or other thermo-mechanical processing and/or heat treatments are used, these regions will remain in the final product where they will act as fracture-initiation sites. It should be noted that these high titanium regions have been observed in powder from a range of sources where a cold-wall technique has been used to melt the solid before atomization. It seems essential therefore to use high temperature processing to remove them. The fact that the highest strength TiAl alloys, which also have reasonable ductilities, have been made via the powder route using high temperature extrusion [2] suggests that the high temperature hot working during extrusion has eliminated these weaker regions.

The strength of the investment-cast alloy was reasonable but microstructural examination (before HIPing) showed significant porosity and incomplete mould filling, especially in the thinner sections. Again it is reasonable to conclude that this is associated in a cold-wall induction furnace with the necessity of maintaining a solid alloy in contact with the walls of the furnace (which is associated with the presence of a mushy zone) and limits the superheat achievable, at least with current designs of furnace.

Microstructures

As is well known, the different thermo-mechanical treatments can result either in an equiaxed structure consisting mainly of γ grains with α_2 at triple points, a duplex structure made up of partially lamellar structure and γ grains, or a fully lamellar structure with very different interlamellar spacings and very different proportions of α_2 which vary with cooling rate from the heat treatment temperatures or with the extrusion conditions. The highest room temperature strength is obtained for alloys extruded in the single phase α field and subsequently cooled reasonably quickly. The lowest room temperature strength in these thermo-mechanically processed samples is obtained in material slowly cooled after being forged. Examination of the structures show that the extruded sample has a fully lamellar structure, and the lamellar spacing is very much finer than the spacings in the other samples.

The interlamellar spacing is the most important factor in determining the yield stress of samples with the lamellar structure, but the precise mechanism which underlies this strengthening effect is not clear. Thus when the interlamellar spacing becomes very small (of the order of tens of nanometres) the proportion of α_2 is inevitably high, and much higher than in the same alloy after furnace cooling (because the samples are not only worked in the single phase α field but are also cooled reasonably rapidly so that excess α is retained). The presence of the much stronger hexagonal phase would by itself be expected to strengthen the alloys as would the reduction in lamellar width of both the γ and α_2 phases. Because fracture occurred at some of the γ/α_2 interfaces after deformation just beyond the yield point it might be suggested that dislocations cannot be easily transmitted across these interfaces (although recent work has shown that α_2 lamellae do deform by the propagation of $1/6\langle 11\bar{2}6\rangle$ and $1/6\langle 11\bar{2}0\rangle$ dislocations [7, 8]).

Thermal stability

Whatever the relationship between the γ/γ spacing, the γ/α_2 spacing, the mean spacing and the inherent higher strength of the α_2 phase in determining the strength of TiAl-based alloys with fine lamellae, it is clear that the microstructure would not be expected to be stable both because of the higher than equilibrium amount of α_2 and because of the enormous interfacial area associated with these microstructures. Hence, it would be expected that heat treatments at temperatures which allow significant diffusion could result in microstructural changes and corresponding changes in properties.

Figure 17.2(*a*) shows the changes induced in a sample of Ti-48.8Al-2Nb-2Cr-1B after 3000 h at 700°C, where it was clear that the α_2 lamellae were

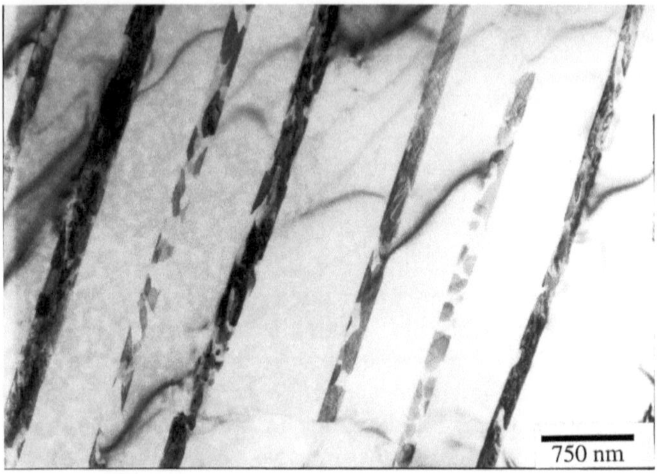

(a)

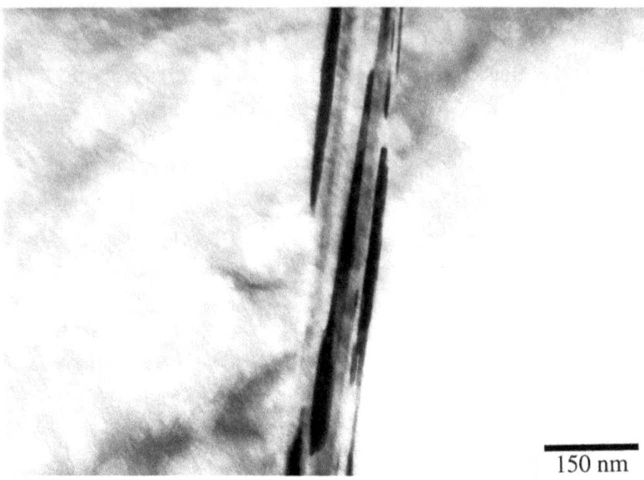

(b)

Figure 17.2. Many-beam bright field micrographs showing the influence of heat treating Ti-48.8Al-2Cr-2Nb-1B at 700°C for 3000 h. (*a*) shows the break up of the α_2 lamellae and (*b*) shows that at some stage extra γ lamellae are formed in an original α_2 lamella.

Table 17.2. The influence of exposure at 700°C on properties of some TiAl-based alloys.

Alloy and condition	Yield stress (MPa)	Elongation (%)
Ti-48Al-2Cr-2Nb-1B		
As extruded	546	1.5
As extruded + 3000h at 700°C	477	1.2
Ti-47Al-2Ta-1Cr-1Mn-0.2Si		
As cast	346	0.3
As cast + 3000h at 700°C	324	0.2
Ti-47Al-12Ta-1Cr-1Mn-0.2Si-1B		
As cast	347	1.44

disappearing. However, there are many examples where the α_2 decomposes to give a number of γ lamellae inside the original α_2 lamellae, suggesting [9] that the initial driving force is associated more with the excess α_2 than with the amount of interface, figure 17.2(*b*). The associated changes in mechanical properties in this alloy, and in others which have been exposed under these conditions, are shown in table 17.2, where it can be seen that the flow stress could be reduced by about 20% and the elongation is also reduced. There were, however, alloys where the strength increased during treatment, e.g. for the tantalum-containing alloy with 1% boron listed in table 17.2. In this case, a significant number of small precipitates were formed during this heat treatment, as shown in figure 17.3, and it is reasonable to consider that the strength increase was associated with this precipitation.

Higher alloy contents

The properties of some alloys with higher alloying additions than the alloys in table 17.1 are shown in table 17.3, where it is clear by comparison with the data in table 17.1 that these higher alloying additions produced materials with a better balance of room temperature yield strength and creep strength. In addition, exposure at 700°C for 3000 h for these alloys resulted in a significant increase in strength, and experiments to date show that this improvement continued for longer times. No obvious microstructural changes were detected accompanying these strength increases. The biggest limitation with these alloys was the limited ductility at room temperature.

Experiments with these higher alloying addition alloys revealed the unexpected result that, although the duplex structure was stronger than the fully lamellar structure in alloys which contained boron additions, the reverse was true for the same alloys without boron. This increase in strength

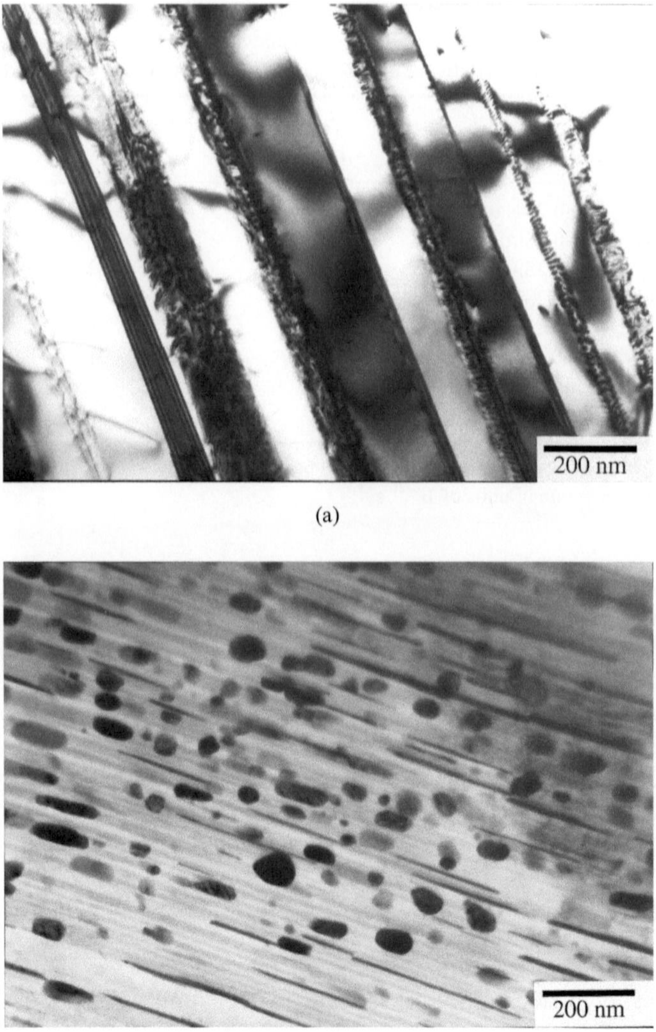

(a)

(b)

Figure 17.3. Bright field many-beam transmission electron micrographs of a sample of Ti-Al-2Ta-1Nb-0.1Mn-1B-2Si alloy (*a*) before and (*b*) after heating at 700°C for 3000 h.

of the boron-containing alloys resulted in a small reduction (probably insignificant in view of the scatter in creep data) in creep properties. These results are also shown in table 17.3.

It is clear that thermo-mechanical processing is capable of producing TiAl-based alloys with a good balance of properties. It is also clear that neither the composition nor the precise thermo-mechanical treatments

Table 17.3 The influence of higher alloying additions on the room temperature strength and creep properties of some TiAl-based alloys. The influence of boron additions on the relative strengths of duplex and lamellar TiAl alloys is also shown.

Alloy composition/structure	Yield stress (MPa)	Primary creep[a]	Secondary creep[a,b]
Ti-44Al-4Nb-4Zr-0.2Si/lamellar	676	0.58% 630 h	7.1
Ti-44Al-4Nb-4Zr-0.2Si/duplex	632	0.56% 725 h	8.5
Ti-44Al-4Nb-4Zr-0.2Si-B/lamellar	644	0.45% 630 h	5.9
Ti-44Al-4Nb-4Zr-0.2Si-B/duplex	732	0.51%	12.9
Ti-44Al-4Nb-4Hf-0.2Si/lamellar	710	0.55% 580 h	6.2
Ti-44Al-4Nb-4Hf-0.2Si/duplex	643	0.69% 515 h	13.4
Ti-44Al-4Nb-4Hf-0.2SiB/lamellar	616	0.78% 710 h	5.9
Ti-44Al-4Nb-4Hf-0.2SiB/duplex	672	0.7% 450 h	15.6

[a] Conditions used for creep were 200 MPa at 700°C.
[b] Secondary creep rate in units of $10^{-10}\ s^{-1}$.

have been fully optimized, although impressive properties have been reported by using high temperature extrusion to produce very fine grains and very finely spaced lamellae [2]. Post-working heat treatments have been used to stabilize the thermo-mechanically processed alloys but the changes in microstructure that give rise to the property changes during post working heat treatments are not generally known and these heat treatments have been developed somewhat empirically. Detailed work on generic TiAl-based alloys will allow a better understanding to be developed of the transformation characteristics of these alloys during cooling from, and holding at, heat treatment temperatures. Thus to optimize TiAl-based alloys it is essential to develop time–temperature transformation and continuous-cooling transformation curves for selected alloys so that their behaviour at high temperatures is fully understood. This is particularly important for the more complex alloys which contain at least three phases (α, α_2 and γ) and which change their properties over long times when heat treated at 900°C.

Increasing thermal stability

Significant property changes caused by high temperature exposure have been reported by a number of workers with the surprising exception of the powder route extruded alloy [2]. In this case, it was reported that no change in microstructure or properties occurred after exposure under conditions under which ingot route material did show a change. The reason for the stability of the powder route alloy is not known, but this may relate to impurities such as oxygen picked up during processing, which may stabilize the very high volume fraction α_2 in these alloys.

Irrespective of the reason for the stability of powder route samples at high temperatures, it is likely that addition of alloying elements such as tantalum or tungsten, which slow down creep slightly (presumably because diffusion is slowed), would be expected to increase the thermal stability. Thus it is possible that TiAl-based alloys which contain tantalum or tungsten would have an acceptable balance of properties after extrusion at high temperatures and would combine this with stability at operating temperatures. Indeed, the tantalum-containing alloy in table 17.2 is an example of an alloy which increased in strength during holding at 700°C.

Cast blades

If it is necessary to add expensive elements such as tantalum and niobium in significant amounts, and then to process the alloy extensively at high temperatures, the cost of the final product will be far higher than would be the case were it possible to manufacture the product by a casting route. However, at this stage there are difficulties in producing cast products without significant porosity (if a cold-wall furnace is used to melt the alloy). The cast materials can of course be HIPed to close the porosity (as would be the case for aerospace applications), but it appears that the porosity is typically so significant that surface dimples are formed, and that extensive machining is necessary to produce the final accurately shaped product.

The applications which have been considered for TiAl-based alloys in aeroengines include low pressure turbine (LPT) blades, high and intermediate pressure (HP/IP) compressor blades, engine casings and stators. The properties of cast alloys are adequate for low risk components such as stators. For virtually all other components, the best properties which can be achieved are adequate, and substitution for nickel alloy components offers a considerable weight saving and associated economies. Nevertheless the processing costs (and thus the final cost of any component) are the main obstacle limiting the introduction of TiAl-based components. The best hope to reduce the final component costs significantly would be to improve casting technology or to combine extrusion with electrochemical machining, which is a mature technology.

The initial engine tests of cast TiAl-based alloys fitted on to the last stage of the low pressure compressor were successful. However, the cast blades used were very expensive because they were cast oversize and very expensive machining operations were required in order to finish the blades. Thus at present the cost of the TiAl components is still too high for aircraft engine manufacturers to realize the potential weight advantage. It is therefore essential that further investigations are performed to improve processing routes to reduce costs without undermining reliability.

Summary

Although it is now possible to produce very strong TiAl-based alloys, which have acceptable fracture toughness, creep and fatigue properties, it is essential that the response of such alloys to exposure at service temperatures and to high temperature heat treatments is put on a sound basis through generation of isothermal (TTT) and continuous cooling (CCT) transformations. Improvements both in melting and in casting technology are required to produce sound pore-free castings of TiAl-based alloys. Despite the weight advantages of TiAl-based alloys over nickel and iron based alloys, the difficulties associated with casting, which at present can be overcome only at an increased cost and the cost of secondary processing, exclude these alloys from use in aeroengines at present.

References

[1] Kim Y W 1994 *J. Metals* **46** 30

[2] Liu C T, Maziasz P J, Clemens D R, Schenibel J H, Sikka V K, Nieb T G, Wright J and Walker L R 1995 *Gamma Titanium Aluminides* ed Y W Kim, R Wagner and M Yamaguchi (Warrendale: TMS) p 688

[3] Godfrey Macha B 1996 PhD thesis (University of Birmingham)

[4] Kuang J P 1998 PhD thesis (University of Birmingham)

[5] Zhang G 1996 PhD thesis (University of Birmingham)

[6] Gouma P I and Loretto M H 1996 *Proc. 8th World Conference on Ti—Ti 95 Science and Technology, Birmingham* ed P A Blenkinsop, W I Evans and N M Flower

[7] Godfrey A, Hu D and Loretto M H 1998 *Phil. Mag. A* **77** 287

[8] Wiezorek J M K, DeLuca P M, Mills M I and Fraser H L 1997 *Phil. Mag. Lett.* **75** 271

[9] Godfrey A B, Hu D and Loretto M H 1997 *International symposium on designing processing and properties of advanced engineering materials, Toyahashi, Japan* (IPS AEM) **37**

Chapter 18

Titanium metal matrix composites

Malcolm Ward-Close, Stuart Godfrey and James Robertson

Introduction

Fibre reinforced titanium alloy composites were first developed about 20 years ago, and have since made substantial progress towards a mature engineering material. The fibre and its protective coating has been refined and optimized. Testing methods have been developed and mechanical properties have been extensively characterized. In broad terms, titanium metal matrix composites (MMCs) have been shown to have an attractive combination of properties, and good stability and durability for extended service. A number of demonstration parts have been manufactured, and tested successfully under service conditions. Cost remains a major concern, and much current effort is aimed at developing cost effective manufacturing methods. This article describes recent progress at DERA, and discusses options for future commercialization of titanium MMCs.

Titanium alloy MMCs, reinforced with continuous ceramic fibre, offer attractive combinations of strength, stiffness and elevated temperature performance and are currently being considered for a range of advanced applications, particularly in the aerospace field. Table 18.1 shows mechanical properties for a typical titanium alloy MMC and its constituent fibre and matrix. The reinforcement is invariably silicon carbide (SiC) monofilament with a carbon-based multilayer coating, of which there are two commercially available types: Textron SCS series (140 μm diameter) and DERA Sigma series (100 μm diameter). Because of the high chemical reactivity of titanium, liquid infiltration cannot be used, and fabrication always involves solid state diffusion bonding.

Figure 18.1 illustrates a number of the lay-up methods which have been used for titanium based MMCs of which the foil–fibre method is the most well established [1]. In the foil–fibre method alternate layers of metal foil and ceramic fibre are hot pressed under inert conditions to give a fully

Table 18.1 Longitudinal strength, stiffness and density of a titanium MMC and its constituents.

	Tensile strength (MPa)	Young's modulus (GPa)	Relative density (g cm^{-3})
Ti-6Al-4V 35 vol% SiC MMC	1690	186	4.0
Titanium alloy Ti-6Al-4V	950	110	4.6
Sigma 100 µm SiC fibre	3500	430	3.0

dense product. The fibres are generally held in place during this process either by a titanium ribbon cross-weave, or an organic binder that is later removed by vacuum degassing. In the powder cloth method [2], alloy powders are mixed with an organic binder and rolled into a cloth-like material as a substitute for the foil. In a variant of this method, tape casting, parallel fibres are coated with a mixture of powder and organic binder to form a precursor tape. Plasma spraying [3], where metal is sprayed on to an array of fibres to form monolayer tapes, now seems unlikely to become a commercially viable route. Experience elsewhere has indicated fibre damage because of thermal shock, and problems because of the high porosity of the matrix and consequently very rough surfaces which interfere with the accurate location of the next layer of fibres. High plasma power tends to result in lower porosity and improved surface quality, but also results in increased fibre damage. In the wire–fibre method SiC fibre and a small diameter titanium

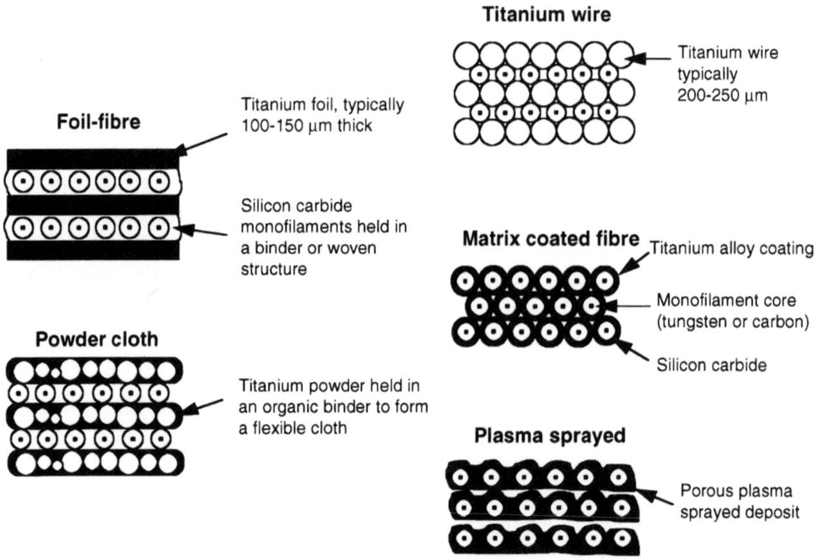

Figure 18.1. Schematic of the different production methods for titanium MMCs.

wire are hot pressed together to form a consolidated MMC. The fibre sits in grooves made by the supporting wire mat, but it is difficult to ensure that the complex lay-up maintains registration during consolidation processing. Also, the fine titanium alloy wires are difficult and expensive to manufacture. In the matrix coated fibre method, the fibre is pre-coated with the titanium matrix alloy by electron beam evaporation to form a precursor material which is then solid-state consolidated.

As yet, no titanium MMC components have reached full production, although a number of prototype titanium MMC parts have been made under various US programmes and several have been tested in aeroengines or have flown on test aircraft. Recently Atlantic Research Corp. (ARC) have used the tungsten wire–fibre route to make a titanium MMC actuator for the F119 engine in the F22 test aircraft [4]. However, it is unclear which of the fabrication methods, if any, will become commercially viable. The choice is likely to be based on overall component cost, availability and product quality. DERA has one of the strongest capabilities in Europe for the development and production of fibre reinforced titanium MMCs, with the DERA Sigma SiC facility (formerly BP Metal Composites), a commercial scale foil–fibre facility, and a pilot scale facility for the matrix coated fibre route.

Foil–fibre process

Processing of titanium MMCs using foil–fibre lay-up techniques is perhaps the oldest and most established approach to the problem of combining the high strength and stiffness of monofilaments with the high temperature capability of reactive titanium alloys. Even so, advances are still being made in this area. High cost, poor fibre distribution, poor foil availability and the difficulty of fabricating complex component shapes are reasons for moving to alternative methods. However, the alternative methods also have problems. For example, the promise of significant cost reduction always comes with proviso of 'if the production volumes are large enough', and this proviso is certainly valid with the foil–fibre techniques.

Foil availability

The production of titanium alloy foil is not a simple process. The $\alpha + \beta$ alloys are difficult to roll, requiring frequent interstage annealing, pack rolling or very controlled processing. The near-α alloys and intermetallics can be produced as foil, but are more difficult to handle. Other problems in using these high temperature, brittle alloys mean their use is likely to be restricted. The β titanium alloys are easier to roll to foil, but have found fewer applications in MMCs.

Many techniques for foil production have been investigated, including plasma spraying, powder processing, chemical milling, and the more conventional thermo-mechanical processing techniques. Foil production volumes are low and, with the demise of the large US spaceplane programme, sheet titanium MMC structures are not required and there is currently little incentive for titanium producers to invest in this small market. Titanium alloy foil will never be cheap, but at increased production volumes it should be an economically viable matrix source. Foil has the advantages of a controlled microstructure, accurate alloy chemistry and low impurity levels.

Matrix coated fibre process

The matrix coated fibre (MCF) process is a new fabrication route for Ti MMCs which uses high rate physical vapour deposition (PVD) to pre-coat continuous fibre with a thick layer of matrix alloy. The coated fibres are then laid-up and hot pressed into the finished MMC, as shown in figure 18.2.

The process has been scaled-up to pilot plant level in both the UK and USA. At DERA, 1000 m lengths of Ti-6Al-4V coated SiC fibre, suitable for a 35 vol% MMC, are now being produced. It is anticipated that further process development will increase the maximum coated fibre length to 5000 m and beyond. Electron beam evaporation from a single alloy bath is possible if the vapour pressures of the constituent elements are relatively close to each other at the bath temperature, which is the case for Ti-6Al-4V. For more complex alloys, containing low vapour pressure refractory elements such as niobium or zirconium, multiple source evaporation is used.

The design of the DERA fibre coater is such that the constituents combine to give a uniform alloy composition in the finished MMC, and the titanium alloy coating has a fine equiaxed microstructure, which adheres well to the fibre without cracking or spalling (figure 18.3). The coated fibres

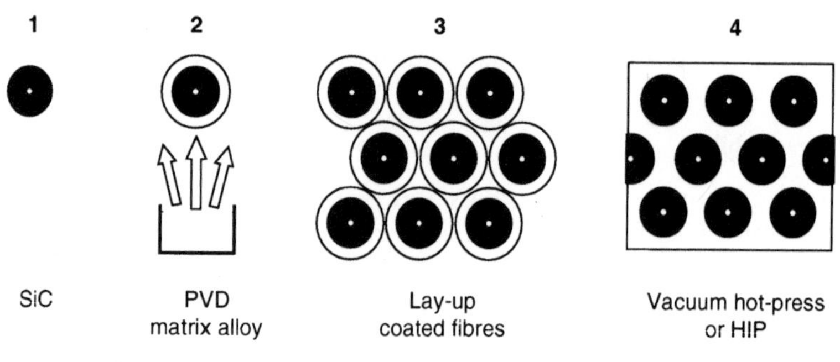

Figure 18.2. Schematic of the matrix coated fibre manufacturing process for titanium MMCs.

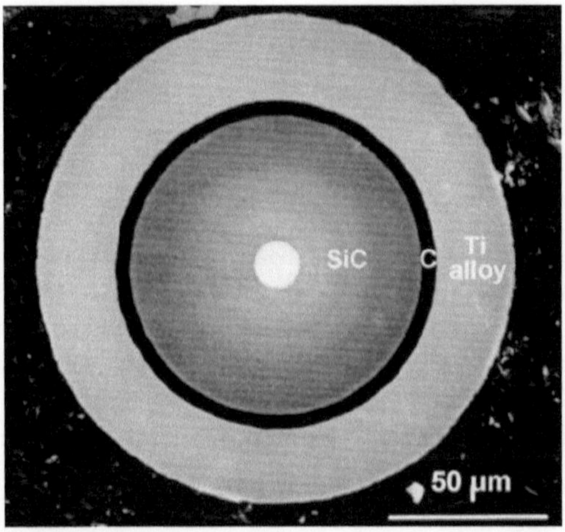

Figure 18.3. A matrix coated silicon carbide fibre.

are consolidated by either vacuum hot pressing or hot isostatic pressing (HIPing), and the finished MMCs have regular fibre distributions. The volume fraction of reinforcing fibre in the finished MMC is determined by the thickness of the matrix coating and examples have been produced with volume fractions in the range 15% to 80%.

The potential advantages of the matrix coated fibre process compared with alternative fabrication methods are summarized as follows:

1. Excellent fibre distribution with no touching fibres. Residual stresses caused by differences in thermal contraction between the fibre and the matrix on cooling from the processing temperature have been shown to give rise to matrix and fibre cracking [1]. For Ti-6Al-4V/140 μm SiC fibre MMC (35% fibre) it has been shown that radial matrix cracking occurs at the point of closest approach of the fibres and that prevalence of cracking increases with decreasing fibre spacing. No cracking occurs if the fibre spacing remains above 30 μm [5]. The inter-fibre spacing for a 50 vol% MMC with 140 μm diameter fibres and the ideal hexagonal packing provided by the matrix coated fibre process will be 35 μm.
2. The consolidation requirements are likely to be less severe for matrix coated fibre than for either foil–fibre or monotapes.
3. Little or no disturbance of the fibre/matrix interface region.
4. Low debulking.
5. Almost any matrix alloy can, in principle, be applied.
6. No foils or powders are required.
7. Matrix coated fibre is suited to filament winding of product forms such as rings, discs, shafts and tubes. For such shapes, the foil–fibre or mono-ply tape methods are likely to be prohibitively expensive and may leave large

numbers of fibre ends embedded in the structure to act as points of chemical attack or stress raisers for crack initiation.

8. Very high volume fractions of fibre are possible (up to 80% has been demonstrated).
9. The metal coating protects the ceramic fibre from damage, both during handling and in the consolidation process.

It should be noted that best practice with other MMC fabrication techniques, such as the foil–fibre method, have also demonstrated very high quality material, with very good fibre distribution, and there remain considerable technical difficulties in implementing the matrix coated fibre process commercially. For example, maintaining consistent alloy chemistry in the coating is difficult, particularly for complex alloys containing refractory elements.

Fibre management

Fibre management is the control of fibre position during MMC processing. The two most common techniques when using foil as the matrix source are weaving and filament winding, although other techniques such as grooved foil and co-winding can be used. Unlike conventional weaving of yarn, the monofilaments must be handled individually, leading to problems of scale and minimum bend radius. The cross weave material must be small enough to allow the target fibre spacing, but strong enough to cope with

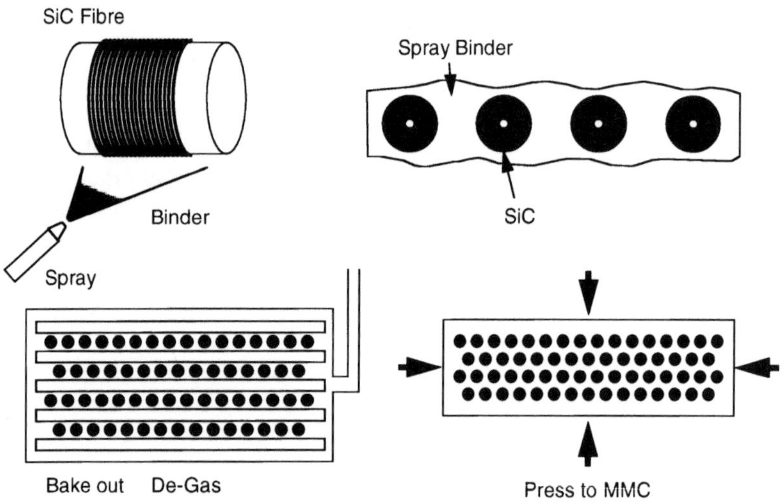

Figure 18.4. Schematic of the foil–fibre filament winding process for the manufacture of titanium MMCs.

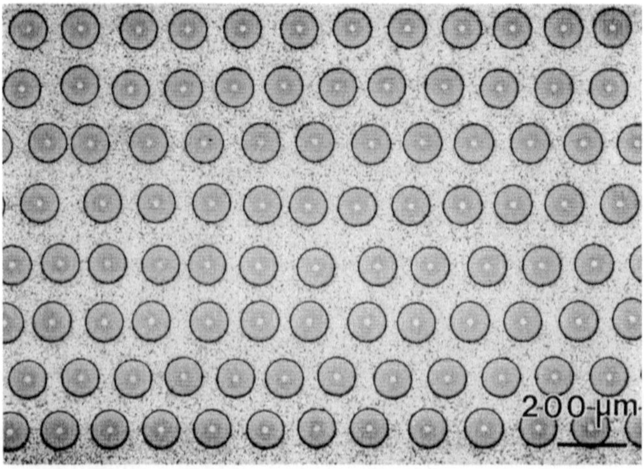

Figure 18.5. Cross-section micrograph of a titanium MMC manufactured by the foil–fibre method.

the stresses during the weaving process. This severely limits the choice and availability of cross weave materials. Various refractory wires have been tried, but the cross weave is usually pure titanium wire or ribbon. Filament winding, illustrated in figure 18.4, offers a cheaper alternative but the

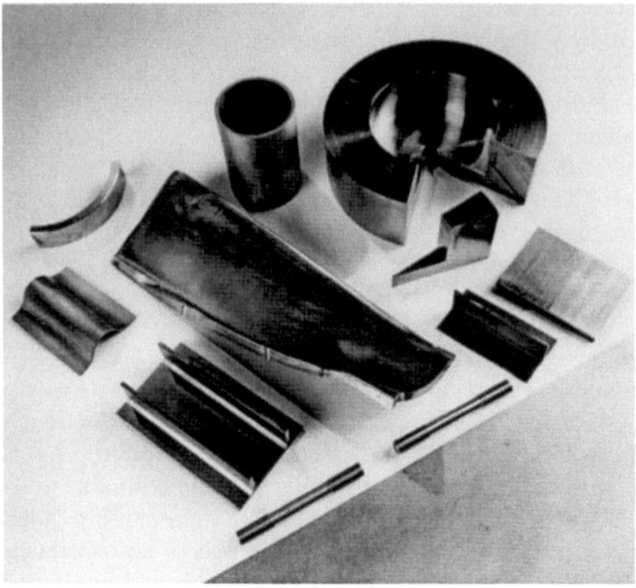

Figure 18.6. A selection of component shapes made by the foil–fibre MMC route.

inclusion of a fugitive binder and its subsequent removal leads to the possibility of contamination and fibre movement. These problems can be largely eliminated with good process control.

The fibre distribution attained when using foil–fibre methods can be very good. Figure 18.5 shows a micrographic section of a panel produced using filament wound foil–fibre. By carefully controlling the process, this quality of fibre distribution can be maintained even for shaped components. Figure 18.6 shows a selection of component shapes produced by DERA using the foil–fibre route. However, the method cannot guarantee that fibres will not touch. The quality of fibre distribution depends to some extent on the diameter of the fibre. Large fibres can be handled more easily, the accuracy of each fibre position is not so critical as neighbouring fibres are farther away. Obtaining acceptable fibre distribution using titanium powder as the source of the matrix is extremely dependent on the relative sizes of the powder and fibre, and good quality, low oxygen titanium alloy powder is not available in the size required for 100 μm SiC fibre.

Selected area reinforcement

Unlike fibre reinforced polymeric composite, where the matrix has relatively poor mechanical properties, MMCs can have relatively good transverse strength. The transverse strength of a typical fibre reinforced titanium MMC is about half the normal matrix strength, and this has important consequences for titanium MMCs. For many potential applications the transverse properties are good enough for titanium MMCs to be used in a unidirectionally reinforced mode, taking maximum advantage of the exceptional strength and stiffness of the fibre. In addition, because of the good mechanical performance of the unreinforced titanium alloy, it is not necessary, or in most cases desirable, for the whole part to be manufactured as an MMC. Thus most applications of titanium MMCs will be as selectively reinforced regions within larger titanium alloy components. In this way, expensive titanium MMC material will be used only in the most critical regions, and damage to fibres will be avoided by carrying out all machining and joining operations, e.g. welding and weld repair, in the unreinforced regions of the component.

The ability of titanium alloys to diffusion bond very easily provides an effective route for small areas of MMC to be incorporated into larger titanium parts and this is illustrated for a reinforced ring structure in figure 18.7. Here, matrix coated fibre is filament wound into a channel in an inner titanium ring, this is then joined to a close fitting outer ring using electron beam welding and the whole assembly is hot isostatically pressed (HIPed) to produce a single solid part. All voids and interfaces are closed by plastic flow and diffusion bonding.

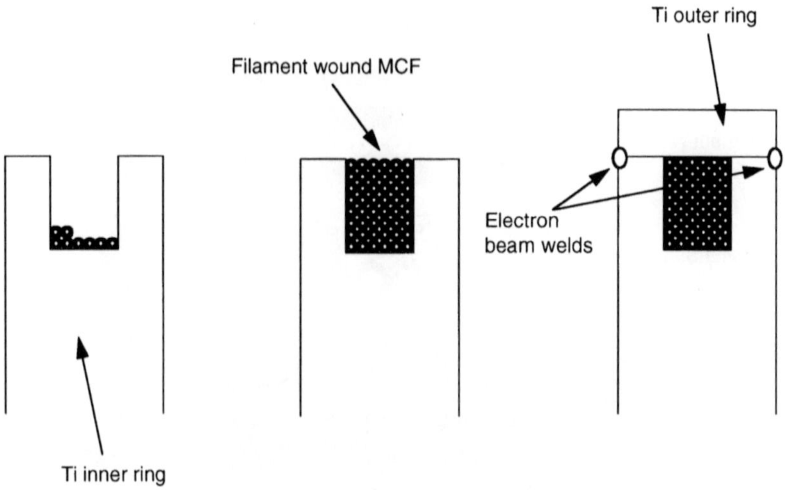

Figure 18.7. Processing ring structures using filament wound metal coated fibres.

Shape control

Debulking, or shrinkage, associated with the hot pressing stage of titanium MMC manufacture is a critical factor in determining both the position of the individual fibres, the position and shape of an area of reinforcement, and the shape and surface finish of the MMC part. The choice of manufacturing route has a major effect on debulking and this can range from over 50% for the foil–fibre method to as little as 10% for filament wound matrix coated fibre or preconsolidated monosheets (figure 18.8). Generally, even with HIPing, shrinkage is not uniform. Proprietary can and tooling designs are used to ensure accurate shape formation without fibre damage. This is illustrated for ring and tube shapes in figure 18.9, which also indicates the most favourable processing routes for the different shapes.

For many component shapes it is essential to restrict consolidation to one direction and at the same time control certain dimensions precisely. For example, in a tube, consolidation must be restricted to the radial direction, as axial movement might result in buckling and severe distortion. But, it may also be necessary to prevent movement of the outside dimension of the tube in order to avoid excessive hoop stressing of the fibre during consolidation, or, depending on the size and shape of the tube, it may be necessary to prevent changes in the inner diameter. In many ways these problems are similar to those found in the manufacture of carbon fibre composite materials, where issues such as dimensional control, fibre uniformity and surface finish are addressed through tool design and choice of curing cycles. For titanium MMC components, the required constraint

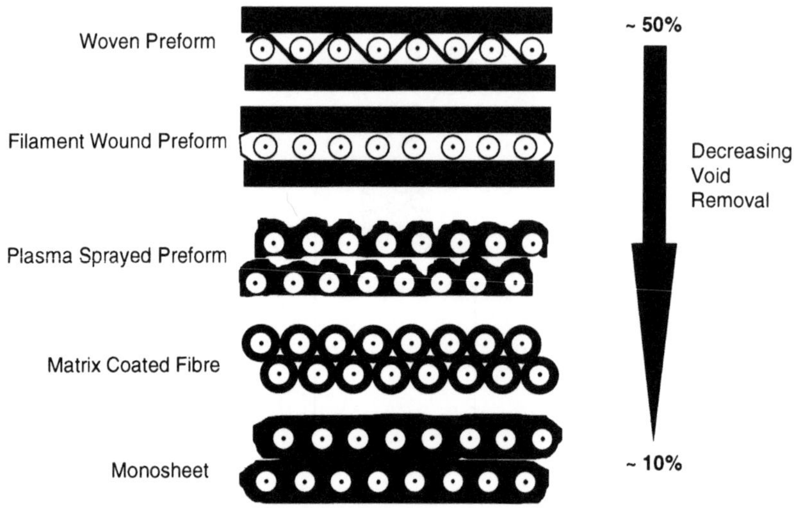

Figure 18.8. Debulking associated with different titanium MMC processing methods.

can often be provided by titanium tooling which is incorporated into the component. In the example shown in figure 18.7, the inner ring will prevent movement in the radial direction and consolidation of the MMC area will be mainly in the axial direction, thus preventing either buckling or overstressing of the fibres. For other component shapes steel tooling can

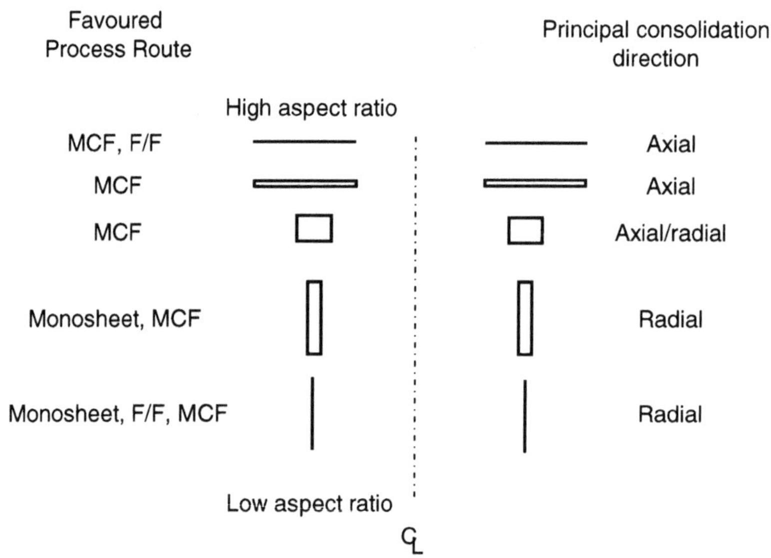

Figure 18.9. Effect of ring aspect ratio on consolidation method.

be used, which will have a high resistance to deformation at the consolidation temperature. Reusable steel tooling can often be utilized, with a special coating to prevent bonding between the steel and the titanium or, if this is not possible, the tooling can be removed by machining or acid pickling.

Choice of fabrication route for titanium MMCs

It is easy to produce flat panels using foil–fibre processing. However, aeroengine applications generally do not require flat panels. Rings, disks, blades, struts and shafts are more likely component shapes. Foil–fibre processing can be used for all of these shapes, as shown in figure 18.3, but may not offer the best alternative in terms of cost, quality and fibre architecture. A significant amount of development is under way evaluating the advantages and disadvantages for the different processing routes.

A major factor in determining the ease of fabrication of each component shape is the volume fraction of voids that must be removed during consolidation. It is clear that the matrix coated fibre method offers the lowest debulking which is most important in ring shaped structures. For the foil–fibre method, grooving the foil reduces the amount of debulking while improving the fibre management, but the cost may not be justified by the improvements for a particular application.

Summary

Used selectively, in relatively small quantities, titanium MMCs can have a dramatic effect on the performance and weight of critical components. The current high cost of titanium fibre reinforced MMCs is the result of a high fibre cost combined with labour and skill intensive fabrication processes. A number of options exist for reduction of fabrication costs and these include the production of robust and convenient intermediate materials, such as matrix coated fibre and fibre/powder-cloth pre-pregs, and automation in fibre handling and placement. Several manufacturing routes have been identified as viable, and choice of route is likely to depend on specific component design. Future use of titanium MMCs will depend on quality, availability and, above all, cost.

References

[1] MacKay R A, Brindley P K and Froes F H 1991 'Continuous fiber-reinforced titanium aluminide composites' *J. Metals* May 23–29

[2] Stephens J R 1988 *AIEE/ASME/SAE/ASEE 24th Joint Propulsion Conference* (Boston: AIAA) pp 1–10

[3] Chou T W, Kelly A and Okura A 1985 'Fiber reinforced metal-matrix composites' *Composites* **16** 187–206

[4] Lavitt M O 1997 'First TMC aircraft part to help power F-22' *Aviation Week and Space Technology* 12 May pp 74–75

[5] Wood M J and Ward-Close C M 1995 'Fibre reinforced intermetallic compounds by vapour deposition' *Materials Sci. Engng. A* **192/193** 590–596

Chapter 19

Anisotropic creep in single crystal superalloys

Malcolm McLean

Introduction

The development to commercial application of a range of nickel-based superalloys, specifically for the production of complex castings in the form of single crystals, has contributed significantly to the progressive improvement in gas turbine efficiency by allowing significant increases in operating temperature. Until now, single crystal superalloys have been predominantly used to produce relatively small rotating turbine blades for aeroengines. However, there is a growing trend to apply this technology to a wider range of aero components which experience more complex loading conditions involving multiaxial stresses and varying stresses and temperatures. In addition, their application is being extended to industrial gas turbines for power generation, where components are expected to have much longer lives and to experience different loading histories from those in aeroengines. Solidification of such large industrial components involves lower temperature gradients and solidification rates than have been previously used, and this leads to difficulties in controlling the crystal orientation and to a propensity for the generation of solidification defects.

Advanced superalloys, particularly those intended for single crystal application, consist of a precipitate of the intermetallic γ' $Ni_3(Al,Ti)$ occupying 65–70% by volume as cube shaped particles (1–2 μm diameter) in a disordered γ austenitic matrix. The philosophy in developing first-generation alloy compositions specifically for use as single crystals was to strip out alloy additions from existing superalloys that were intended to strengthen grain boundaries. This increased the solidus temperature to allow higher γ' volume fractions and solution temperatures to be achieved, and opened a temperature window between the γ' solvus and the alloy solidus to facilitate precipitate control through solution/ageing heat treatments. Second- and third-generation alloys have explored the additional benefits to be obtained through novel solid solution strengtheners, such as tungsten and rhenium [1]. However, these highly segregating refractory additions have extended the melting range of the alloys, exacerbating the problems of orientation and defect control.

The benefits of single crystal over polycrystalline superalloys derive from a combination of retention of creep strength to higher temperatures, greatly improved creep ductility (around 25% rather than around 2%) and alignment of the elastically soft $\langle 001 \rangle$ direction parallel to the solidification direction (and the turbine blade axis) leading to improved thermal fatigue resistance. In order to take maximum advantage of the properties of single crystal superalloys, it is important both to control the orientation and to ensure the absence of defects such as stray grains. However, orientation control is proving to be difficult in the more advanced superalloys, and the low temperature gradients and solidification rates achieved in large industrial castings increase the probability of casting defects. Consequently, it is important to understand how deviations from the ideal orientation will degrade the properties in order to define acceptability criteria for blade castings and/or to modify the design data to accommodate the degraded performance.

The creep behaviour of single crystal superalloys exhibits a significant anisotropy that is a complex function of stress, temperature and alloy system. Current design practice appears to depend on an anisotropic elastic stress analysis with creep lives being estimated from conventional creep data. This is an inherently conservative approach in which large design margins are introduced. If a fully time-dependent plastic analysis is to be developed, it will be necessary to provide reliable constitutive equations that account for both anisotropic creep and the time independent plasticity/elastic stress that develops during service. These equations must be capable of being incorporated in finite element calculations. The experimental effort that would be required in providing a comprehensive database to account for all possibilities encountered by a turbine blade (varying stress/temperature, multiaxial stresses, crystal anisotropy, defect density) is prohibitive. Rather, it is necessary to model the process to allow the complex service to be deduced from a limited database. This chapter describes the current status of a model of anisotropic creep that is consistent with current knowledge of mechanisms of creep deformation, and the experimental validation of this model.

Creep mechanisms

The high temperature strength of nickel-based superalloys results from the coherent precipitation of the ordered γ' phase in the disordered nickel-based solid solution γ matrix. In modern single crystal superalloys the γ' constitutes about 70% by volume of the alloy. The superalloy has superior creep performance to either of the constituent phases in monolithic form and is capable of supporting stresses to homologous temperatures (T/T_M) of above 0.8. The advanced single crystal superalloys show little improvement over the earlier equiaxed and directionally solidified versions in low temperature performance. However, the creep behaviour at 850°C and above is significantly improved leading to increases in use temperature of about 100 and 50 K respectively.

There have been many electron microscope studies which show that there are quite clear differences in the micromechanisms occurring during creep in different temperature regimes. Figures 19.1 and 19.2 illustrate these differences for the single crystal alloys SRR99 and CMSX4. It is noteworthy that the shapes of creep curves in these three temperature regimes are quite different, suggesting that the change in mechanisms are controlling the creep behaviour.

Regime 1 ($\leq$ 750°C)

The creep curves at low temperatures have a large primary creep component for both $\langle 001 \rangle$ and $\langle 111 \rangle$ oriented specimens; indeed the $\langle 111 \rangle$ specimens also show significant primary creep at 850°C. Deformation occurs in both the γ and γ' phases, and is consistent with $\{111\}\langle 2\bar{1}\bar{1} \rangle$ slip. For certain orientations with high resolved shear stresses for this slip, there is considerable mechanical twinning, although the strain associated with this twinning is much less than the observed creep strains [2].

Regime 2 (750 to 1000°C)

Tertiary creep dominates the creep curves at these temperatures, with a small amount of primary creep being observed in some conditions. The stresses to cause creep at these higher temperatures are significantly less than in regime 1, and are too small to cause cutting of the γ'. Consequently, deformation is restricted to the minority γ matrix phase. The microstructure is essentially stable in regimes 1 and 2, consisting of aligned, cuboidal γ' particles.

Dislocations in the γ matrix have $\frac{1}{2}\{111\}\langle 1\bar{1}0 \rangle$ character. However, because of the low matrix volume fraction, the γ' particles cannot be bypassed by the mechanisms commonly proposed for materials with low volume fraction reinforcement. Rather, the dislocations interact with the γ–γ' interface, which coincides with (001) and overall deformation through the superalloy requires combined deformation on the $\{111\}\langle 1\bar{1}0 \rangle$ and $\{001\}\langle 110 \rangle$ slip systems. The dislocations bowing in the γ channels and interacting with the γ–γ' interface transfer stress to the γ', and dislocation motion occurs by a combination of glide in the γ and climb along the γ–γ' interface. The finite dimensions of the γ' particles limit the transferred stress.

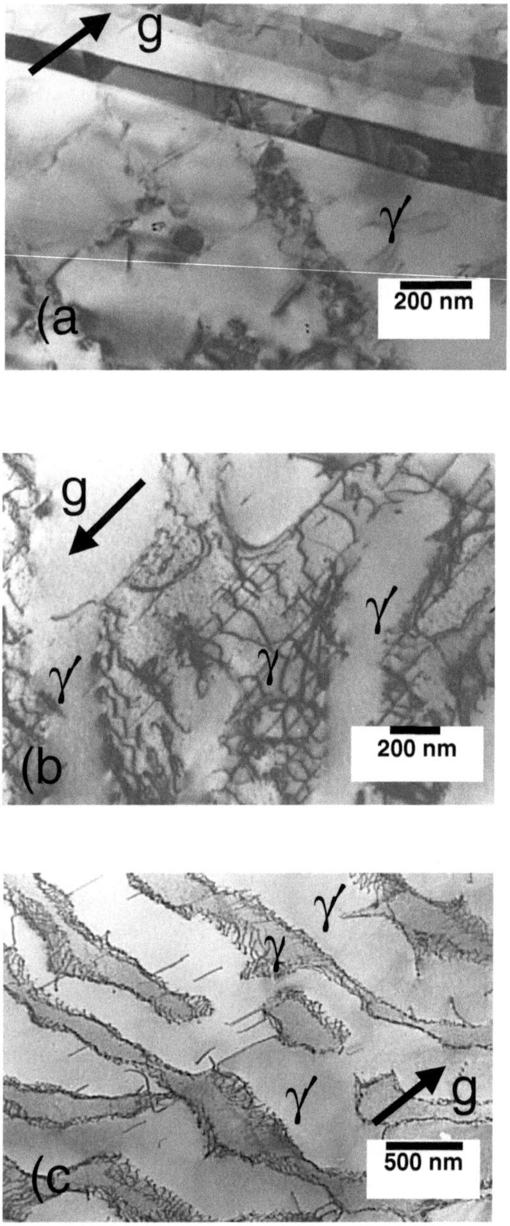

Figure 19.1. (*a*) SRR99 ⟨011⟩ crystal creep tested at 750°C/850 MPa, showing microtwins. Beam parallel to [112], $g = [402]$. (*b*) SRR99 ⟨001⟩ crystal creep tested at 950°C/175 MPa, showing deformation restricted to γ. Beam parallel to [011], $g = [200]$. (*c*) SRR99 ⟨001⟩ crystal creep tested at 1050°C/150 MPa, showing rafts and dislocation pairs in γ'. Beam parallel to [011], $g = [111]$.

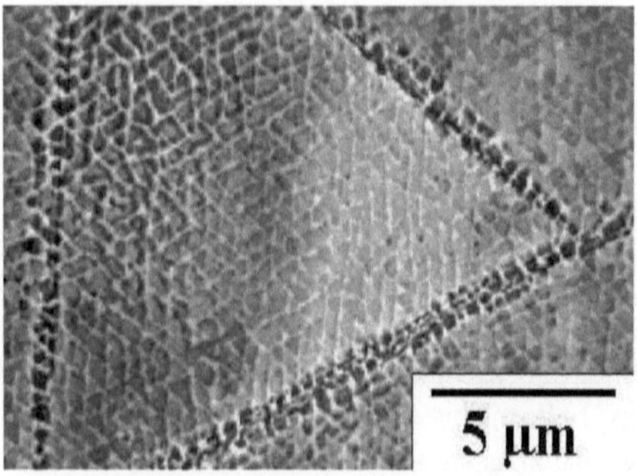

Figure 19.2. Scanning electron micrograph showing the localized slipped bands observed on a section perpendicular to the tensile axis in CMSX-4 [111].

Regime 3 ($>$**1000°C**)

At the highest temperatures, the creep curves show a very low creep rate over most of the life and a sudden onset of a very steep tertiary creep leading to fracture. In this temperature regime, creep is accompanied by a significant change in γ' morphology with a rafted structure developing normal to the applied tensile stress. The low creep rates appear to be due to the entrapment of dislocations in extended γ channels. However, this allows higher stresses to be transferred to the γ' which is eventually cut by dislocation pairs of the type $\frac{1}{2}\{111\}\langle 1\bar{1}0\rangle$.

Modelling of creep

A number of anisotropic creep models have been proposed to account for behaviour of single crystal superalloys. These range from the empirical which provide a fit of creep data to a mathematical function having the required crystal symmetry, to approaches which attempt to capture elements of the micromechanisms that are known to operate during high temperature deformation of these materials. For example, Cailletaud and co-workers [3] have extended an isotropic visco-plastic model incorporating isotropic and kinematic hardening terms by restricting deformation to specific slip systems. Ghosh *et al* [4] have also proposed a crystallographic based model of anisotropic creep, later extended by Pan *et al* [5], that uses the formalism of continuum damage mechanics to incorporate features of both primary and tertiary creep. This again is an extension of a previous isotropic model of creep that has been successful in accounting for the creep behaviour of a range of materials. This chapter will give a more detailed description of the last of these models and of the results of experiments devised to test predictions of the model.

The underlying assumptions of the model are:

1. Deformation occurs by viscous creep on a restricted number of slip systems. For nickel-based superalloys slip is assumed to be confined to $\{111\}\langle 1\bar{1}0\rangle$. This has some validity in regime 2, described in the previous section, but does not take into account the $\{111\}\langle 2\bar{1}\bar{1}\rangle$ slip that occurs at 750°C and below. Matan *et al* [6] have recently modified the model to include this low temperature behaviour.
2. The creep rate increases with increasing creep strain due to an accumulation of mobile dislocations. This feature accounts for the dominant tertiary creep behaviour that characterizes creep curves obtained in most temperature/stress conditions.
3. Primary creep is a consequence of stress transfer between soft and hard regions of the duplex microstructure and is restricted to a relatively short initial period.

The overall creep deformation is obtained by summing the shear strains that occur on each active slip system, and these depend on the resolved shear stress on that slip system caused by the applied stress system. In addition, the shear creep rate depends on two state variables: S^k, accounting for the state of kinematic hardening due to internal stress redistribution, and ω^k, determined by the mobile dislocation density. This shear creep is represented by coupled differential equations:

$$\dot{\gamma}^k = \dot{\gamma}_i^k(1 - S^k)(1 + \omega^k),$$

$$\dot{S}^k = H^k \dot{\gamma}_i^k \left(1 - \frac{S^k}{S_{ss}^k}\right),$$

$$\dot{\omega}^k = \beta^k \dot{\gamma}^k,$$

where $\dot{\gamma}^k$ is the initial shear strain rate on the active slip system (designated by the superscript k) which is a sensitive function of stress and temperature, and $\dot{\gamma}_i^k$, H^k, S_{ss}^k and β^k are constants. If these constants are known as functions of stress and temperature, the shear displacements on each active slip system can be calculated and the general deformation is obtained by summing all possible shear displacements:

$$\varepsilon_{ij} = \sum_{k=1}^{N} \gamma^k b_i^k n_j^k.$$

where b_i^k and n_j^k are direction cosines of the slip direction, k, relative to the i and j directions. This leads to both a change in orientation of an arbitrary direction from $[x_1 \;\; x_2 \;\; x_3]$ to $[X_1 \;\; X_2 \;\; X_3]$:

$$\begin{bmatrix} X_1 \\ X_2 \\ X_3 \end{bmatrix} = \begin{bmatrix} 1+\varepsilon_{11} & \varepsilon_{12} & \varepsilon_{13} \\ \varepsilon_{21} & 1+\varepsilon_{22} & \varepsilon_{23} \\ \varepsilon_{31} & \varepsilon_{32} & 1+\varepsilon_{33} \end{bmatrix} \begin{bmatrix} x_1 \\ x_2 \\ x_3 \end{bmatrix}$$

and a linear strain in that direction of $(\bar{X} - \bar{x})/\bar{x}$. With this formulation it is quite straightforward to compute the creep deformation for arbitrary loading in any crystallographic direction.

The constants ($\dot{\gamma}_i^k$, H^k, S_{ss}^k, β^k) completely define the creep curve. Analysis of a series of creep curves for $\langle 001\rangle$ and $\langle 111\rangle$ orientations allow these constants to be determined for both sets of slip systems as functions of stress and temperature, and it is found that all can be represented as exponential functions. Only $\dot{\gamma}_i^k$ varies significantly with both stress and temperature. β^k is virtually independent of temperature and has a moderate dependence on stress. There is a large scatter in the values of H^k and S_{ss}^k determined by analysis of the creep curves, because there was little observable primary creep for most of the test conditions, so that these constants which represent the primary creep component were subject to a large uncertainty. Indeed they can essentially be regarded as stress and temperature independent constants. Consequently a set of 18 constants completely describes all of the creep behaviour at all stresses and temperature. Having established these constants it is possible, in principle, to calculate the creep deformation for arbitrary orientations and loading history (including temporally varying stress/temperature and multiaxial stresses).

Validation

The minimum requirement of the model is that it should be capable of representing adequately the data from which the optimized parameter set is derived. It is unrealistic to expect individual model-generated creep curves to provide a better fit to the experimental creep curves than the intrinsic scatter of the input data. It is more important to test the accuracy of the model predictions over the entire range of temperature and stress for which data are available. Figure 19.3 compares the model predictions and measurements of times to achieve 1% strain for all the creep conditions for which data are available for SRR99. A similar comparison has been made for CMSX4. The agreement between model representation and experimental measurement is of the same order as the intrinsic scatter in the experimental data.

Representation of the data leading to the quantification of the model is not a particularly severe test of the model and the same level of agreement could probably be achieved using empirical approaches, such as those based on parametric approximations (e.g. Larson–Miller). It is more important to be able to predict reliably the response in conditions that differ from the input data. Using the model parameters derived from analysis of uniaxial creep curves for $\langle 001\rangle$ and $\langle 111\rangle$ crystal orientations it is quite straightforward to predict the creep response for a range of different orientations and loading conditions:

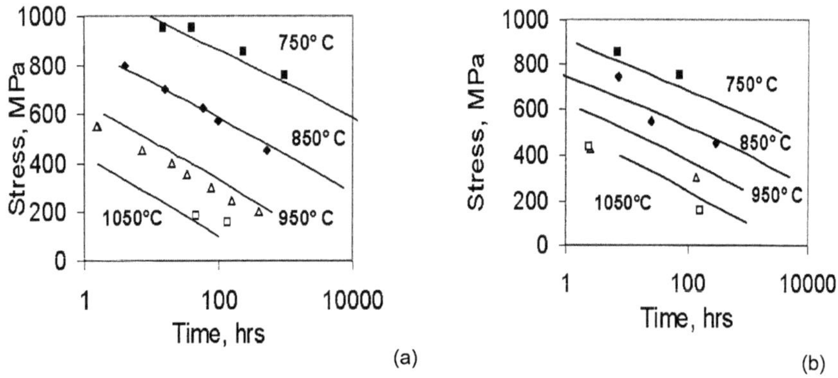

Figure 19.3. Comparison of model predictions and creep data of times to achieve arbitrary creep strains in SRR99 at various temperatures and stresses (*a*) ⟨001⟩, time to 5% strain, and (*b*) ⟨111⟩, time to 2% strain.

1. Creep under a constant axial stress for axial loading in an arbitrary crystal direction is obtained by determining the resolved shear stresses on the allowed slip systems and by calculating each shear strain in each computational time element. Thus the axial creep curve for any orientation can be predicted. For high symmetry orientations, such as ⟨001⟩ and ⟨111⟩, the initial orientation of the specimen should be stable. However, for non-symmetrical orientations the crystallographic orientation of the axial load changes during deformation and the cross-sectional shape of the deforming specimen changes. These features provide a powerful means of testing the predictions of the model which is described in detail below.
2. The effects of changing stress and/or temperature under stress controlled loading conditions are computed in a straightforward manner through continuity of the state variables S^k and ω^k on changing the loading conditions. Other empirical approaches based on representing creep strain as a function of time (rather than strain rate as a function of state variables) can only deal with changing loads by introducing an additional rule (such as strain hardening, time hardening, life fraction) to determine the strain rate immediately on a change in stress and/or temperature.
3. Extension of the calculations to strain rather than load controlled deformation requires additional information on the anisotropic elastic constants as functions of temperature to allow the relaxation of elastic stresses due to creep deformation to be calculated. Pan *et al* [5] have described this development of the model in some detail. Figure 19.4 compares the observed and model generated response of SRR99 with strain controlled low cycle fatigue. The predictions are made using the model parameters derived from creep tests supplemented by the Young's modulus and consequently figure 19.4 is a true prediction.

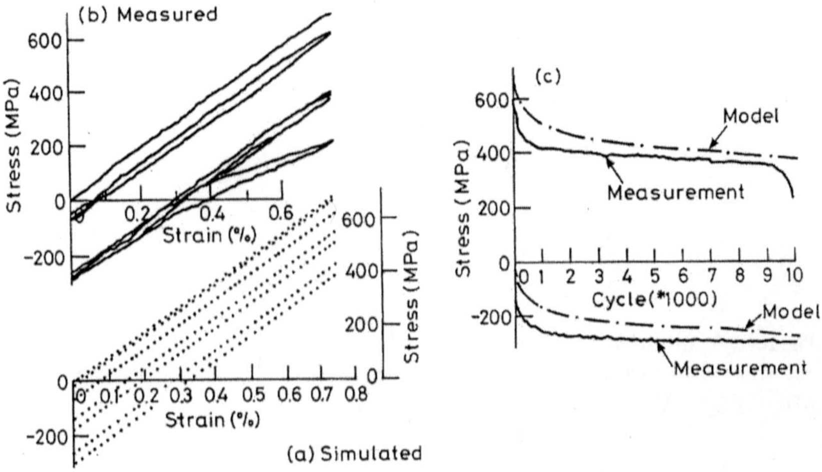

Figure 19.4. Simulated and measured low cycle fatigue behaviour of SRR99 at 950°C for cycling between 0 and 0.75% strain. (*a*) and (*b*) are measured and simulated stress/strain curves, (*c*) compares predicted and calculated maximum and minimum stresses versus cycle number.

It is also important to assess the validity of the underlying assumptions of the model. As discussed above, anisotropic deformation leads to changes in the orientation and shape of creep specimens with complex orientations. Comparison of model predictions with experimental measurements of these two types of variation provides a very sensitive test of the model.

Orientation changes

Crystal orientations have been determined along the length of ruptured creep specimens using the technique of electron backscatter diffraction (EBSD) which has a lateral resolution of around 1 μm [2]. For most data in the intermediate temperature regime the crystal rotates continuously and with increasing scatter of orientation. For orientations within about 45° of $\langle 001\rangle$ the rotations are consistent with slip on $\{111\}\langle 1\bar{1}0\rangle$ and for orientations close to $\langle 111\rangle$, $\{001\}\langle 110\rangle$ slip dominates. For orientations close to $\langle 111\rangle$, $\langle 011\rangle$ and $\langle 112\rangle$ there are occasional large deviations in orientation, consistent with mechanical twinning leading to a $\{111\}\langle 2\bar{1}\bar{1}\rangle$ displacement. Figures 19.5 and 19.7 show examples of crystal rotations and twinning orientations. The main continuous rotation occurs by $\{111\}\langle 1\bar{1}0\rangle$ slip, whereas mechanical twinning, which dominates at low temperatures and for tensile stresses remote from $\langle 001\rangle$, is associated with shearing of the γ' particles in the $\{111\}\langle 2\bar{1}\bar{1}\rangle$ direction. The unambiguous identification of the operating slip systems from crystal rotations is often difficult, but when combined with an analysis of shape changes, the analysis is more certain.

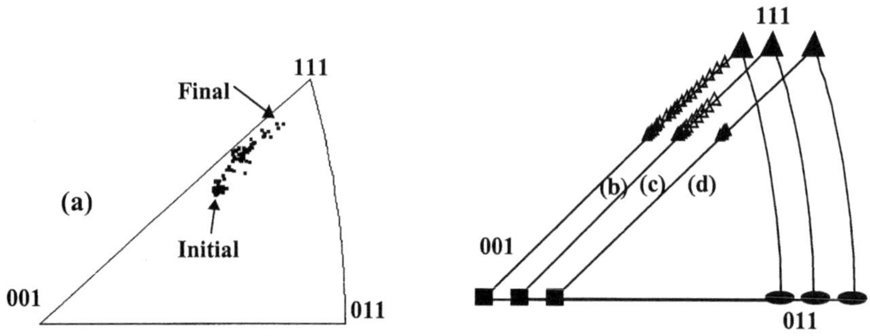

Figure 19.5. (*a*) Inverse pole figure showing continuous rotations during creep at 950°C/200 MPa of a single crystal of SRR99 initially close to ⟨211⟩, and predictions for (*b*) (001)⟨110⟩, (*c*) (111)⟨112⟩ and (*d*) (111)⟨011⟩ slip.

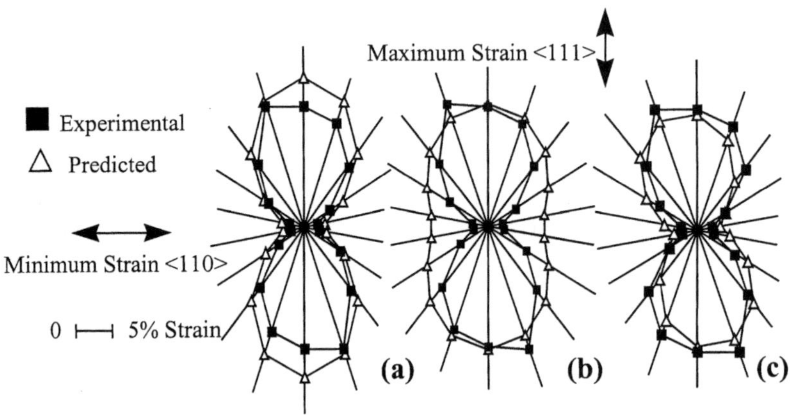

Figure 19.6. Measured specimen shape changes in a near ⟨112⟩ SRR99 specimen creep tested at 950°C/200 MPa compared with predictions for (*a*) (001)⟨110⟩, (*b*) (111)⟨112⟩ and (*c*) (111)⟨011⟩ slip.

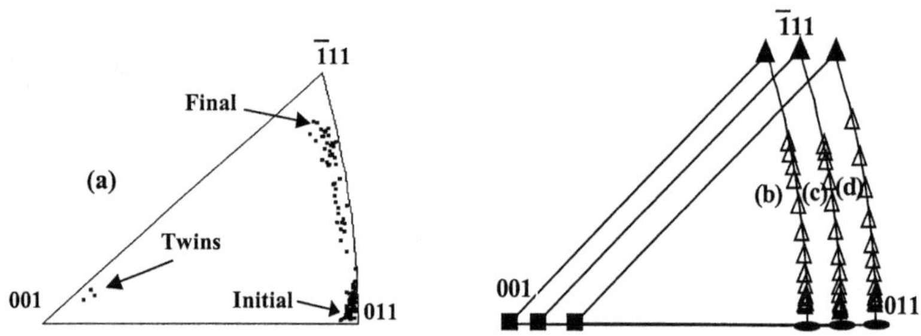

Figure 19.7. (*a*) Inverse pole figure showing continuous rotation and discontinuous twinning rotations during creep at 750°C/850 MPa of a single crystal of SRR99 initially close to ⟨011⟩, and predictions for (*b*) (001)⟨110⟩, (*c*) (111)⟨112⟩ and (*d*) (111)⟨011⟩ slip.

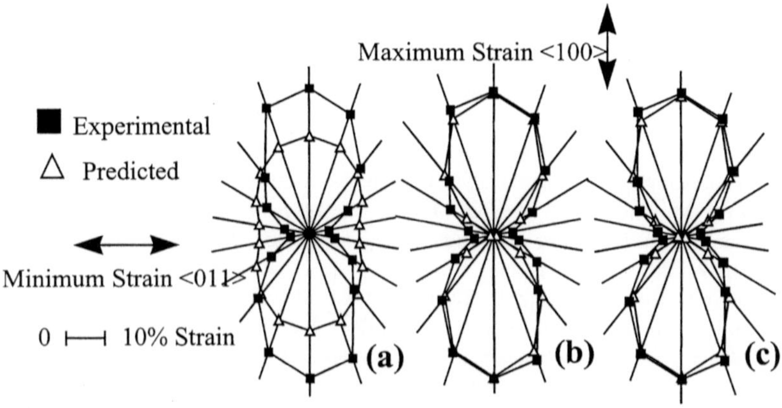

Figure 19.8. Measured specimen shape changes in a near ⟨011⟩ SRR99 specimen creep tested at 750°C/850 MPa compared with predictions for (*a*) (001)⟨110⟩, (*b*) (111)⟨112⟩ and (*c*) (111)⟨011⟩ slip.

Shape changes

The change in the cross-sectional shape of initially cylindrical creep specimens has been determined as a function of local reduction in area and compared with the model prediction [6]. Examples are shown in figures 19.6 and 19.8 for the same specimens analysed in figures 19.5 and 19.7. These results confirm the orientation measurements which show octahedral slip to dominate for orientations close to ⟨001⟩, and cube slip to become important for tensile orientations close to ⟨111⟩.

Summary

The described model is consistent with observed deformation mechanisms in the intermediate temperature regime where viscous glide is largely confined to the γ phase. Nevertheless, the model gives an adequate description of creep strain as a function of time for all of the obtained creep data. Further development of the model will have to account for shearing of γ' at low temperatures and slowing the deformation kinetics through γ' rafting at high temperatures.

References

[1] Duhl D N 1987 *Superalloys II* ed C T Sims, W Hagel and N S Stoloff (New York: Wiley)

[2] Ardakani M G, Ghosh R N, Brien V, Shollock B A and McLean M 2001 'Implications of dislocation micromechanisms for changes of orientation and shape of single crystal superalloys' *Scripta Mater* in press

[3] Meric L, Poubanne P and Cailletaud G 1991 *Trans. ASME J. Eng. Mater. Tech.* **113** 162–170

[4] Ghosh R N, Curtis R V and McLean M 1990 'Creep deformation of a single crystal superalloy—modelling the crystallographic anisotropy' *Acta Metall. Mater.* **38** 1977

[5] Pan L-M, Shollock B A and McLean M 1997 'Modelling of high-temperature mechanical behaviour of a single crystal superalloy' *Proc. Roy. Soc. London* **453A** 1689–1715

[6] Matan N, Cox D C, Carter P, Rist M A, Rae C M F and Reed R C 1999 'Rationalisation of the creep performance of CMSX4 superalloy single crystals' *Modelling of Microstructural Evolution in Creep Resistant Materials* ed A Strang and M Mclean (London: The Institute of Materials)

[7] Shollock B A, Buffiere J Y, Curtis R V, Henderson M B and McLean M 1997 'Characterisation of crystallographic evolution during creep deformation of a single crystal superalloy' *Scripta Mater* **36** 1471–1478

[8] Pan L M, Scheibli I, Henderson M B, Shollock B A and McLean M 1995 'Asymmetric creep deformation of a single crystal superalloy' *Acta Metall. Mater.* **43** 1375–1383

Chapter 20

Microstructural evolution in single crystal nickel-based superalloys during high temperature creep

Tetsuya Ohashi

Introduction

Nickel-based superalloys are widely used as a gas turbine blade material and their integrity during service is important. When these alloys are creep deformed at high temperature, cuboidal shaped strengthening precipitates of γ' phase usually grow in the direction perpendicular to the loading axis [1–3]. In this chapter, a model for the mechanism of this rafting phenomenon is presented in terms of the free energies in narrow channels of the γ phase and the local equilibrium concentrations of the component elements. Anisotropic distribution of the free energy is attributed to the difference in stress state in the γ channels. The stresses caused by lattice misfit between the γ and γ' phases and dislocations formed at the γ/γ' interface during the initial stage of creep deformation are then evaluated. Finally, the rate of rafting may be calculated from the difference in the local equilibrium concentration of component elements, and may be compared with experimental results.

The rafting phenomenon

Figure 20.1(*a*) shows a scanning electron microscope image of the microstructure of nickel-based superalloy CMSX4. Cuboidal shaped regions with dark contrast are precipitates of the γ' phase, while the matrix is the γ phase. After a creep deformation at high temperature (figure 20.1(*b*)), the γ' phase grows in an anisotropic manner, resulting in plate-shaped precipitates which align perpendicular to the loading axis.

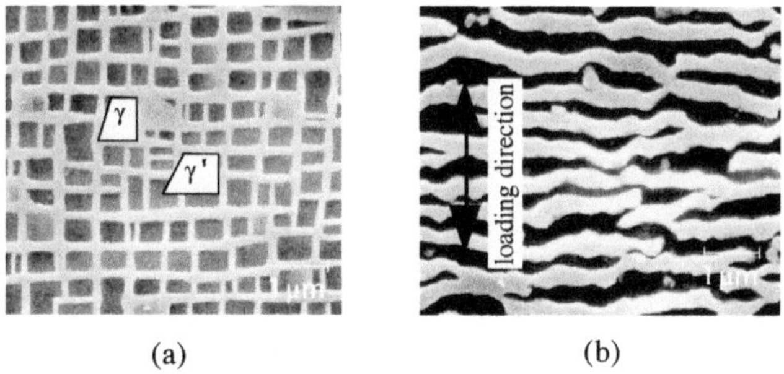

Figure 20.1. Microstructure of the nickel-based superalloy before (*a*) and after (*b*) creep deformation at 1213 K/343 MPa for 52 hours. The loading axis lies vertical in (*b*).

As is well known, the equilibrium concentration of chemical elements in an alloy is determined by points of common tangency to the free energy-concentration curves of each phase (figure 20.2(*a*)). Then, let us suppose that under an application of external or centrifugal force, free energy in the vertical channel of the γ phase is higher than that in the horizontal one (figure 20.2(*b*)). The equilibrium concentration is modified by the local fluctuation of the free energy. The flow of aluminium atoms from the horizontal to the vertical channel takes place because the aluminium concentration C_2 in the horizontal channel of the γ phase is larger than C_1 in the vertical channel. Similar atom flow in the γ' phase takes place since C_4 is greater than C_3. As a

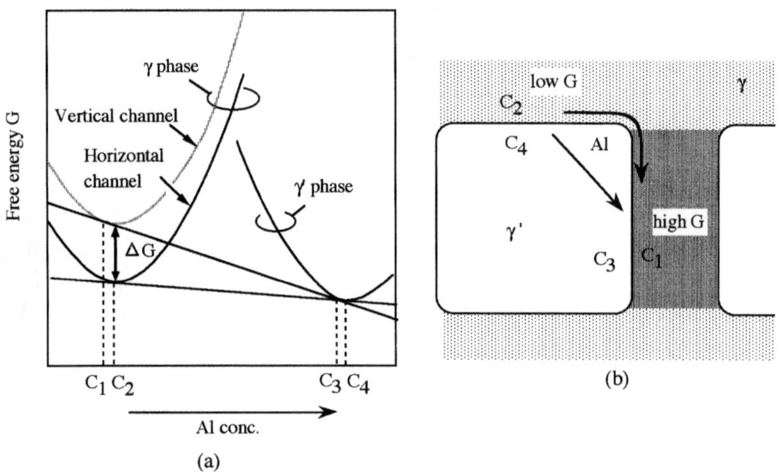

Figure 20.2. Thermodynamic mechanism for rafting. (*a*) Schematic illustration of free energy in the vertical and horizontal channels of the γ matrix and local equilibrium concentration C_1–C_4. (*b*) When the free energy G in the vertical channel of γ phase is higher than that in the horizontal channel, atomic flow takes place because of the shift of local equilibrium concentrations, and results in anisotropic growth of the γ' precipitate.

result of this atomic flow, the local equilibrium of atomic concentration near the γ/γ' interface is broken. To maintain the equilibrium, the γ' which is facing the horizontal channel will shrink, while the γ' facing the vertical channel will expand, thus changing the shape of the cuboidal precipitates into platelets. The importance of this proposed mechanism of rafting in real material systems depends on the magnitude of the difference in the free energy in the horizontal and vertical channels, and the ease of atomic diffusion in the γ and γ' phases.

Stress generation by lattice misfit, external loading and dislocation

The stresses which are generated by lattice misfit and external loading are now evaluated using a three-dimensional finite element technique. Deformation by the lattice misfit and by an application of external load are assumed to be elastic. The effect of the plastic relaxation during creep deformation is then taken into account by introducing dislocations in the narrow channels of the γ phase. Stresses generated by the dislocations in the γ channels are evaluated from elasticity theory. The total stress state can be evaluated by simply adding the stresses caused by these three effects.

Model γ/γ' system for numerical analysis

Figure 20.3 shows the geometry of a model γ/γ' system used for the subsequent numerical analysis. The γ' phase is given by square-shaped plates

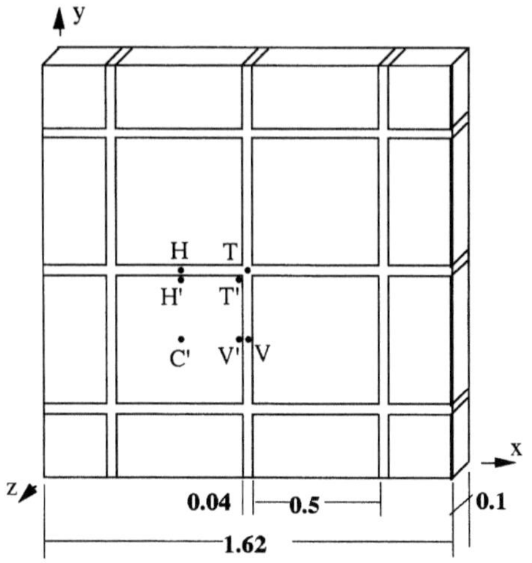

Figure 20.3. Geometry and dimensions of the model γ/γ'system used for the finite element analysis. Dimensions are in μm.

Table 20.1. Calculated elastic constants for γ with various aluminium concentrations (data 1–3) and published [6] constants for γ' (data 4). (All at room temperature).

		Elastic compliances, 10^{-11} Pa^{-1}		
No.	Material	s_{11}	s_{12}	s_{44}
1	γ ($f = 0$)	0.734	−0.274	0.802
2	γ ($f = 0.25$)	0.943	−0.348	1.479
3	γ ($f = 0.5$)	1.152	−0.421	2.156
4	γ'	0.954	−0.377	0.795

with a lateral dimension of 0.5 μm and thickness of 0.1 μm. The width of the γ channels is 0.04 μm and the volume fraction of the γ' phase is about 0.8. The model is divided into 1600 finite elements. The elements are compound type with eight nodes [4] and the software used for the finite element deformation analysis is the same as that used for crystal plasticity analyses [5].

Elastic compliances for the γ phase are determined in the following way. The γ phase is approximated as an alloy of nickel and aluminium and its compliances are evaluated by linear interpolation of compliances for nickel and aluminium:

$$s_{ij}^{\gamma} = f s_{ij}^{\mathrm{Al}} + (1 - f) s_{ij}^{\mathrm{Ni}}, \tag{1}$$

where f, s_{ij}^{Al} and s_{ij}^{Ni} denote aluminium concentration, and elastic compliances for aluminium and nickel, respectively, and the indices ij are 11, 12 or 44. For the compliance of γ', experimental data are used [6]. Table 20.1 shows the compliances for γ' and interpolation of data for nickel and aluminium [7] at room temperature with volume fractions $f = 0$, 0.25 and 0.5.

Stress because of lattice misfit

The magnitude of the lattice misfit is usually discussed in terms of the misfit parameter δ, which is defined as

$$\delta = 2(a^{\gamma'} - a^{\gamma})/(a^{\gamma'} + a^{\gamma}), \tag{2}$$

where a^{γ} and $a^{\gamma'}$ denote lattice constants for the γ and γ' phases, respectively. In this case, $\delta = 0.1\%$. The strain tensor $\{\varepsilon^*\} = \{-\delta, -\delta, -\delta, 0, 0, 0\}^T$ is given as an eigen strain for the elements which make up the γ phase. Figure 20.4 shows the numerical result for the distribution of the normal stress in the y direction, σ_{yy} for $f = 0$ along the line through the sites C', V' and V, as shown in figure 20.3. The stress is 170–180 MPa in the γ phase and 40–10 MPa in the γ' phase. Stresses obtained at the seven sites

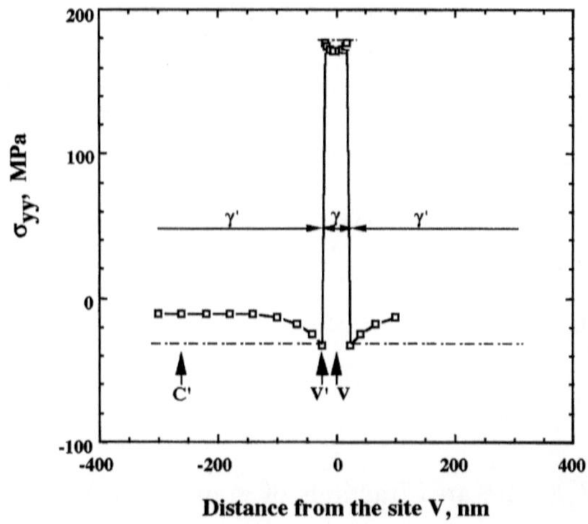

Figure 20.4. Distribution of the normal stress in the y direction caused by the elastic component of the lattice misfit of 0.1% in the specimen for $f = 0$. The results obtained by the finite element analysis are given by the solid lines, while those obtained by the approximate analytical expression are shown by the dashed lines.

shown in figure 20.3 for the model with $f = 0, 0.25$ and 0.5 are summarized in table 20.2. At site V, major components of the stress are σ_{yy} and σ_{zz}, while at site H the major components are σ_{xx} and σ_{zz}. The major stress components in the γ channels are the normal components whose directions are lying parallel to the γ/γ' interfaces.

Table 20.2. Finite element analysis results for the normal stress components at the seven sites shown in figure 20.3 for a lattice misfit of 0.1% between γ and γ' phases. Stresses are in MPa.

		Sites							
f		V	H	T	V'	H'	T'	C'	Remarks
0	σ_{xx}	−10.9	171.0	26.3	−12.3	−32.7	−46.0	−10.8	γ: 100%Ni
	σ_{yy}	171.0	−10.9	26.2	−32.7	−12.3	−46.0	−10.8	
	σ_{zz}	117.2	117.2	68.7	−60.2	−60.2	−90.9	−1.9	
0.25	σ_{xx}	−8.0	135.0	26.4	−9.0	−25.8	−40.5	−8.6	γ: Ni + 25%Al
	σ_{yy}	135.0	−8.0	26.4	25.8	−9.0	−40.5	−8.6	
	σ_{zz}	94.5	94.5	59.2	−48.4	−48.4	−78.9	−1.7	
0.5	σ_{xx}	−6.2	111.6	23.3	−7.0	−21.3	−36.9	−7.2	γ: Ni + 50%Al
	σ_{yy}	111.6	−6.2	23.3	−21.3	−7.0	−36.9	−7.2	
	σ_{zz}	79.3	79.3	50.9	−40.5	−40.5	−70.2	−1.5	

If it is assumed that the stress is constant in the γ' phase and also in the vertical and horizontal channels of the γ phase, approximate analytical expressions for the stresses in γ and γ' may be introduced. First consider the stresses and strains in the vertical channels of γ. On the vertical interface of γ and γ', the following conditions of strain continuity hold:

$$\varepsilon_{yy}^{\gamma'} + \delta = \varepsilon_{yy}^{\gamma}, \qquad (3a)$$

$$\varepsilon_{zz}^{\gamma'} + \delta = \varepsilon_{zz}^{\gamma}. \qquad (3b)$$

The stresses in γ and γ' balance since there is no external force:

$$A^{\gamma}\sigma_{yy}^{\gamma} + A^{\gamma'}\sigma_{yy}^{\gamma'} = 0, \qquad (4a)$$

$$A^{\gamma}\sigma_{zz}^{\gamma} + A^{\gamma'}\sigma_{zz}^{\gamma'} = 0, \qquad (4b)$$

where A^{γ} and $A^{\gamma'}$ are area fractions of γ and γ' in the cross-section. If the shape of the γ' phase is assumed to be cuboidal and surrounded by a γ phase of a uniform thickness,

$$A^{\gamma'} = (V_f^{\gamma'})^{2/3}, \qquad (5a)$$

$$A^{\gamma} = 1 - (V_f^{\gamma'})^{2/3}, \qquad (5b)$$

where $V_f^{\gamma'}$ denotes the volume fraction of the γ' phase in the alloy. Hooke's law holds for the stresses and strains:

$$\varepsilon_{yy}^{\gamma} = (\sigma_{yy}^{\gamma} - \nu^{\gamma} \cdot \sigma_{zz}^{g})/E^{\gamma}, \qquad (6a)$$

$$\varepsilon_{zz}^{\gamma} = (\sigma_{zz}^{\gamma} - \nu^{\gamma} \cdot \sigma_{yy}^{\gamma})/E^{\gamma}, \qquad (6b)$$

$$\varepsilon_{yy}^{\gamma'} = (\sigma_{yy}^{\gamma'} - \nu^{\gamma'} \cdot \sigma_{zz}^{\gamma'})/E^{\gamma'}, \qquad (6c)$$

$$\varepsilon_{zz}^{\gamma'} = (\sigma_{zz}^{\gamma'} - \nu^{\gamma'} \cdot \sigma_{yy}^{\gamma'})/E^{\gamma'}. \qquad (6d)$$

Solutions for equations (3)–(6) are

$$\sigma_{yy}^{\gamma} = \sigma_{zz}^{\gamma} = \beta\delta, \qquad (7a)$$

$$\sigma_{yy}^{\gamma'} = \sigma_{zz}^{\gamma'} = -\frac{1-(V_f^{\gamma'})^{2/3}}{(V_f^{\gamma'})^{2/3}}\beta\delta, \qquad (7b)$$

$$\beta = \frac{(V_f^{\gamma'})^{2/3}\dfrac{E^{\gamma}}{1-\nu^{\gamma}}\dfrac{E^{\gamma'}}{1-\nu^{\gamma'}}}{[1-(V_f^{\gamma'})^{2/3}]\dfrac{E^{\gamma}}{1-\nu^{\gamma}} + (V_f^{\gamma'})^{2/3}\dfrac{E^{\gamma'}}{1-\nu^{\gamma'}}}, \qquad (7c)$$

where E^{γ}, $E^{\gamma'}$, ν^{γ} and $\nu^{\gamma'}$ are Young's moduli and Poisson's ratios for γ and γ', respectively and may be approximated by $1/s_{11}^{\gamma}$, $1/s_{11}^{\gamma'}$, $-s_{12}^{\gamma}/s_{11}^{\gamma}$ and $-s_{12}^{\gamma'}/s_{11}^{\gamma'}$, respectively. The stresses obtained from (7) are indicated in figure 20.4 by the dashed lines. The finite element analysis results and the approximate results show good agreement. Applying a similar discussion to the stress state in the horizontal channels of γ, the following equation is obtained:

$$\sigma_{xx}^{\gamma} = \sigma_{zz}^{\gamma} = \beta\delta. \tag{8}$$

Stress because of external forces

Now elastic stress caused by external or centrifugal forces is considered. It assumed that uniform tensile displacement of 0.0162 µm in the y direction is imposed on the upper surface of the specimen while the bottom surface is fixed, which results in an average strain in the loading direction of 0.1%. The model includes no lattice misfit in this analysis. Distribution of the normal stress component σ_{yy} in the specimen with $f = 0$ is shown in figure 20.5. The stress in the vertical channel of the γ phase is about 30 MPa larger compared with that in the γ' phase. Table 20.3 summarizes the normal component of stresses and strains at the seven sites indicated in figure 20.3 for $f = 0$, 0.25 and 0.5: (1) σ_{yy} changes with the aluminium concentration f at sites V and T, while it does not change significantly at the other sites; (2) σ_{xx} and σ_{zz} are 0–17% of the average stress $\bar{\sigma}$ at sites H and T; (3) ε_{yy} is nearly equal to the average strain at all sites except H where the strain changes with f; and (4) variations in ε_{xx} and ε_{zz} are negligible.

On the basis of the results obtained above, the Reuss hypothesis for site H states that the local stress state is equal to the average:

$$\sigma_{yy}^{\gamma} = \bar{\sigma}, \qquad \text{at site } H. \tag{9}$$

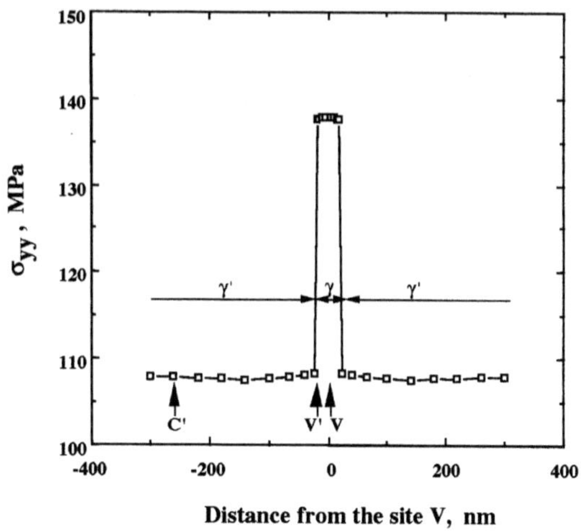

Figure 20.5. Finite element analysis of the stress distribution due to an external force. The elastic constants obtained for $f = 0$ are used and the average stress, σ, in the loading direction is 110 MPa.

Table 20.3. The finite element analysis results of normal components of stress and strain at seven sites in the specimen to which a uniaxial tensile displacement of 0.1% is given in y direction. The stresses are normalized by the average stress $\bar{\sigma}$ and denoted as $\tilde{\sigma}_{xx}$, $\tilde{\sigma}_{yy}$ and $\tilde{\sigma}_{zz}$. Strains have units of 10^{-3}.

		Sites							
f		V	H	T	V'	H'	T'	C'	Remarks
0	$\tilde{\sigma}_{xx}$	0.014	−0.170	−0.145	0.014	0.003	0.008	0.008	γ: 100%Ni
	$\tilde{\sigma}_{yy}$	1.253	0.980	1.160	0.985	0.982	0.991	0.981	$\bar{\sigma} = 110.0$ MPa
	$\tilde{\sigma}_{zz}$	−0.010	−0.127	−0.070	0.018	0.054	0.040	0.001	
	$\tilde{\varepsilon}_{xx}$	−0.36	−0.39	−0.45	−0.40	−0.39	−0.42	−0.40	
	$\tilde{\varepsilon}_{yy}$	1.01	0.88	1.00	1.01	0.98	1.01	1.02	
	$\tilde{\varepsilon}_{zz}$	−0.39	−0.35	−0.36	−0.39	−0.37	−0.37	−0.41	
0.25	$\tilde{\sigma}_{xx}$	0.003	−0.040	−0.016	0.003	0.008	0.008	0.002	γ: Ni + 25%Al
	$\tilde{\sigma}_{yy}$	0.994	1.000	1.008	1.004	1.000	1.007	1.000	$\bar{\sigma} = 105.9$ MPa
	$\tilde{\sigma}_{zz}$	−0.019	−0.028	−0.017	0.010	0.014	0.019	0.000	
	$\tilde{\varepsilon}_{xx}$	−0.36	−0.40	−0.38	−0.40	−0.40	−0.40	−0.40	
	$\tilde{\varepsilon}_{yy}$	1.00	1.02	1.02	1.00	0.99	1.00	1.00	
	$\tilde{\varepsilon}_{zz}$	−0.39	−0.38	−0.38	−0.39	−0.39	−0.39	−0.40	
0.5	$\tilde{\sigma}_{xx}$	−0.005	0.005	0.064	−0.005	−0.011	0.009	−0.001	γ: Ni + 50%Al
	$\tilde{\sigma}_{yy}$	0.823	1.013	0.906	1.017	1.013	1.022	1.013	$\bar{\sigma} = 102.9$ MPa
	$\tilde{\sigma}_{zz}$	−0.022	0.039	0.017	0.006	−0.016	0.003	0.000	
	$\tilde{\varepsilon}_{xx}$	−0.35	−0.40	−0.32	−0.40	−0.40	−0.39	−0.39	
	$\tilde{\varepsilon}_{yy}$	0.99	1.16	1.04	0.99	0.99	0.99	0.99	
	$\tilde{\varepsilon}_{zz}$	−0.38	−0.41	−0.40	−0.39	−0.40	−0.40	−0.39	

Contrary to this, the strain ε_{yy} at the sites C', V' and V is approximately equal to the average strain. Then, applying the Voigt hypothesis to these points, i.e. that the strain is a constant and equal to the average strain, an expression for the stress state at site V can be derived. In approximating the uniaxial stress condition, the condition of uniform strain on the cross section through sites C', V' and V is written:

$$\frac{\sigma_{yy}^{\gamma}}{E^{\gamma}} = \frac{\sigma_{yy}^{\gamma'}}{E^{\gamma'}}, \qquad \text{at sites } C', V' \text{ and } V, \tag{10}$$

where σ_{yy}^{γ}, E^{γ}, $\sigma_{yy}^{\gamma'}$ and $E^{\gamma'}$ denote the tensile stress and the Young's modulus in the loading direction in γ and γ', respectively. Using σ_{yy}^{γ} and $\sigma_{yy}^{\gamma'}$, the average stress is given by:

$$\bar{\sigma} = A^{\gamma}\sigma_{yy}^{\gamma} + A^{\gamma'}\sigma_{yy}^{\gamma'}. \tag{11}$$

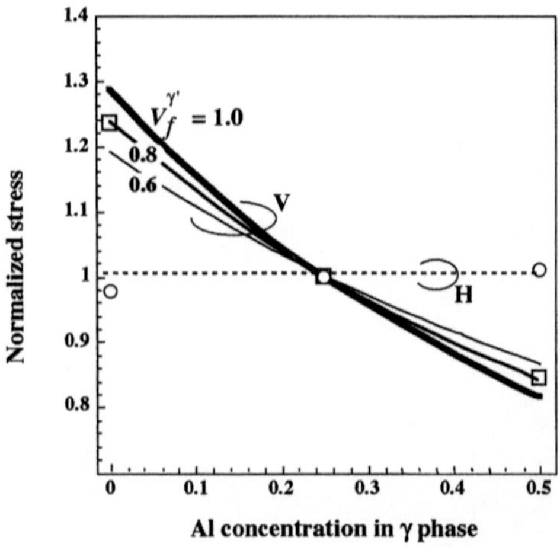

Figure 20.6. Normalized by the stresses at sites V and H in the specimen which is subjected to a uniaxial load. Stresses are plotted as a function of the aluminium concentration in the γ phase. Approximate analytical results are given by solid and dashed lines while the finite element result is given by □ for site V and ○ for site H.

By equations (5) and (9)–(11), the stress in the vertical channels of the γ phase can be obtained:

$$\sigma_{yy}^{\gamma} = \alpha \cdot \bar{\sigma}, \qquad \text{at site } V, \tag{12a}$$

$$\alpha = \frac{E^{\gamma}}{[1-(V_f^{\gamma'})^{2/3}]E^{\gamma} + (V_f^{\gamma'})^{2/3}E^{\gamma'}}. \tag{12b}$$

The normalized stress, which is defined as the stress divided by $\bar{\sigma}$, is equal to α in the vertical channel of γ. In the horizontal channel of γ, the stress is approximated as $\bar{\sigma}$ and then the normalized value in the approximation is 1. The effect of aluminium concentration in the γ phase and the volume fraction $V^{\gamma'}$ on the normalized stress is shown in figure 20.6. Equation (12) for $V_f^{\gamma'} = 0.8$ provides excellent agreement with the finite element analysis.

Stress because of dislocations

Creep deformation accompanies movement of dislocations, and the dislocations accumulated in the microstructure generate stresses, which should be added to the stresses caused by the lattice misfit and the external force. Accumulation of dislocations is evaluated by assuming a distribution of plastic distortion in the γ channels. Assuming that during the initial stage of creep deformation the γ' phase remains elastic, plastic slip occurs in the γ channels and the maximum plastic distortion is given by $f_{\max}$. Figure 20.7 shows a simplified profile of the plastic distortion under these constraints.

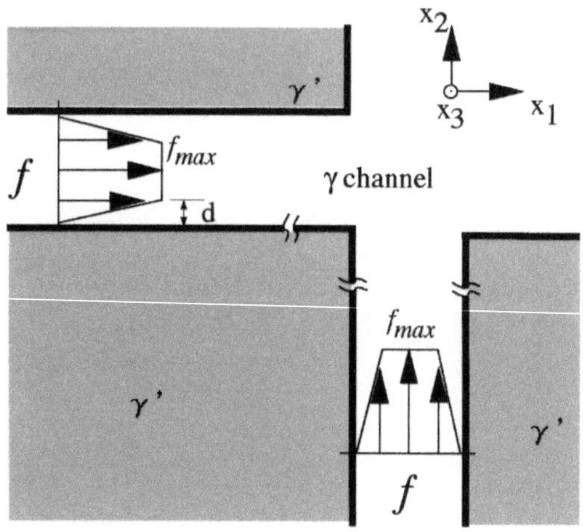

Figure 20.7. Assumed distribution of the plastic distortion in the γ channels, represented by a piecewise linear function. The plastic distortion is uniform in the direction parallel to the γ/γ' interface, while it is zero on the interface and increases linearly in the transition region of thickness *d*.

The profile is given by a piecewise linear function with a transition region of thickness d at the γ/γ' interface and a plateau of the height $f_{\max}$ in the central region of the channels. Variations in the plastic distortion in the direction parallel to the interface near the crossing area of the horizontal and vertical channels are neglected.

To evaluate the density and character of dislocations, we utilize the concept of the geometrically necessary dislocations [8]. The components of the plastic distortion β^*_{ij} at a point in the horizontal and vertical γ channels and far away from the edge of γ' assumed to be given as follows:

$$\beta^*_{ij} = \begin{pmatrix} -f/2 & 0 & 0 \\ 0 & f & 0 \\ 0 & 0 & -f/2 \end{pmatrix}, \tag{13}$$

where f is a scalar function given as a piecewise linear function and illustrated in figure 20.7. Physically, equation (13) implies volumetric change does not take place by plastic distortion. The density of the then geometrically necessary dislocations which accompany the gradient of plastic distortion is [9]

$$\alpha_{hi} = -\varepsilon_{hlj}\beta^*_{ji,l}, \tag{14}$$

where α_{hi} gives the sum of the Burgers vectors of the dislocations whose Burgers vector and line direction are directed parallel to h and i axes $(h, i = x, y, z)$, respectively. $\beta^*_{ji,l}$ is the gradient of β^*_{ji} in the l direction and ε_{hlj} denotes the permutation tensor. Since the function f is piecewise linear, the gradient of f and the dislocation density are constants in the transition region near the γ/γ' interfaces and zero in the remaining part of the channel. In other words, dislocations lie close to the γ/γ' interfaces. Applying equations (14) and (13) and referring to figure 20.7, the dislocation density in the horizontal γ channels is

$$\alpha_{ij} = \begin{pmatrix} 0 & 0 & f_{\max}/2d \\ 0 & 0 & 0 \\ -f_{\max}/2d & 0 & 0 \end{pmatrix}. \tag{15}$$

The non-zero components of the dislocation density tensors, α_{13} and α_{31}, correspond to the edge dislocations lying in the x_1 direction with the Burgers vector in the x_3 direction, and dislocations lying in the x_3 direction with the Burgers vector in the x_1 direction, respectively. In the γ channels perpendicular to the x_1 direction, the gradient of the plastic distortion in the x_1 direction is non-zero and the dislocation density tensor is

$$\alpha_{ij} = \begin{pmatrix} 0 & 0 & 0 \\ 0 & 0 & -f_{\max}/2d \\ 0 & -f_{\max}/d & 0 \end{pmatrix}. \tag{16}$$

In these γ channels, there are edge dislocations lying in the x_2 direction with Burgers vector in the x_3 direction and edge dislocations lying in the x_3 direction with Burgers vector in the x_2 direction. A similar calculation for the dislocation densities in the γ channels perpendicular to the x_3 axis, generates edge dislocations lying in the x_2 direction with Burgers vector in the x_1 direction and edge dislocations lying in the x_1 direction with Burgers vector in the x_2 direction.

Denoting the unit vectors parallel to the dislocation line and the Burgers vector as $\vec{\xi}$ and $\vec{b}$ respectively, the slip plane normal is $\vec{\xi} \otimes \vec{b}$ and the extra half planes of edge dislocations with a density >0 are on the positive side of $\vec{\xi} \otimes \vec{b}$. The distribution of edge dislocations obtained from equations (15) and (16) in the γ channels perpendicular to the x_3 axis is shown in figure 20.8. The dislocation density in the γ channels parallel to the loading axis and lying perpendicular to the loading axis, given by the α_{32} component, is double that of other dislocations.

Figure 20.9 shows a transmission electron microscope image of dislocations which are lying in the horizontal channels. The direction of the dislocation lines is parallel to $\langle 110 \rangle$, making an angle of $\pm\pi/4$ to the x axis, because they are formed as a result of crystallographic slip on $\{111\}$ slip planes [10]. For the same reason, dislocation lines in the vertical channels of γ phase make an angle of $\pm\pi/4$ to the loading angle. Then, components of the dislocation density tensors given in equations (15) and (16) do not represent the characters of real dislocations. This discrepancy arises because the expression for the components of the dislocation density tensor is dependent on the coordinate system. On the other hand, the dislocation tensor in the

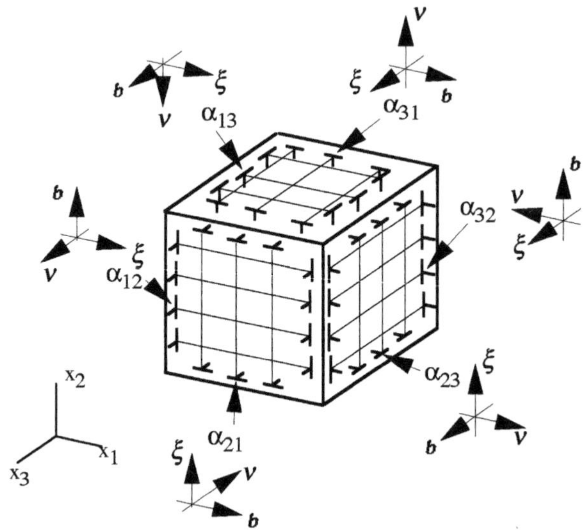

Figure 20.8. A schematic illustration of the dislocations in the γ channels, which are geometrically necessary to accommodate the spatial gradient of plastic strain.

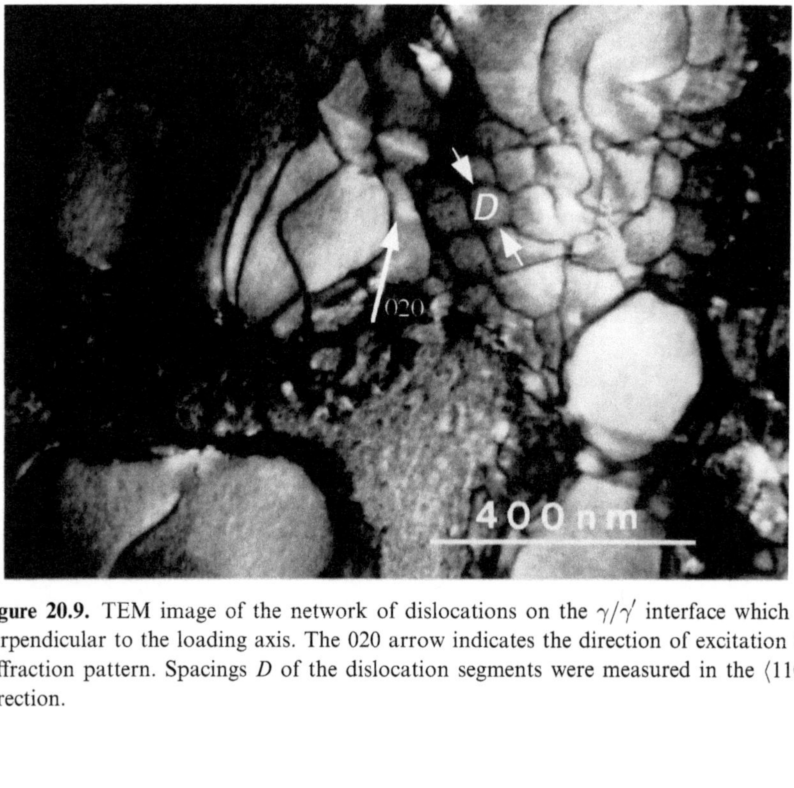

Figure 20.9. TEM image of the network of dislocations on the γ/γ' interface which is perpendicular to the loading axis. The 020 arrow indicates the direction of excitation in diffraction pattern. Spacings D of the dislocation segments were measured in the $\langle 110 \rangle$ direction.

horizontal channels and given in equation (15) is antisymmetric. An antisymmetric tensor remains antisymmetric under any coordinate transformation. Introducing in the horizontal channels of the γ phase, a new coordinate system which is rotated $\pi/4$ around the y axis, the dislocation density components under the new coordinate system are again pure edge in character. This conclusion is in accordance with the discussion made from the viewpoint of glide dislocations and their reaction [11]. However, in the vertical channels, the components of the dislocation density tensor are neither symmetric nor antisymmetric. Introducing a coordinate system whose coordinate axes lies along the dislocation lines, the dislocations have edge and screw characters. Real dislocations in the vertical channels of the γ phase have a mixed character of edge and screw components [12]. Details of the dislocation character in the vertical channels of the γ phase are points for further research and discussion.

To calculate the internal stresses introduced by the dislocations, it is convenient to use the expressions given in equations (15) and (16), and to count the number of dislocations in the γ channels. The density of edge dislocations in the γ channels perpendicular to the loading axis and lying in the x_3 direction with Burgers vector in the x_1 direction is $-f_{\max}/2d$. The number N_{31} of dislocations which penetrate an area of width 1 in the x_1 direction and height d in the x_2 direction is

$$N_{31} = \alpha_{31} \cdot 1 \cdot d/\tilde{b} = -f_{\max}/2\tilde{b}, \tag{17}$$

where $\tilde{b}$ is the magnitude of the Burgers vector of the dislocations. At the limit of $d \to 0$, all the dislocations are placed on the γ/γ' interface and their mean spacing D_{31} is

$$D_{31} = 1/N_{31} = -2\tilde{b}/f_{\max}. \tag{18a}$$

Similar calculations for the mean spacings D_{13}, D_{23}, and D_{32}, for dislocations of α_{13}, α_{23}, and α_{32} give

$$D_{13} = 2\tilde{b}/f_{\max}, \tag{18b}$$

$$D_{23} = -2\tilde{b}/f_{\max}, \tag{18c}$$

$$D_{32} = \tilde{b}/f_{\max}. \tag{18d}$$

As shown in figure 20.9, the measured distance of dislocation segments ranges from 50 to 150 nm, while published experimental results (for example, figure 20.19 [13]) also indicate dislocation spacings close to these values. Substitution of these values, together with a typical value for $\tilde{b}$ as 0.25 nm, into equation (18a) gives the range of the maximum plastic distortion as $0.3\% < f_{\max} < 1\%$, which seems to be reasonable.

Assuming that all the dislocations are on the γ/γ' interface, the stress field generated by the dislocations may be estimated. First, straight and parallel edge dislocations in a periodic array between $-a < x < a$ (figure 20.10) are

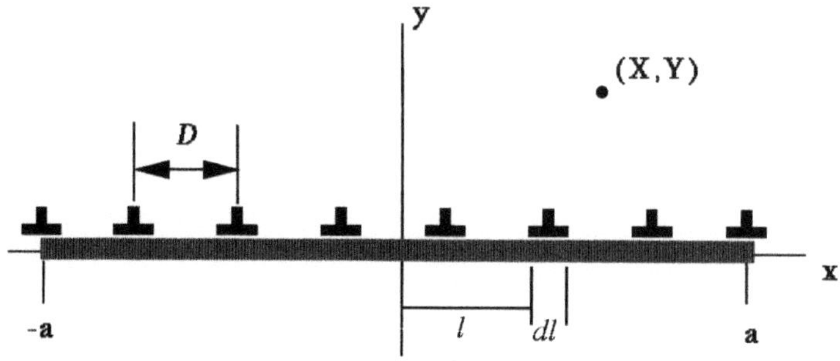

Figure 20.10. Calculation of stresses generated by an array of edge dislocations on the γ/γ' interface.

assumed. To simplify the calculation of the stresses made by this array of dislocations, it is further assumed that infinitesimal and continuously distributed dislocations replace the finite number of discrete dislocations with finite magnitudes of the Burgers vector. Denoting the mean distance of discrete dislocations as D and the magnitude of their Burgers vector as $\tilde{b}$, the magnitude of the Burgers vector divided by a unit length of the dislocation distribution is

$$B = \tilde{b}/D, \tag{19}$$

and the periodic array of discrete dislocations is replaced by the continuous distribution of infinitesimal dislocations with line density of B. The normal stress in the x direction at a point (X, Y) in the section dl is

$$\mathrm{d}\sigma_{xx} = -K\,\mathrm{d}l\,\frac{Y\{3(X-l)^2 + Y^2\}}{\{(X-l)^2 + Y^2\}^2}, \tag{20}$$

where

$$K = \frac{\mu B}{2\pi(1-\nu)}, \tag{21}$$

and μ and ν denote the elastic shear modulus and Poisson's ratio, respectively. The stress generated by the dislocations in $-a < x < a$ is

$$\begin{aligned} \sigma_{xx} &= \int_{-a}^{a} \mathrm{d}\sigma_{xx} \\ &= KY\left(\frac{X+a}{(X+a)^2 + Y^2} - \frac{X-a}{(X-a)^2 + Y^2}\right) \\ &\quad - 2K\,\mathrm{sgn}(Y)\left(\arctan\frac{X+a}{|Y|} - \arctan\frac{X-a}{|Y|}\right), \end{aligned} \tag{22}$$

where $\mathrm{sgn}(Y) = 1$ when $Y > 0$ and $\mathrm{sgn}(Y) = -1$ when $Y < 0$. Using a similar procedure to that for equations (20)–(22),

$$\sigma_{yy} = KY\left(\frac{X+a}{(X+a)^2+Y^2} - \frac{X-a}{(X-a)^2+Y^2}\right), \tag{23}$$

$$\sigma_{zz} = -2\nu K\,\mathrm{sgn}(Y)\left(\arctan\frac{X+a}{|Y|} - \arctan\frac{X-a}{|Y|}\right), \tag{24}$$

$$\sigma_{xy} = \sigma_{yz} = \sigma_{zx} = 0. \tag{25}$$

In the vicinity of the γ/γ' interface, $Y \to 0$ and

$$\sigma_{xx} \approx -\frac{\mu}{1-\nu}\frac{\tilde{b}}{D}\,\mathrm{sgn}(Y), \tag{26}$$

$$\sigma_{yy} \approx 0, \tag{27}$$

$$\sigma_{zz} \approx -\nu\frac{\mu}{1-\nu}\frac{\tilde{b}}{D}\,\mathrm{sgn}(Y), \tag{28}$$

$$\sigma_{xy} = \sigma_{yz} = \sigma_{zx} = 0. \tag{25}$$

Using the expressions obtained above, the internal stress field in the γ channels generated by the dislocations can be calculated. In the γ channels perpendicular to the loading axis, the contribution from α_{31} is

$$\sigma_{xx} \approx -\frac{\mu}{1-\nu}\frac{\tilde{b}}{D_{31}} = \frac{\mu}{2(1-\nu)}f_{\mathrm{max}}, \tag{29}$$

$$\sigma_{zz} \approx -\nu\frac{\mu}{1-\nu}\frac{\tilde{b}}{D_{31}} = \frac{\nu\mu}{2(1-\nu)}f_{\mathrm{max}}. \tag{30}$$

The contribution from α_{13} is

$$\sigma_{xx} \approx -\nu\frac{\mu}{1-\nu}\frac{\tilde{b}}{D_{13}}(-1) = \frac{\nu\mu}{2(1-\nu)}f_{\mathrm{max}}, \tag{31}$$

$$\sigma_{x} \approx -\frac{\mu}{1-\nu}\frac{\tilde{b}}{D_{13}}(-1) = \frac{\mu}{2(1-\nu)}f_{\mathrm{max}}. \tag{32}$$

The coefficient -1 enters here since the γ phase is on the negative side of the normal direction of the slip plane. Using equations (29)–(32), the total contribution from α_{31} and α_{13} is

$$\sigma_{xx} = \sigma_{zz} \approx \frac{(1+\nu)}{2(1-\nu)}\mu f_{\mathrm{max}} = (1+\nu)\Gamma, \tag{33}$$

where

$$\Gamma = \frac{1}{2(1-\nu)}\mu f_{\mathrm{max}}. \tag{34}$$

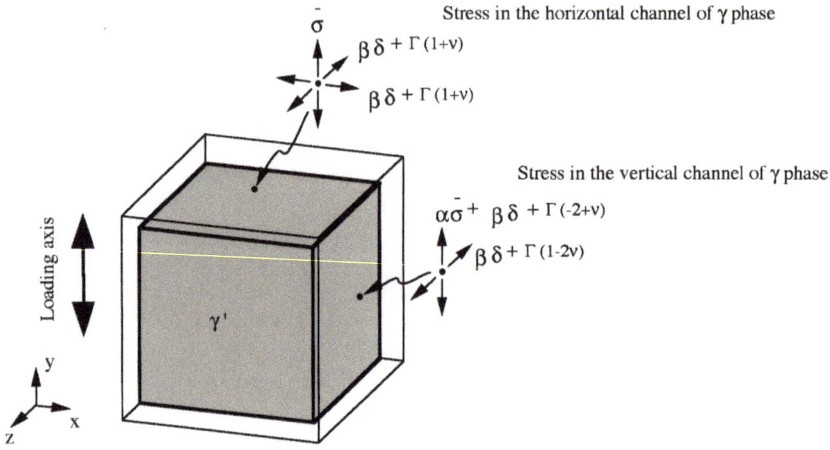

Figure 20.11. A γ' precipitate surrounded with horizontal and vertical channels of the γ phase. Stresses in the horizontal and vertical channels of the γ phase consist of terms from the external or centrifugal force with coefficient α, lattice misfit with coefficient β, and dislocation coefficient Γ. Definitions for α, β, Γ and δ are given in equations (12*b*), (7*c*), (34) and (2) respectively.

That is, the internal stress field resulting from glide dislocations in the horizontal γ channels perpendicular to the loading axis is a bi-axial tensile stress. The two stress components are equal in magnitude and lying parallel to the interface. Similarly for the γ channels perpendicular to the x axis,

$$\sigma_{yy} \approx (-2+\nu)\Gamma, \tag{35a}$$

$$\sigma_{zz} \approx (1-2\nu)\Gamma. \tag{35b}$$

In these channels there is again a bi-axial stress field but with compressive stress in the loading direction. The stresses in the γ channels perpendicular to the z axis are calculated similarly and obtained as equation (35*a*) and

$$\sigma_{xx} \approx (1-2\nu)\Gamma. \tag{36}$$

The stresses in the γ channels obtained in equations (33)–(36) are shown in figure 20.11, together with the elastic stresses caused by the centrifugal force and lattice misfits.

Rate of rafting

In order to evaluate the magnitude of stresses with some typical data for materials and loading conditions, the following values are assumed:

$$
\begin{aligned}
\bar{\sigma} &= 100\,\text{MPa} \\
\delta &= -0.2\% \\
\tilde{b} &= 2.5 \times 10^{-10}\,\text{m} \\
E^{\gamma} &= \frac{1}{1.465} \times 10^{11}\,\text{MPa} \\
E^{\gamma'} &= \frac{1}{1.284} \times 10^{11}\,\text{MPa} \\
\mu &= \frac{1}{1.395} \times 10^{11}\,\text{MPa} \\
\nu^{\gamma} &= \nu^{\gamma'} = 0.35 \\
f_{\text{max}} &= 0.5\%.
\end{aligned}
\tag{37}
$$

Applying equations (37) to the equations for the stresses shown in figure 20.11, the contributions from the external force, lattice misfit and dislocations and the total are obtained, with the results shown in table 20.4.

The calculated strain energy density from the total stresses, given in table 20.4, are approximately 0.5 and 2.6 MJ/m^3 in the horizontal and vertical γ channels, respectively and the difference in the strain energy density is about 2.1 MJ/m^3. In the actual slip process, the movement of dislocations is governed by the elastic stress field which is not the same in the horizontal and vertical γ channels. Then, different amounts of plastic slip and dislocation accumulation will take place in the horizontal and vertical γ channels. If it is assumed that there is 1.5 times wider spacing between dislocations in the vertical channels compared with that in the horizontal channels [14], the energy difference between channels is about 1 MJ/m^3.

After a change of the free energy distribution in the γ/γ' system, local concentrations of aluminium will change to develop local equilibrium. A

Table 20.4. Estimated stresses in the γ channels perpendicular to the loading axis (denoted as site *H*) and the *x* axis (denoted as site *V*). Stresses in MPa.

		Contribution from			
Site	Stress	External force	Lattice misfit	Dislocations	Total
H	σ_{xx}	0	–209	374	165
	σ_{yy}	100	0	0	100
	σ_{zz}	0	–209	374	165
V	σ_{xx}	0	0	0	0
	σ_{yy}	87	–209	–457	–579
	σ_{zz}	0	–209	87	–122

difference $\Delta C \equiv C_1 - C_2$ is developed between the local concentrations in the horizontal and vertical channels of the γ phase (figure 20.2). The difference in aluminium concentration corresponding to the free energy gap is evaluated by the thermodynamics software system Thermo-Calc [15] and its data base [16]. For the calculation, an Ni-Al binary system at 1273 K with the regular solution approximation for the γ phase and the sub-lattice model for γ' phase is assumed. The aluminium concentration difference which corresponds to the energy gap of 1 MJ/m^3 is determined to be on the order of 10^{-4}.

Assuming that this difference is maintained during the rafting, the growth rate of the γ' phase can be estimated as follows. Consider the diffusion of aluminium in the γ/γ' system of figure 20.2(*b*) where the aluminium concentrations at the horizontal and the vertical channels of γ phase differ by ΔC. According to Fick's first law, the flux J of aluminium from the horizontal to the vertical channel is

$$J = -D \frac{V_m \Delta C}{d} \qquad (\text{mol/m}^2\,\text{s}), \tag{38}$$

where D, d and V_m denote diffusion coefficient, diffusion distance and number of moles per unit volume in the γ' phase, respectively. This atomic movement causes an advance of the γ/γ' interface at the vertical channel to the γ side. At the same time, the horizontal channel of the γ phase loses aluminium and local equilibrium will be destroyed. To maintain the local equilibrium, a γ' phase region near the horizontal channel changes into γ phase and emits aluminium atoms into the horizontal channel, thus causing a retreat of the interface to the γ' side.

Experimental studies of the early stage of rafting have shown the formation of humps at the edges of γ' particles, an increase of aluminium and tantalum or decrease of chromium and molybdenum in the vertical channels of γ [17], and an increase of aluminium concentration at the γ' side near the interface which is perpendicular to the loading axis [18]. These facts seem to demonstrate atom flows between the vertical and horizontal channels of the γ phase and a dynamic aspect of concentration profiles during the movement of the γ/γ' interface.

To estimate the migration rate R_G of the interface, suppose that the vertical γ channels receive aluminium atoms from the horizontal γ channels and change into the γ' phase when the aluminium concentration reaches a nickel : aluminium ratio of 3 : 1. If we suppose C_1 and C_2 to be negligibly small compared with C_3 or C_4, then

$$R_G = -\frac{J}{\frac{1}{4} V_m}. \tag{39}$$

Using equation (38), equation (39) becomes

$$R_G = 4D \frac{\Delta C}{d}. \tag{40}$$

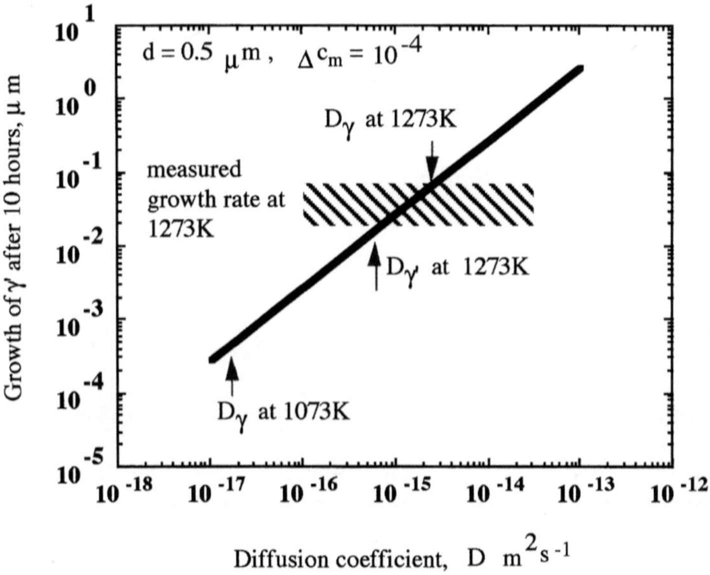

Figure 20.12. Estimated rafting rate as a function of diffusion coefficient and the actual rate at 1273 K. Published data for diffusion coefficients [19] are indicated by arrows.

Figure 20.12 shows the growth rate R_c as a function of the diffusion coefficient D when $\Delta c = 10^{-4}$ and the diffusion distance $d = 0.5\,\mu m$, which is a typical value for γ' precipitates in Ni-based superalloys. The magnitude of the measured growth rate at 1273 K, which is indicated in the same figure, is a rough estimate made from the fact that a γ channel with a thickness of about 0.1 μm disappears in a few tens of hours. Published data [19] for the diffusion coefficients for γ and γ' at 1273 K and 1073 K are also indicated by arrows.

This result shows that an estimated energy difference of 1 MJ/m^3 causes rafting at a realistic rate. To describe the plastic deformation and accumulation of strain energy in the γ channels under a non-uniform distribution of elastic stresses, not only experimental but also analytical or numerical studies [12, 20] are needed.

Summary

A mechanism for rafting in Ni superalloys containing γ' precipitates has been proposed and elastic deformation described by a finite element analysis and analytical expressions for the elastic stress distribution. The accumulation of dislocations in γ channels during rafting may be described by geometrically necessary dislocations which accompany gradients of plastic distortion, and lead to the generation of internal stresses. The total elastic strain energy during service is the sum due to externally applied, lattice misfit and dislocation induced stresses, and can account for the local shift of the equilibrium concentration of component atoms. The rate of rafting can be estimated by consideration of the diffusion of component atoms and agrees reasonably well with observed rafting rates.

References

[1] Tien J K and Copley S M 1971 *Metall. Trans.* **2** 215
[2] Tien J K and Copley S M 1971 *Metall. Trans.* **2** 543
[3] Nathal M V and Ebert L 1983 *J. Scripta Metall.* **17** 1151
[4] Zienkiewicz O C 1977 *The Finite Element Method* 3rd ed (London: McGraw-Hill) p 141
[5] Ohashi T 1994 *Phil. Mag.* **A70** 793; Ohashi T 1987 *Trans. JIM* **28** 906
[6] Kayser F X and Stassis C 1981 *Phys. Stat. Sol.* **A64** 335
[7] Hirth J P and Lothe J 1982 *Theory of Dislocations* 2nd ed (New York: Wiley) p 835
[8] Ashby M F 1970 *Phil. Mag.* **21** 399
[9] T Mura 1982 *Micromechanics of Defects in Solids* (The Hague: Martinus Nijhoff) p 45
[10] Scheunemann-Frerker *et al* 1993 *Phil. Mag.*
[11] Ohashi T, Hidaka K and Saito M 1997 *Mat. Sci. Eng.* **A238** 42
[12] Ohashi T and Busso E P 1997 'Physics and mechanics of finite plastic and viscoplastic deformation' *Proc. Plasticity 97* (Fulton, MD: Juneau Neat) p 269
[13] Pollock T M and Argon A S 1992 *Acta Metall. Mater.* **40** 1
[14] Feller-Kniepmeier M and Link T 1989 *Met. Trans.* **20A** 1233
[15] Thermo-Calc Division of Physical Metallurgy KTH, S-10044 Stockholm, Sweden
[16] Ishida K, Ohtani H and Oikawa K private communication
[17] Svetlov I L, Golovko B A, Epishin A I and Abalakin N P 1992 *Scripta Metall. Mater.* **26** 1353
[18] Saito M, Aoyama T, Hidaka H, Tamaki H, Ohashi T, Nakamura S and Suzuki T 1996 *Scripta Metall. Mater.* **34** 1189
[19] Watanabe M, Horita Z, Sano T and Nemoto M 1994 *Acta Metall. Mater.* **42** 3389
[20] Meissonnier F T, Busso E P and Odowd N P private communication

Chapter 21

Effects of tantalum and rhenium on creep in single crystals of nickel–20% chromium

Takashi Matsuo, Takayuki Kashiwa, Masanori Nijyo and Yoshihiro Terada

Introduction

It has been shown that enhancement of creep resistance in advanced single crystal superalloys is achieved by increasing the amount of tantalum and rhenium [1, 2]. Tantalum dissolves preferentially in γ' (Ni_3Al) and acts as a solid solution strengthener of the γ' precipitates, while rhenium dissolves preferentially in the γ matrix, and acts as a solid solution strengthener of the γ matrix [3, 4]. In a γ single phase Ni-20Cr polycrystalline alloy, it has been shown further that tantalum acts as the most effective solid solution strengthener [5, 6]. Solid solution strengthening effects are divided into two features: tantalum decreases the creep rate, and suppresses the onset of accelerating creep [6]. In comparison with reports concerning the effect of tantalum additions, reports of rhenium additions are somewhat limited. Because the effect of solute additions on the creep rate and the onset of accelerating creep becomes more prominent at the lower stresses, creep tests must be performed at lower stresses [6].

In this chapter, the effect of tantalum additions on the creep rate–time curve in a single phase Ni-20Cr single crystal is investigated over a wide stress range. Because a single crystal containing rhenium could not be obtained the effect of solute rhenium additions on creep rate–time curves in Ni-20Cr is investigated using unidirectionally solidified specimens (designated UDS hereafter). The results on the UDS specimens are compared with those of polycrystalline specimens.

Model nickel-based alloys for creep studies

Ni-20Cr, Ni-20Cr-3at%Ta and Ni-20Cr-3at%Re alloys were melted in a vacuum induction furnace and cast to 4 kg ingots in argon, and then hot forged into bars with a diameter of 13 mm. Using the bars, single crystals or UDS samples were produced in a modified Bridgman furnace. Polycrystalline specimens in the three alloys were also prepared, and had an average grain diameter of 200 μm. Constant stress tensile creep tests were conducted at 1173 K in the stress range from 19.4 to 98 MPa. Creep strain was automatically recorded through a differential transformer attached to

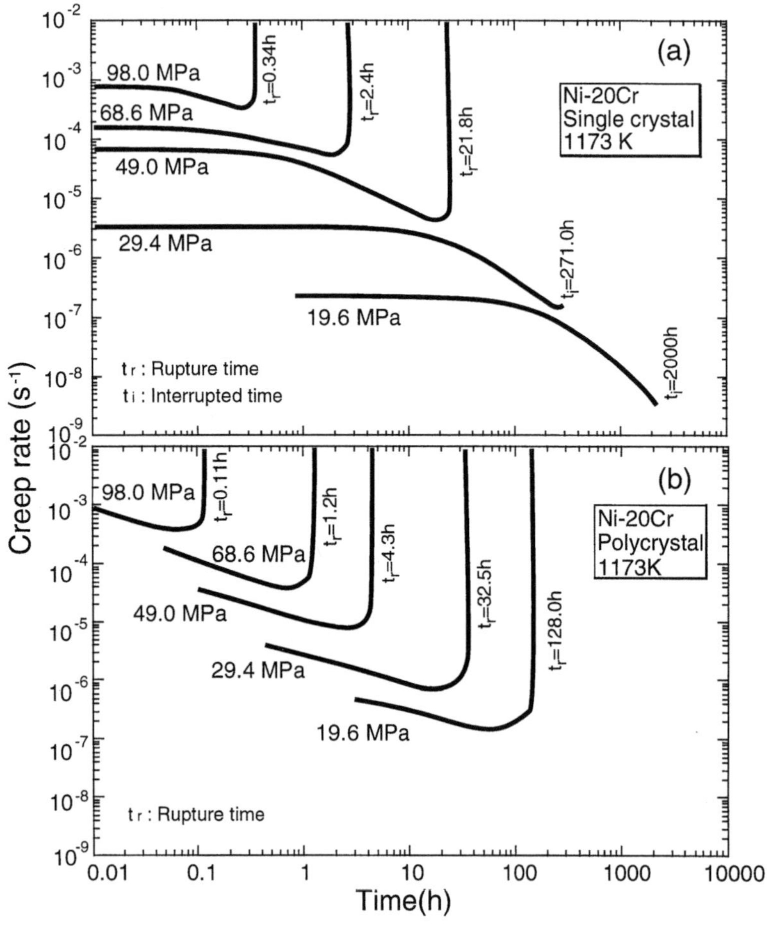

Figure 21.1. Creep rate–time curves from Ni-20Cr single crystals at 1173 K in the stress range from 19.6 to 98 MPa (*a*), and corresponding curves from Ni-20Cr polycrystals (*b*).

an extensometer and the stress axis of the single crystals was always located within the standard triangle of the (001) projection.

Stress dependence of creep rate–time curves of γ single phase Ni-20Cr single crystal

The creep rate–time curves of single crystal Ni-20Cr alloy creep tested in the stress range from 19.6 to 98 MPa are shown in figure 21.1(*a*), with corresponding creep rate–time curves for polycrystalline equivalents shown in figure 21.1(*b*) for the same stresses. The shape of the creep rate–time curve for the single crystal is similar to that of the polycrystalline specimen at the highest stress of 98 MPa, while the difference in creep rate–time curve between single crystals and polycrystals becomes more pronounced with decreasing stress. At lower stresses, the transient creep region in the single crystal was markedly extended, giving lower creep rates and longer rupture times than those of the polycrystal. These features can be well understood by using the creep rate–strain curves, i.e. by plotting the data from figure 21.1 against strain, as shown in figure 21.2. Smaller creep rate and larger rupture strain are attained in the single crystal, not by extending the steady state or accelerating creep stage, but by extending the transient creep stage.

Effect of tantalum on creep rate

Creep rate–time curves of Ni-20Cr-3at%Ta single crystals at 1173 K in the stress range from 29.4 to 98 MPa are shown in figure 21.3, accompanied by the creep rate–time curves of the Ni-20Cr single crystals at 98 and 29.4 MPa. Comparison of the creep rate–time curve of Ni-20Cr-3at%Ta single crystals with that of Ni-20Cr single crystals shows that the difference in creep rate between these two alloys begins just after loading when the creep rate in the tantalum-containing alloy is smaller than that in the base alloy. Then, at higher stresses, the creep rate in Ni-20Cr-3at%Ta single crystal decreases gradually to the minimum creep rate. With decreasing stress level, there is then a more pronounced decrease in the creep rate. In this region the rate of decrease in the creep rate increases with decreasing stress. This feature cannot be detected in the creep rate–time curves for the polycrystalline specimen because the failure occurs relatively rapidly.

Stress–minimum creep rate curves for Ni-20Cr and Ni-20Cr-3at%Ta single crystals at 1173 K are shown in figure 21.4, accompanied by corresponding curves for the matching polycrystals. At higher stress levels, alloys with and without tantalum show similar minimum creep rates, with single crystal alloys in each case exhibiting approximately an order of magnitude lower creep rate. At low stress levels, however, the extension of

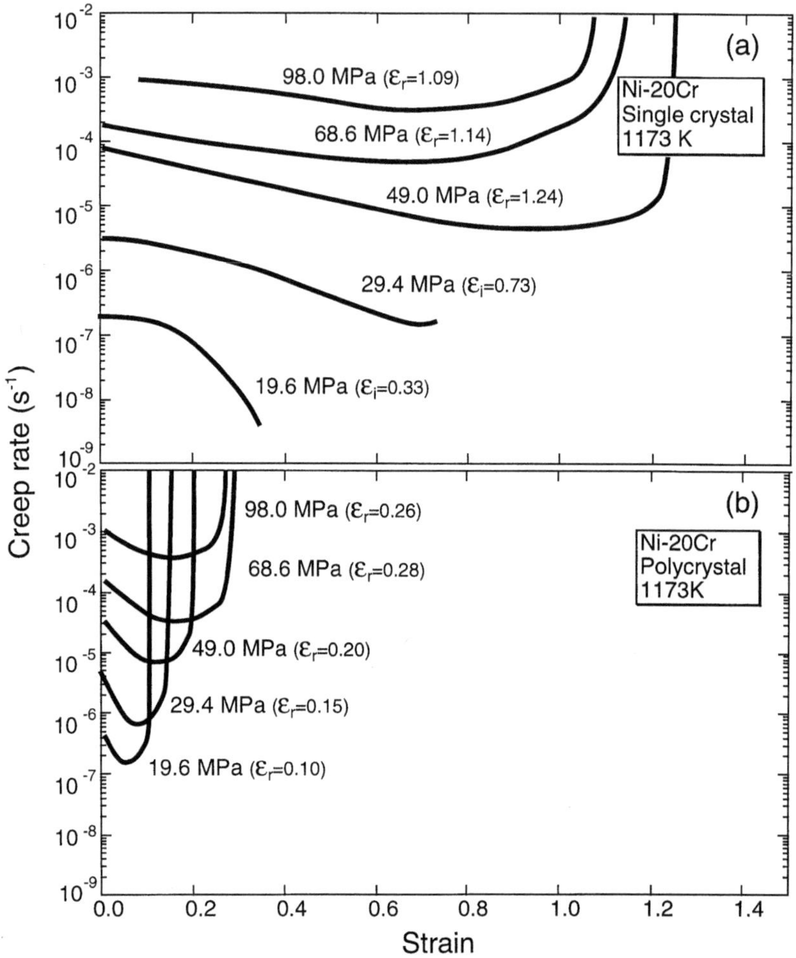

Figure 21.2. Creep rate–strain curves from Ni-20Cr single crystals at 1173 K in the stress range from 19.6 to 98 MPa (*a*), and corresponding curves from Ni-20Cr polycrystals (*b*).

the transient creep stage in tantalum-containing alloys leads to a more marked decrease in creep rate.

Effect of rhenium on creep rate

Creep rate–time curves of Ni-20Cr-3at%Re UDS specimens at 1173 K and 49 MPa are shown in figure 21.5(*a*), accompanied by the corresponding curves for the polycrystalline alloy in figure 21.5(*b*)). The previous results for Ni-20Cr and Ni-20Cr-3at%Ta at 49 MPa are also shown. In the case

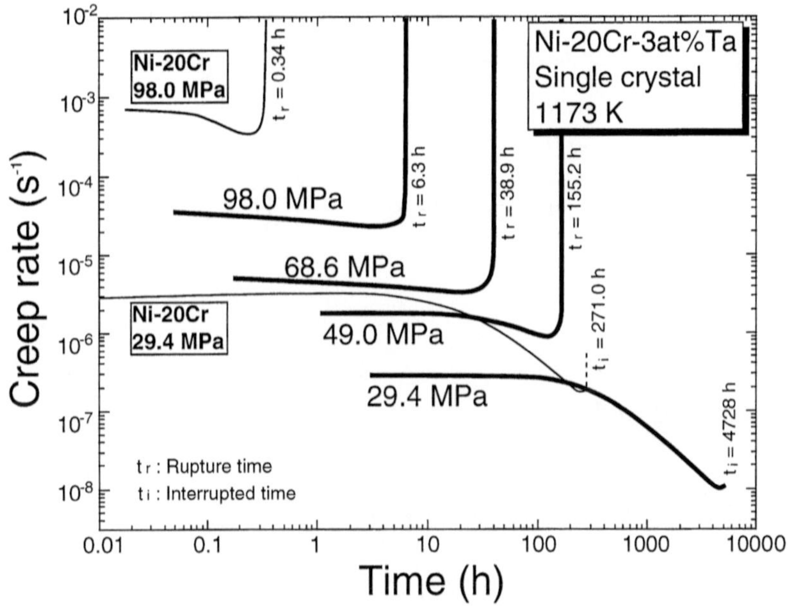

Figure 21.3. Creep rate–time curves from Ni-20Cr-3at%Ta single crystals at 1173 K in the stress range from 29.4 to 98 MPa, accompanied by curves from Ni-20Cr single crystal at 1173 K under the stress of 98 and 29.4 MPa shown by a thin line.

of polycrystalline specimens, the creep rate in Ni-20Cr-3at%Re is smaller than that in Ni-20Cr-3at%Ta, while the creep rupture life in Ni-20Cr-3at%Re is shorter than that in Ni-20Cr-3at%Ta. Additionally, the rupture elongation in the rhenium-containing alloy is very small. In contrast, the creep rupture life in UDS Ni-20Cr-3at%Re is longer than that in Ni-20Cr-3at%Ta, corresponding to the smaller minimum creep rate. Furthermore, the rupture elongation in the rhenium containing alloy, which was shorter in polycrystal alloys, becomes larger in the UDS alloy.

Microstructural factors leading to onset of accelerating creep

As previously described, the difference in creep rate in Ni-20Cr between single crystal and polycrystals is emphasized at lower stresses. With decreasing stress, the minimum creep rate in the single crystal alloy becomes smaller than that of the polycrystals by suppressing the onset of accelerating creep. Hence, the origin of the onset of accelerating creep must be elucidated. The change in dislocation substructure with increasing creep strain is shown in the micrographs of figure 21.6 for creep interrupted at 1173 K and 29.4 MPa. At a strain of 0.1, straight sub-boundaries appear parallel to the slip plane. With

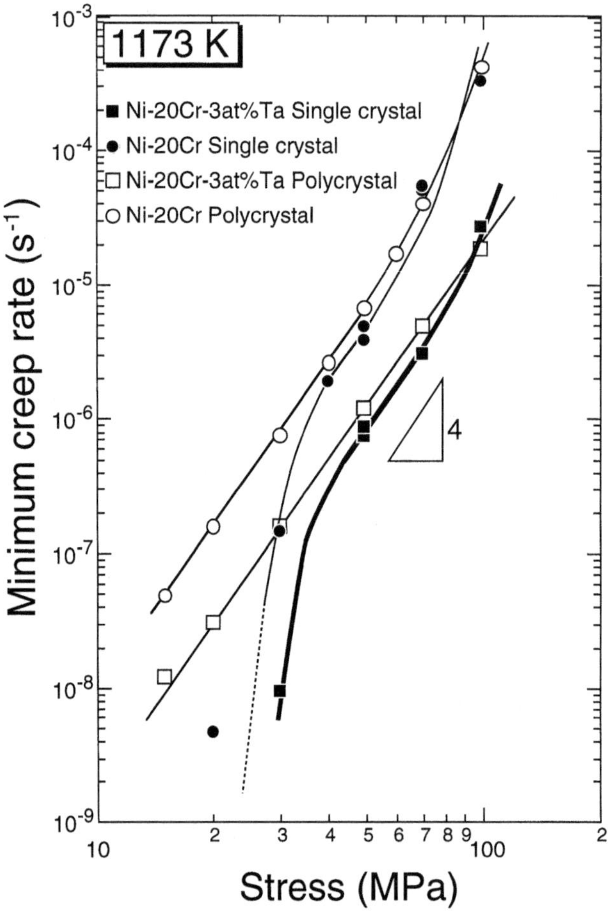

Figure 21.4. Stress–minimum creep rate curves from Ni-20Cr and Ni-20Cr-3at%Ta single crystals at 1173 K, accompanied by curves from Ni-20Cr polycrystals.

an increase in strain from 0.1 to 0.73, sub-boundaries become wavy, and dynamically recrystallized grain boundaries appear. Evolution of dynamic recrystallization occurs during accelerating creep, progressively transforming the microstructure to equiaxed grains. These equiaxed grains are uniformly evolved within the gauge portion. From these microstructural observations, the onset of accelerating creep can be interpreted as the appearance of dynamic recrystallization, i.e. the appearance of high-angled grain boundaries. In dislocation creep, it is generally suggested that the creep rate is independent of the grain size. In contrast, an experimentally observed grain size dependence of creep rate has been termed the core–mantle model and has been proposed previously by Terada *et al* [7].

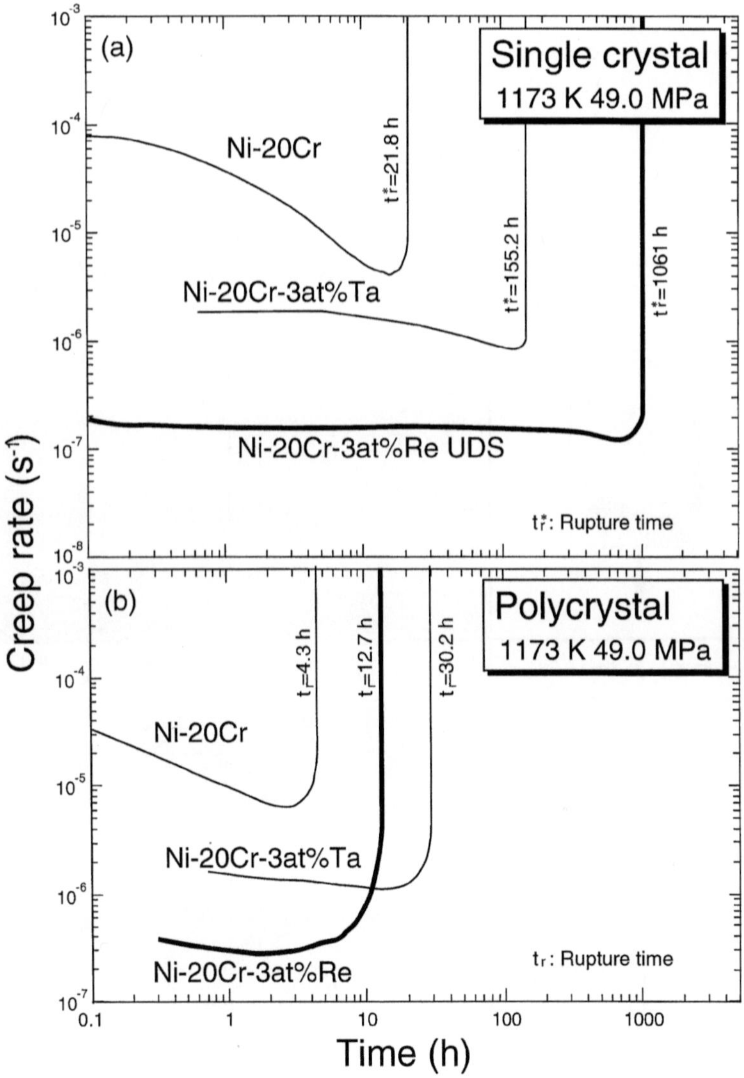

Figure 21.5. Creep rate–time curves from Ni-20Cr-3at%Re UDS specimens at 1173 K/49 MPa, accompanied by curves from Ni-20Cr and Ni-20Cr-3at%Ta single crystals shown by thin lines (*a*), and corresponding curves for polycrystals from the three alloys (*b*).

Effect of tantalum and rhenium on dynamic recrystallization

Similar transmission electron microscope investigations of the change in dislocation sub-structure with increasing creep strain in creep tested Ni-20Cr-3at%Ta single crystals confirm that the formation of straight sub-boundaries

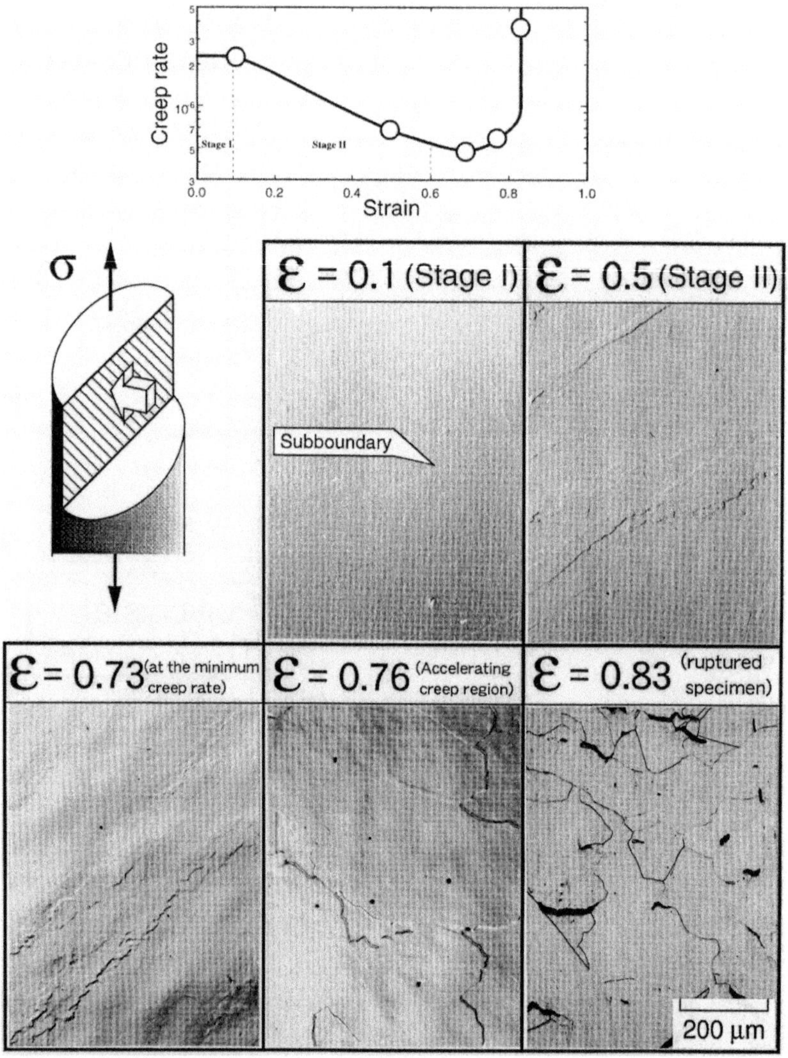

Figure 21.6. Optical micrographs from Ni-20Cr single crystals subjected to creep strains of 0.1, 0.5, 0.73, 0.76 and 0.83 at 1173 K/29.4 MPa.

is suppressed and that dynamic recrystallization occurs along the primary sub-boundaries, rather than uniform evolution of dynamic recrystallization across the gauge length. Rhenium fully suppresses homogeneous dynamic recrystallization and leads to brittle fracture at the grain boundaries. In the case of single crystals, without grain boundaries, brittle fracture at the grain boundaries is of course fully suppressed, and the solute rhenium then acts as an efficient suppressor of the onset of accelerating creep.

Summary

In single phase Ni-20Cr alloys, the difference in creep rate–time curves between single crystal and polycrystalline samples is emphasized with decreasing stress. The minimum creep rate in single crystal alloys is lower than that of the corresponding polycrystalline alloys, and the onset of accelerating creep in the single crystals is comparatively delayed. Ni-20Cr-3at%Ta alloy shows a lower creep rate just after loading than a similar Ni-20Cr alloy, and there is a gradual decrease in creep rate in the transient region. At lower stresses in single crystals there is a relatively large decrease in creep rate. At these lower stresses the creep rate of single crystals becomes lower than that of polycrystals, because of the suppression of accelerating creep. The creep rate of Ni-20Cr-3at%Re polycrystals is smaller than that of Ni-20Cr-3at%Ta. The creep rupture life of the rhenium-containing alloy, however, is shorter than that of the tantalum-containing alloy. In contrast to polycrystalline results, the rupture life of unidirectionally solidified specimens of Ni-20Cr-3at%Re is longer than that of Ni-20Cr-3at%Ta single crystal specimens.

The onset of accelerating creep in Ni-20Cr and Ni-20Cr-3at%Ta corresponds to dynamic recrystallization through the evolution of subboundaries. The suppressing effect of dynamic recrystallization through additions of rhenium and tantalum was confirmed. Rhenium is more effective in suppressing dynamic recrystallization than tantalum.

References

[1] Harada H 1997 *Report of the 123rd Committee on Heat-Resisting Metals and Alloys* **28** pp 185–196

[2] Kondo Y and Matsuo T 1997 *Report of the 123rd Committee on Heat-Resisting Metals and Alloys* **28** pp 269–286

[3] Decker R F 1969 'Strengthening mechanisms in nickel-base superalloy' Climax Molybdenum Company Symposium, Zurich

[4] Hoppin III G S and Danesi W P 1987 in *Superalloy II* ed C T Sims (New York: Wiley) pp 549–561

[5] Matsuo T, Nakajima K, Terada Y and Kikuchi M 1991 *Mater. Sci. Eng.* **A146** 261–272

[6] Nakajima K 1992 Doctoral Thesis, Tokyo Institute of Technology

[7] Terada Y, Matsuo T and Kikuchi M 1993 *Proceedings of the 7th JIM International Symposium on Aspects of High Temperature Deformation and Fracture in Crystalline Materials* ed Y Hosoi *et al* (Sendai: Japan Institute of Metals) pp 27–34

Chapter 22

Thermal barrier coatings

Patrick Grant

Introduction

Aeroengines make extensive use of coatings for a wide range of crucial applications, where they allow optimum bulk component properties, e.g. creep resistance, fatigue life, etc., to be combined with enhanced surface properties, e.g. wear and corrosion resistance. Coating applications include anti-fretting coatings on casing flanges, wear protection (WC/Co) on shafts, abrasive coatings (Al_2O_3, Cr_3C_2) on compressor blade tips, and anti-fretting/wear coatings on the fan disk, fan blade root and fan snubbers (WC/Co, NiCr + Cr_3C_2). Abradable coatings are used on compressor and other casing linings where they are combined with abrasive coatings on blade tips to produce conformal seals with minimal clearance when the blade is rotating. These seals are essential to minimize leakage of intake air between adjacent sections of the compressor which are operating at different pressures. Minimizing circumferential pressure leaks results in significant increases in efficiency.

In combustor and turbine sections of the aeroengine, thermal barrier coatings (TBCs) are used to insulate nickel superalloy components from ever-increasing combustion gas temperatures which in some cases may exceed the melting point of the nickel superalloy itself. As well as structural failure by melting, the thermal barrier coating also confers additional protection from the effects of creep and thermal fatigue [1] and the oxidizing effect of sulphates and chlorides [2]. Thermal insulation is usually provided by a partially stabilized ZrO_2 (PSZ) ceramic coating. However, in order for the partially stabilized zirconia to adhere to the nickel superalloy component and withstand the interfacial shear stresses generated during thermal cycling, an intermediate MCrAlY (M = Co, Ni or Co + Ni) bond coat is used. The bond coat additionally provides oxidation and corrosion resistance for the nickel superalloy component. To add further complexity, the nickel superalloy component may itself have a initial oxidation resistant aluminized layer formed at the surface by a

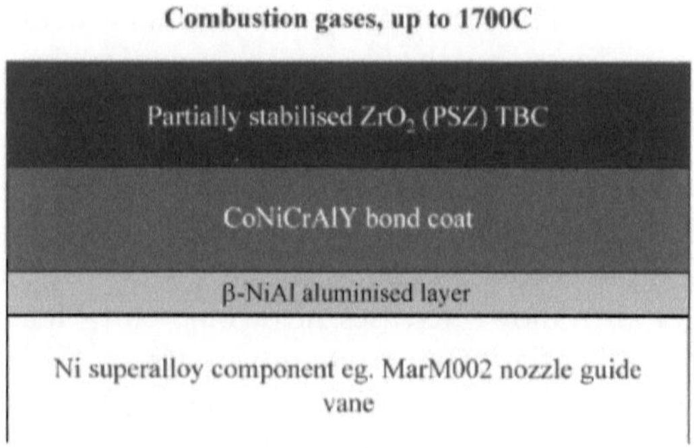

Figure 22.1. Schematic of a typical TBC arrangement.

solid or vapour based diffusion treatment. A typical thermal barrier coating arrangement is shown schematically in figure 22.1. The demanding mix of property requirements of coating systems for gas turbine aerofoil applications has been described in detail [2], and is summarized in table 22.1.

Table 22.1. Property requirements of coating systems for gas turbine aerofoil applications [2].

Property	Requirement
Corrosion and oxidation resistance	Rapid formation of a continuous, adherent protective oxide film Slow subsequent growth rate of oxide Highly stable and adherent scale High concentration of scale forming elements Acceptable oxidation/corrosion rate
Erosion resistance	Ductile and adherent scale Moderate coating ductility
Interfacial stability	Low rates of diffusion across the interface Minimum brittle phase formation
Mechanical properties	Ability to withstand thermal exposure Appropriate ductility Minimum effect of substrate
Aerodynamic properties	Best possible surface finish Acceptable thickness and uniformity Minimum loss of surface finish in service
Coating process	Ability to coat complex shapes Optimized for composition, structure, thickness, thickness and uniformity Cost effectiveness
Adhesion	Similar/matched coating and substrate properties

The use of thermal barrier coatings for thermal insulation in the turbine section has been driven by the desire to minimize the parasitic use of compressor discharge air for: (1) component surface boundary layer cooling; and (2) dilution of hot combustion gases entering the turbine first stage. Reduced use of cooling compressor air enables the recovery of maximum work from the compressor and promotes thermodynamic efficiency. As well as reducing the need for cooling air for a given temperature, thermal barrier coatings also allow increases in turbine entry temperature [2] and currently offer a potential temperature advantage of up to 170°C over the base superalloy, as shown schematically in figure 22.2. It is estimated that successful thermal barrier coating application to all high temperature aerofoils in a typical modern gas turbine engine could result in annual fuel savings as high as 38 million litres for a 250 aircraft fleet [1].

Aluminized layers

Aluminized layers on nickel superalloy components are formed by exposing the clean and selectively masked surface of the component to a solid or vapour aluminium source at high temperature in an inert atmosphere. Diffusion of the aluminium into the component (high activity coating) or outward diffusion of nickel (low activity coating) leads to the formation of a 30–50 μm wide layer of 1–5 μm grains of β-NiAl (B2 ordered bcc) phase, as well as some Ni_2Al_3 which is transformed to β-NiAl on subsequent heat treatment [2]. Chemical vapour deposition (CVD) aluminizing is preferred for the coating of complex shaped components. The β-NiAl provides surface oxidation resistance by the ready formation of a tenacious 50–200 nm grained α-Al_2O_3 scale. However, bulk β-NiAl has little ductility below about 750°C,

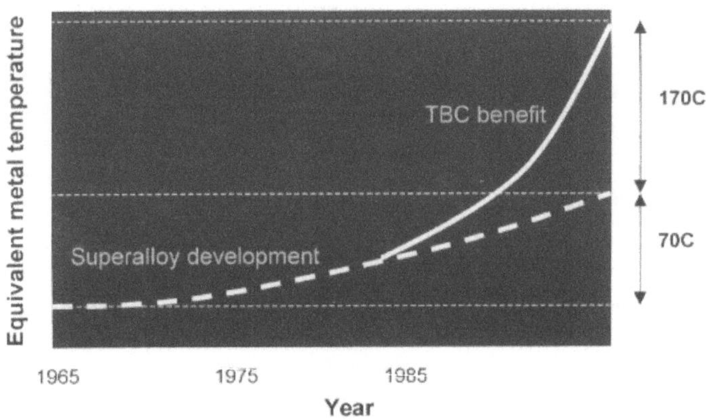

Figure 22.2. Operating temperature advantage offered by TBCs [1].

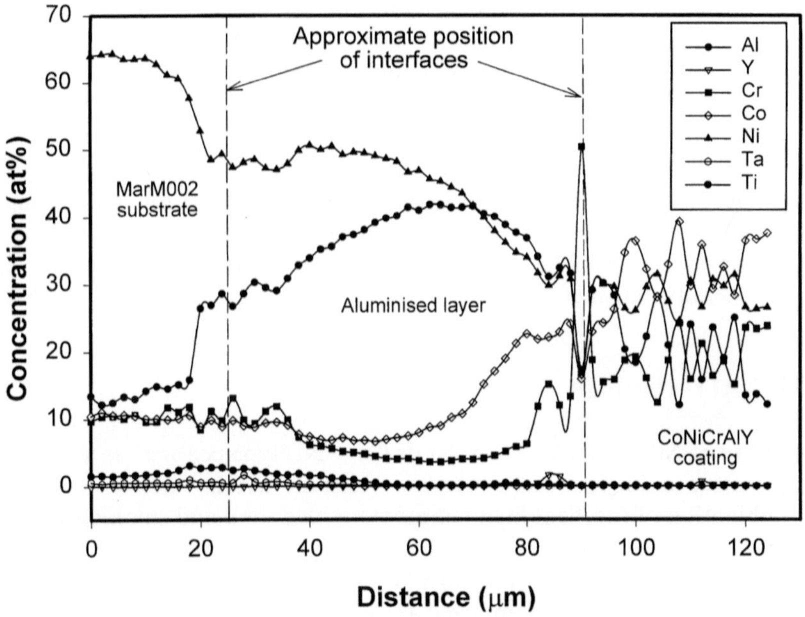

Figure 22.3. EPMA line scan across a MarM002/aluminized layer/CoNiCrAlY coating arrangement.

and excessive aluminized layer thickness can lead to coating cracking, and the propagation of these cracks into the underlying component.

The usually complex chemistry of the underlying nickel superalloy component leads to the formation of a range of minority phases within the aluminized layer after manufacture and in-service. For example, γ'-$Ni_3(Al,Ti)$ (ordered fcc structure) precipitates, $M_{23}C_6$ phases where M is predominantly chromium but may include molybdenum, tungsten or tantalum, and intragranular α-chromium rich precipitates are frequently formed [3–8]. Figure 22.3 shows an electron probe microanalysis (EPMA) line scan for cobalt, nickel, chromium, aluminium, yttrium, tantalum and titanium across a heat treated MarM002/aluminized layer/Co-32Ni-21Cr-8Al-0.5Y coating arrangement shown schematically in figure 22.1. Figure 22.3 shows the aluminized layer was about 60 μm in width and consisted primarily of β-NiAl with cobalt and chromium in solid solution. Close to the MarM002 substrate, peaks in the tantalum trace suggested the formation of tantalum carbide. Close to the interface with the CoNiCrAlY coating, there was diffusion of cobalt, chromium and to a lesser extent yttrium from the CoNiCrAlY into the aluminized layer. The aluminized layer/CoNiCrAlY interface was chromium rich because of the formation of σ-chromium rich phases [9].

Additions of platinum, chromium, tantalum, silicon and other active elements to aluminized layers have been used to improve high temperature oxidation resistance, and the most common addition is platinum. A platinum layer is deposited by electroplating or sputtering, followed by aluminizing and heat treatment [2]. Platinum is relatively concentrated at the surface of the aluminized layer as $PtAl_2$, as well as dissolved in the NiAl [8]. The presence of platinum improves the diffusional stability of the coating by restricting the outward transport of substrate elements, and inhibits the formation of a complex series of internal and surface precipitates which are generally detrimental to the retention of the protective Al_2O_3 scale [8].

MCrAlY coatings

The use of MCrAlY (M = Co, Ni or Co + Ni) type coatings as a further or alternative oxidation and corrosion resistant overlay derives from their combination of excellent high temperature oxidation and sulphidation resistance, and mechanical compatibility with nickel superalloys [10]. The detailed role of MCrAlY constituents on coating properties is described elsewhere [11] and may be summarized as: (i) nickel forms a series of high melting point phases including γ-nickel (disordered fcc structure), γ'-Ni_3Al and oxidation resistant β-NiAl; (ii) cobalt readily substitutes for nickel and provides improved sulphidation resistance; (iii) aluminium alloys with nickel and cobalt to form β-(Ni,Co)Al which readily forms a protective Al_2O_3 scale; (iv) chromium forms a Cr_2O_3 scale, improves hot corrosion resistance, and reduces the level of aluminium additions needed to form the Al_2O_3 scale; and (v) yttrium additions segregate preferentially to Al_2O_3 scale grain boundaries to form yttrium-based phases which pin the 50–200 nm Al_2O_3 grains (pegging), fill intergranular voids and inhibit oxygen diffusion and the resulting growth of complex oxide and spinel phases which generate detrimental MCrAlY/Al_2O_3 interfacial stresses [12–14]. A wide range of MCrAlY compositions have been formulated to optimize mechanical and protective properties and have typical contents of (wt%) 15–38Cr, 8–13Al, 0.5–1Y with cobalt and/or nickel and small additions of other active elements comprising the remainder. The effect of active element additions in slowing oxidation kinetics is temperature dependent, hence multiple additions have been suggested e.g. Ni-20Co-18Cr-12.5Al-0.6Y-0.25Hf-0.4Si [2]. The benefit of an aluminizing heat treatment to enrich the surface of the MCrAlY coating with β-(Ni,Co)Al has also been demonstrated [15, 16].

MCrAlY coatings consist primarily of a two phase β-(Ni,Co)Al and γ-(Ni,Co) mixture. Figure 22.4 shows a scanning electron micrograph of a typical MCrAlY β/γ microstructure after heat treatment employed to promote MCrAlY adhesion by interdiffusion with the substrate. This interdiffusion leads to the formation of a range of minority phases. The

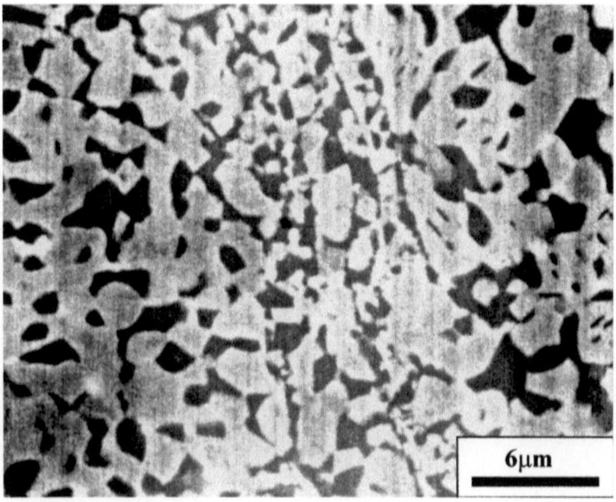

Figure 22.4. SEM micrograph of typical CoNiCrAlY γ/β microstructure after heat treatment.

EPMA line scan across the MarM002/aluminized layer/CoNiCrAlY arrangement in figure 22.2 shows that aluminium diffuses from the aluminized layer into the CoNiCrAlY coating after heat treatment, and figure 22.5 shows a transmission electron microscope (TEM) micrograph of the resulting fine scale γ'-$(Ni,Co)_3(Al,Ti)$ precipitation in 1–3 μm grains in regions close to the CoNiCrAlY/aluminized layer interface [17]. Figure 22.6 shows a TEM micro-

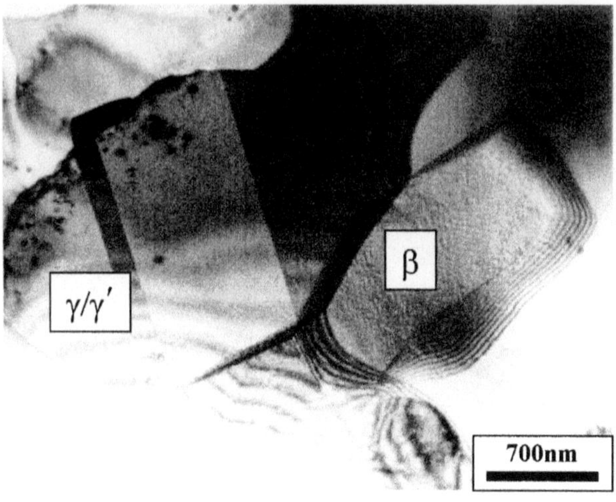

Figure 22.5. TEM micrograph of the formation of γ' in CoNiCrAlY coatings after heat treatment.

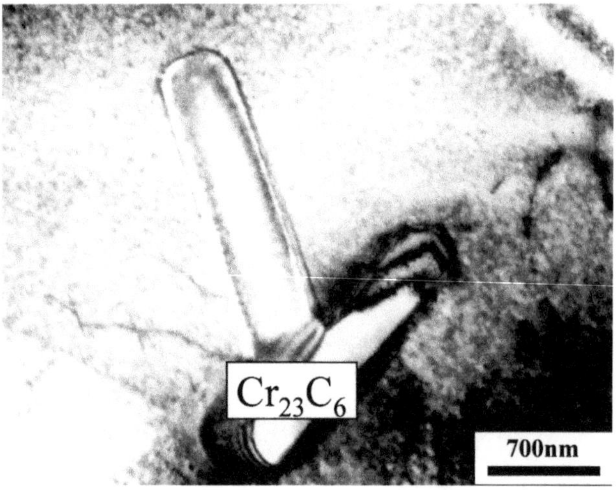

Figure 22.6. TEM micrograph of the formation of $Cr_{23}C_6$ in CoNiCrAlY coatings after heat treatment.

graph of blocky 200 nm × 700 nm $Cr_{23}C_6$ particles formed in the CoNiCrAlY coating because of carbon diffusion from the base MarM002 superalloy. Figure 22.7 shows that as well as γ' and carbide formation, as deposited M_3Y or M_5Y phases transform by internal oxidation to 100 nm β/γ' grain boundary Y_2O_3 particles on heat treatment. Intra-granular 50–100 nm Y_2O_3 particles are also formed [18].

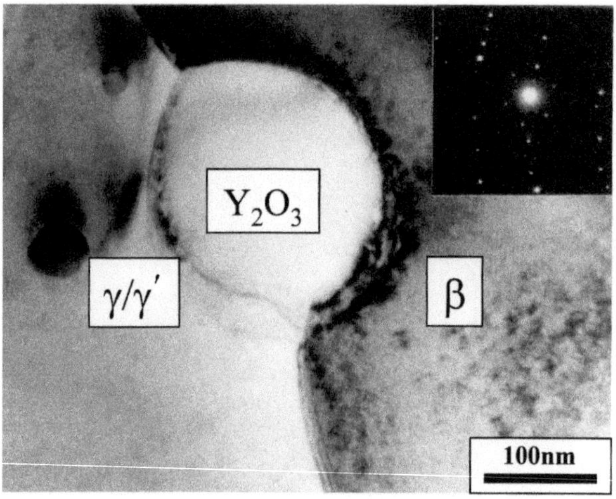

Figure 22.7. TEM micrograph of the formation of Y_2O_3 in CoNiCrAlY coatings after heat treatment.

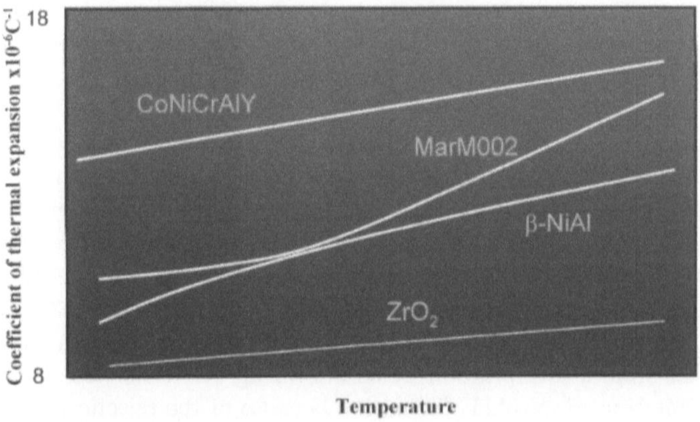

Figure 22.8. Typical variation of CTE variation with temperature for TBC components.

Ceramic coatings

MCrAlY coatings can be used without any subsequent top coat as an oxidation and hot corrosion resistant overlay. However, in the most demanding applications, an insulating ceramic top coat is also used and the MCrAlY coating must then additionally serve as a bond coat. Because of its relatively high thermal expansion coefficient ($7\text{–}10 \times 10^{-6}\,K^{-1}$), low thermal conductivity ($0.8\text{–}1.5\,W\,m^{-1}\,K^{-1}$) and fully oxidized ground state, ZrO_2 is used widely in thermal barrier coating applications. Typical variations of the coefficient of thermal expansion (CTE) for thermal barrier coating materials as a function of temperature are shown in figure 22.8. The difference in coefficient of thermal expansion between substrate, bond coat and ceramic leads to the generation of significant thermally induced stresses during thermal cycling of the engine, and can lead to cracking and spalling of the ceramic coating. Zirconia also undergoes a solid state transformation on cooling from tetragonal to monoclinic at ~950°C. This transformation is accompanied by a 3–10% volume change which may cause cracking. Therefore, partially stabilized ZrO_2 in which 6–8% Y_2O_3 additions are used to stabilize the tetragonal phase [2]. However, sulphation of Y_2O_3 in service may reduce the Y_2O_3 stabilization effect, and the resulting phase changes may induce cracking [2].

Degradation of thermal barrier coatings

The in-service degradation of thermal barrier coatings stems from the synergistic effects of: (1) high temperature interdiffusion and reaction of environment, coating and substrate constituents, because of steep or discontinuous through-thickness compositional gradients resulting in internal stresses because of associated volume changes; (2) the generation of additional internal and/or interfacial stresses because of coefficient of thermal expansion mismatch; and (3) the progressive sintering and associated increase in stiffness of the ceramic top coat [19].

In the case of the aluminized layers, microstructural change is dominated by the oxidation of the coating surface and interdiffusion with the substrate material. Both of these processes lead to the loss of aluminium [3], which in turn destabilizes β-NiAl and promotes the progressive formation of γ'-Ni_3Al [4]. Interdiffusion of coating and substrate constituents at high temperature also leads to the formation of minority phases (depending on underlying superalloy substrate composition and temperature) including further $M_{23}C_6$ and α-chromium precipitation, re-solutionizing of γ' followed by subsequent re-precipitation, β'-Ni_2AlTi formation because of the rejection of Ti from β-NiAl, Cr_2Ti Laves phase and Mo rich M_6X, as well as χ (chromium and molybdenum rich intermetallic $M_{18}C$) and σ-chromium phases [4]. Similarly to aluminized layers, MCrAlY coatings also show an increase in γ'-Ni_3Al volume fraction on thermal exposure, again because of the progressive loss of aluminium, but also because of interdiffusion leading to the loss of cobalt and chromium which tend to favour the formation of γ over γ' [10].

For thermal barrier coating arrangements incorporating a partially stabilized zirconia coating, the volume changes associated with the continued Al_2O_3 scale growth at the PSZ/MCrAlY interface because of oxygen diffusion and ZrO_2 reduction by aluminium, as well as $Ni(Al,Cr)_2O_4$ spinel oxide and $Y_3Al_5O_{12}$ garnet growth [14], leads to interfacial stresses which are superimposed on the thermally induced stresses because of coefficient of thermal expansion differences within the thermal barrier coating arrangement. The evolution of the resulting complex stress field eventually leads to thermal barrier coating failure by spalling at or close to the MCrAlY/PSZ interface [15, 20, 21] and a critical scale thickness for coating spalling of 2–4 μm has been suggested [2].

The synergistic and evolving effect of stresses arising from microstructural change and coefficient of thermal expansion differences, and their relaxation at high temperature, is highly complex and makes thermal barrier coating lifetime prediction problematical. For example, the use of a bond coat with relatively poor oxidation resistance may still provide excellent thermal barrier coating life if it has appropriately matched mechanical and thermo-physical properties to the thermal barrier coating system as a whole [22].

Processing of thermal barrier coatings

Plasma spraying

The majority of thermal barrier coatings in a commercial gas turbine engine are manufactured by plasma spraying of a MCrAlY bond coat followed by a

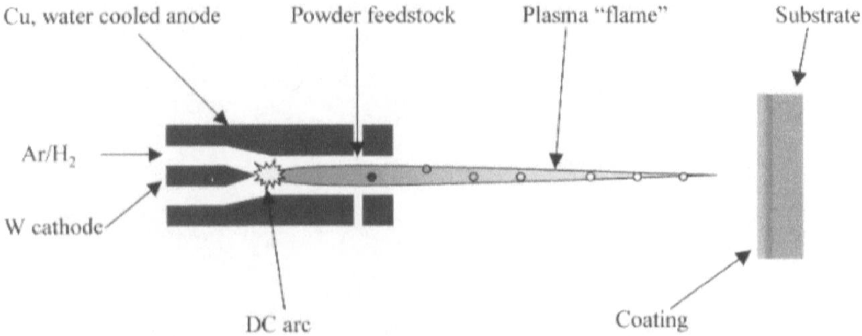

Figure 22.9. A schematic of plasma spraying.

partially stabilized zirconia top coat. The principles of plasma spraying are shown schematically in figure 22.9. A dc arc is established between a water cooled copper anode and a concentrically located tungsten cathode which causes dissociation and ionization of argon and hydrogen gases to form a plasma. The partially ionized 'flame' exits from the end of the plasma gun at velocities up to $500\,m\,s^{-1}$ and at a temperature of up to 10 000 K [23, 24]. Particulate feedstock material is introduced radially into the axial flow by an argon carrier gas, downstream of the arc region. Particulate is entrained into the flow where it is accelerated, heated, melted and projected at high velocity onto the substrate.

Atmospheric plasma spraying (APS) in air is used for the partially stabilized zirconia coating, whereas an argon shroud is employed for the MCrAlY coating to restrict oxidation. Alternatively, low pressure or vacuum plasma spraying (LPPS, VPS) may be used, in which MCrAlY plasma spraying is carried out in a sealed chamber which is evacuated of air and then back-filled with argon to around 0.1 atm. Figure 22.10 shows a vacuum plasma spraying system for MCrAlY coatings employing five-axis manipulation of the plasma gun in order to coat complex shaped surfaces evenly. Figure 22.11 shows a typical plasma sprayed thermal barrier coating arrangement. Figure 22.11 shows that the 100 μm thick atmospheric plasma spraying partially stabilized zirconia typically has porosity of up to 15%, with extensive micro-cracking which promotes strain tolerance and toughness [25]. The 100 μm thick vacuum plasma spraying MCrAlY has typical porosity of less then 1% arising from the higher droplet velocities in vacuum plasma spraying, which also improves adhesion. The MCrAlY coating is relatively rough and promotes mechanical locking of the partially stabilized zirconia. However, the rough interface also leads to complex residual stress patterns and stress concentration effects at or close to the PSZ/MCrAlY interface [21]. To enhance vacuum plasma spraying MCrAlY adhesion to the nickel superalloy component, an arc is struck

Figure 22.10. A VPS system for the manufacture MCrAlY coatings.

between the plasma gun and the metallic workpiece during manufacture. This transferred arc is used to sputter clean and then heat the component prior to and during plasma spraying. The heated nickel superalloy surface is free from oxide and a diffusion bond is readily formed on deposition of

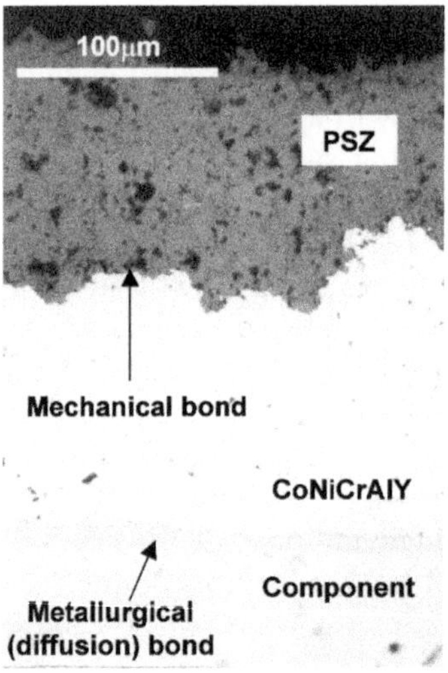

Figure 22.11. A typical plasma sprayed TBC arrangement.

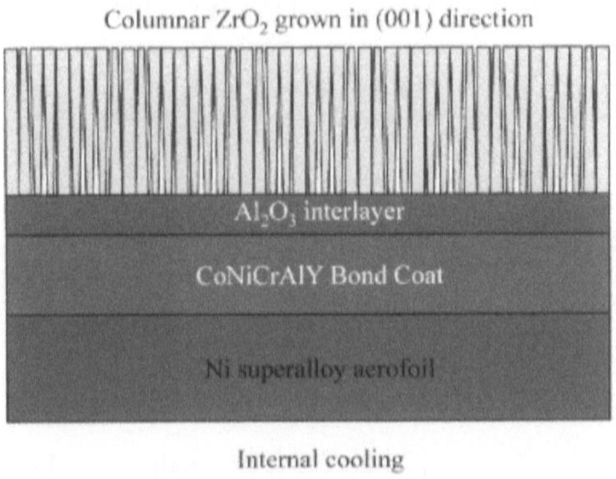

Figure 22.12. Schematic of a typical EB-PVD processed TBC microstructure.

the first MCrAlY layer. Figure 22.11 shows that MCrAlY/substrate interdiffusion after heat treatment forms a 5 μm thick reaction layer.

Electron beam–physical vapour deposition

Electron beam–physical vapour deposited thermal barrier coatings have been developed in which improved strain tolerance of the partially stabilized zirconia is conferred by the formation of columnar grain boundaries arranged perpendicularly to the substrate surface. The characteristic electron beam–physical vapour deposited columnar microstructure is shown schematically in figure 22.12, and the principle of electron beam–physical vapour deposition (EB-PVD) is shown schematically in figure 22.13. A

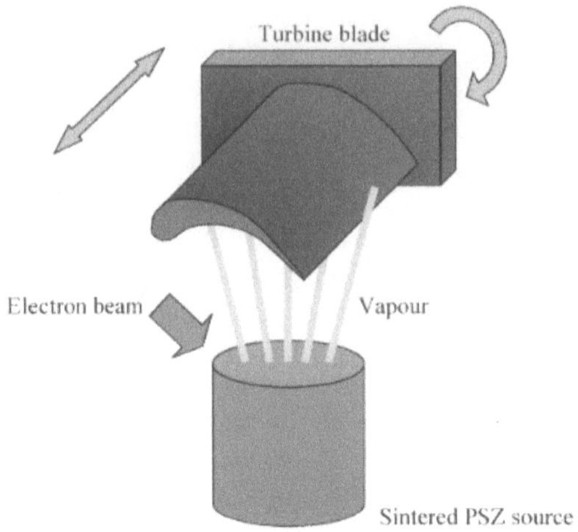

Figure 22.13. Schematic of EB-PVD processing of TBCs on aerofoil sections.

Table 22.2. Properties of plasma sprayed and EB-PVD TBCS at room temperature [15, 26].

	Plasma spray	EB-PVD
Thermal conductivity (W/mK) (bulk ZrO_2 ~2.2)	0.8	1.5
Surface roughness (μm)	10	1
Adhesive strength (MPa)	20–40	400
Young's modulus (GPa)	200	90
Erosion rate (normalized to PVD)	7	1
Hole closure	Excessive on small diameters	Little or none
Thickness	Thick coatings economically produced	Good control for thin coatings
Coating distribution	Excellent although line of sight process	Excellent on single aerofoils; poor structure and thickness between aerofoil pairs

nickel superalloy component is rotated and translated in the vapour of an evaporating partially stabilized zirconia source heated to up to 4000°C [26] by electron beam guns under a low vacuum of 10^{-4}–10^{-6} Torr [2]. The partially stabilized zirconia vapour condenses onto the component to progressively grow a coating up to 100 μm thickness or more.

A comparison of plasma spraying and EB-PVD techniques is shown in table 22.2 [15, 27]. Most importantly, the improved strain tolerance of EB-PVD partially stabilized zirconia shown in table 22.2 allows the application of thermal barrier coatings to curved aerofoil sections, whereas less strain tolerant vacuum plasma sprayed coatings are largely restricted to planar or low radius of curvature surfaces such as nozzle guide vane platforms and combustor cans. EB-PVD thermal barrier coatings also provide improved surface finish, resistance to erosion, hole definition and adhesion, but have a higher thermal conductivity and therefore reduced thermal barrier effect, and high capital and processing costs. The incremental and controllable nature of the EB-PVD process allows the manipulation of the through thickness microstructure of either the MCrAlY or zirconia layers. For example, an Al_2O_3 layer is formed on the substrate to improve zirconia adhesion, or a transition from columnar to equiaxed fine grain may be engineered to reduce zirconia thermal conductivity [27].

Future challenges

Future developments in thermal barrier coatings will focus on improvements in materials, processing, design and lifetime prediction, and their interrelationships. Material developments will include improved mechanical and chemical matching of the substrate, aluminized layers and MCrAlY coatings, accounting for the inevitably of the interdiffusion. Consequently, developments in the chemistry of the underlying component will be taken into consideration in coating formulation. An improved understanding of the role of superalloy and coating active elements, e.g. yttrium, cerium, rhenium and platinum, on oxidation behaviour will lead to more complex, expensive and effective oxidation resistant coatings. The maintenance of titanium alloy bulk properties to higher operating temperatures [28] will lead to new coatings for surface protection [29, 30].

For the insulating ceramic coating, further reductions in thermal conductivity will be realized by alloying additions, controlled microcracking, and the manipulation of grain morphology through the coating thickness by improvements in both plasma sprayed and EB-PVD processing. Graded coatings in which the MCrAlY and partially stabilized zirconia contents of a composite coating are changed continuously as a function of thickness successfully reduce metal/ceramic interfacial shear stresses generated during thermal cycling [31], but require complex processing, cause difficulties for repair, and have complex and sometimes inferior oxidation performance because of increased MCrAlY/PSZ interfacial area.

All thermal barrier coating processing techniques will be optimized for further cost reduction in order to increase margins. For more established plasma spraying based processes, cost reduction will be derived from a reduction in component cycle time, improved process yield (particularly of expensive MCrAlY + active element powders), improved quality (adhesion, porosity and surface roughness), extension of the component range that can be processed, and ease of repair. Process reproducibility and robustness will be improved by progressive implementation of feedback control during processing, e.g. measurement of coating temperature during manufacture and coating thickness distribution [32]. For less mature EB-PVD processing, reductions in capital and running costs will be paramount. Ultimately, the choice of atmospheric plasma spraying, vacuum plasma spraying, EB-PVD or other processing will be increasingly component specific and depend upon the precise combination of cost and benefit. The boundaries between component and coating and their respective processing technologies will become less well defined as traditional coating processes such as plasma spraying and EB-PVD are also used for manufacturing advanced components such as Ti alloys continuously reinforced with SiC fibres for compressor applications [33, 34], graded structures and ultrathin wall, near-net turbine aerofoil components.

As with all aeroengine components, an improved fundamental understanding of the complex factors controlling thermal barrier coating life, and their incorporation into a model to predict with confidence coating lifetimes, will be increasingly important. As part of improved predictions of coating life, non-destructive inspection techniques such as eddy current techniques, thermography and ultrasonics [32] and finite element and other modelling techniques will be used increasingly [35, 36]. Finally, an understanding of the factors controlling coating lifetime and the effect of coating performance on underlying component life will allow the advantages and restrictions conferred by thermal barrier and other coatings to be integrated into the earliest stages of component design.

Summary

Coatings are used extensively in aeroengines to provide a wide range of enhanced surface properties. Thermal barrier coatings have been particularly successful in improving the thermodynamic efficiency by reducing the need for cooling air and increasing operating temperatures. Current thermal barrier coating arrangements are complex and lead to steep through-thickness compositional gradients. During high temperature service, thermal barrier coatings degrade to some extent by continued interdiffusion, oxidation and other environmental attack. The resulting microstructural change and stresses derived from the coefficient of thermal expansion mismatch across the coating arrangement usually result in failure by spalling. A range of coating compositions and processing techniques have been optimized to extend coating life, and to extend the range of coating components. However, no single thermal barrier coating solution is used on all components. The particular coating/process arrangement employed depends upon a cost–benefit analysis for each component. This trend is likely to increase with progressively customized solutions for different components. The distinction between bulk and coating technologies will become blurred through the introduction of graded or composite structures and ultrathin wall and hollow aerofoil sections.

References

[1] Manning Meier S, Gupta D K and Sheffler K D 1991 'Ceramic barrier coatings for commercial gas turbine engines' *JOM* **43**(3) 50–53

[2] Rhys-Jones T N 1989 'Coatings for blade and vane applications in gas turbines' *Corrosion Science* **29** 623–646

[3] Gale W F and King J E 1992 'Microstructural study of aluminide surface coatings on single crystal nickel base superalloy substrates' *Mat. Sci. Tech.* **8** 673

[4] Gale W F and King J E 1993 'Microstructural degradation of aluminide coatings on single crystal nickel based superalloys during high temperature exposure' *Mat. Sci. Tech.* **9** 793

[5] Gale W F and King J E 1992 'Decomposition of the B2-type matrix of aluminide diffusion coatings on single-crystal nickel-base superalloy substrates' *Met. Trans.* **23A** 2657–2665

[6] Gale W F and King J E 1993 'Precipitation of Cr containing phases in aluminide coated nickel superalloy single crystals' *J. Mat. Sci.* **28** 4347–4354

[7] Gale W F, Totemeier T C and King J E 1995 'Modification of the B2-type matrix of aluminide diffusion coatings on nickel base superalloys—bulk aluminide analogues' *Met. Mat. Trans.* **26A** 949

[8] Tawancy H M, Sridhar N and Abbas N M 1995 'Comparative thermal stability characteristics and isothermal oxidation behaviour of an aluminized and Pt aluminized Ni base superalloy' *Scripta Met. Mat.* **33** 1431–1438

[9] Swindells N 1986 'Structural stability of coatings' *Mat. Sci. Tech.* **2** 250–255

[10] Totemeier T C, Gale W F and King J E 1994 'Microstructural evolution of an overlay coating on a single-crystal nickel-base superalloy' *Met. Mat. Trans.* **25A** 2837–2840

[11] Goward G W 1986 'Protective coatings—purpose role and design' *Mat. Sci. Tech.* **2** 194–199

[12] Tawancy H M 1991 'On the role of yttrium during high-temperature oxidation of an NiCrAlFeY alloy' *Met. Trans.* **22A** 1463–1465

[13] Sacré S, Wienstroth U, Feller H-G and Thomas L K 1993 'Influence of the phase compositions on the transient high-temperature oxidation behaviour of an NiCoCrAlY coating material' *J. Mat. Sci.* **28** 1843–1848

[14] Alperine S and Lelait L 1994 'Microstructural investigations of plasma sprayed yttria stabilised thermal barrier coating' *Trans. ASME: J. Eng. Gas Turbines and Power* **116** 258–265

[15] Wortman D J, Nagaraj B A and Duderstadt E C 1989 'Thermal barrier coatings for gas turbine use' *Mat. Sci. Eng.* **A121** 433–440

[16] Wu B C, Chao C H, Chang E and Chang T C 1990 'Effects of bond coat pre-aluminizing treatment on the properties of ZrO_2-8wt%Y_2O_3/Co-29Cr-6Al-1Y thermal barrier coatings' *Mat. Sci. Eng.* **A124** 215–221

[17] Noguchi K, Nishida M and Chiba A 1996 'Transmission electron microscopy of low pressure plasma sprayed CoNiCrAlY coating' *Scripta Mat.* **35** 1359–1354

[18] Gudmundsson B and Jacobson B E 1988 'Structure formation and interdiffusion in vacuum plasma sprayed CoNiCrAlY coatings on IN738LC' *Mat. Sci. Eng.* **100** 207–217

[19] Thompson J A and Clyne T W 1999 'The stiffness of plasma sprayed zirconia top coats in thermal barrier coatings to appear' in *Proc. 2nd United Thermal Spray Conference and Exposition* 17–19 March, Dusseldorf, Germany

[20] Meier S M and Gupta D K 1994 'The evolution of thermal barrier coatings in gas turbine applications' *Trans. ASME: J. Engineering for Gas Turbines and Power* **116** 250–257

[21] Bennett A 1986 'Properties of thermal barrier coatings' *Mat. Sci. Tech.* **2** 257–261

[22] Brindley W J and Whittenberger J D 1993 'Stress relaxation of low pressure plasma sprayed NiCrAlY alloys' *Mat. Sci. Eng.* **A163** 33–41

[23] Rahmane M, Soucy G, Boulos M I and Henne R 1998 'Fluid dynamics study of direct current plasma jets for plasma spraying applications' *J. Thermal Spray & Tech.* **7** 349–356

[24] Pfender E and Chang C H 1998 'Plasma spray jets and plasma–particulate interaction: modelling and experiments' *Proc. 15th Int. Thermal Spray Conf.* ed C Coddet (Ohio: ASM Int) pp 315–327

[25] Pawlowski L, Lombard D and Fauchais P 1985 'Structure-thermal properties relationship in plasma sprayed zirconia coatings' *J. Vac. Sci. Tech.* **A3** 2494

[26] Rigney D V, Vigue R, Wortman D J and Skelly D W 1997 'PVD thermal barrier coating applications and process development for aircraft engines' *J. Thermal Spray & Tech.* **6** 193–198

[27] Nicholls J R, Lawson K J, Rickerby D S and Morrell P 1998 *Proc AGARD SMP Meeting 'Thermal Barrier Coatings'* AGARD report 823 (Neuilly, France: AGARD) pp 6-1

[28] Bartolotta P, Barret J, Kelly T and Smashey R 1997 'The use of cast Ti-48Al-2Cr-2Nb in jet engines' *JOM* **49**(5) 48

[29] Brady M P, Brindley W J J, Smialek L and Locci I E 1996 'Oxidation and protection of gamma titanium aluminides' *JOM* **48**(11) 46–50

[30] Jha S K, Khanna A S and Harendranath C S 1997 'Oxidation characteristics of Ti_3Al-Nb alloys and improvement in the oxidation resistance by pack aluminizing' *Oxidation of Metals* **47** 465–493

[31] Sampath S, Herman H, Shimoda N and Saito T 1995 'Thermal spray processing of FGMs' *MRS Bulletin* **20** 27–31

[32] Bettridge D F and Ubank R G 1986 'Quality control of high temperature protective coatings' *Mat. Sci. Tech.* **2** 232–242

[33] Wood M and Ward-Close M 1995 'Fibre reinforced intermetallic compounds by physical vapour deposition' *Mat. Sci. Eng.* **A192/193** 590–596

[34] Grant P S 1997 'Titanium composites: ready for flight' *Materials World* **5** 77–78

[35] Tsui Y C and Clyne T W 1997 'An analytical model for predicting residual stresses in progressively deposited coatings. Part 1: Planar geometry; Part 2: Cylindrical geometry; Part 3: Further developments and applications' *Thin Solid Films* **306** 23–33, 34–51, 52–61

[36] Freborg A M, Ferguson B L, Brindley W J and Petrus G J 1998 'Modeling oxidation induced stresses in thermal barrier coatings' *Mat. Sci. Eng.* **A245** 182–190

Index